ENERGY AUDIT
OF
BUILDING SYSTEMS

An Engineering Approach

Mechanical Engineering Series
Frank Kreith - Series Editor

Published Titles

Energy Audit of Building Systems: An Engineering Approach
Moncef Krarti
Finite Element Method Using MATLAB, 2nd Edition
Young W. Kwon & Hyochoong Bang
Principles of Solid Mechanics
Rowland Richards, Jr.
Entropy Generation Minimization
Adrian Bejan
Finite Element Method Using MATLAB
Young W. Kwon & Hyochoong Bang
Fundamentals of Environmental Discharge Modeling
Lorin R. Davis
Intelligent Transportation Systems: New Principles and Architectures
Sumit Ghosh & Tony Lee
Mathematical & Physical Modeling of Materials Processing Operations
Olusegun Johnson Ileghus, Manabu Iguchi & Walter E. Wahnsiedler
Mechanics of Composite Materials
Autar K. Kaw
Mechanics of Fatigue
Vladimir V. Bolotin
Mechanism Design: Enumeration of Kinematic Structures According to Function
Lung-Wen Tsai
Nonlinear Analysis of Structures
M. Sathyamoorthy
Practical Inverse Analysis in Engineering
David M. Trujillo & Henry R. Busby
Thermodynamics for Engineers
Kau-Fui Wong
Viscoelastic Solids
Roderic S. Lakes

Forthcoming Titles

Distributed Generation: The Power Paradigm for the New Millennium
Anne-Marie Borbely & Jan F. Kreider
Engineering Experimentation
Euan Somerscales
Introductory Finite Element Method
Chandrakant S. Desai & Tribikram Kundu
Mechanics of Solids & Shells
Gerald Wempner & Demosthenes Talaslidis

ENERGY AUDIT
OF
BUILDING SYSTEMS

An Engineering Approach

Moncef Krarti, Ph.D., P.E.
University of Colorado
Boulder

CRC Press
Boca Raton   London   New York   Washington, D.C.

Library of Congress Cataloging-in-Publication Data

Krarti, Moncef.
    Energy audit of building systems : an engineering approach / Moncef Krarti.
        p. cm.—(Mechanical engineering series)
    Includes bibliographical references and index.
    ISBN 0-8493-9587-9
      1. Building—Energy conversation. 2. Energy auditing. I. Title. II. Advanced topics in
mechanical engineering series.

    TJ163.5.B84 K73 2000
    696—dc21   00-050751

Cover Design: Adapted from the artwork of Hajer Tnani Krarti.

About the Author

Moncef Krarti, Ph.D., P.E., is currently associate Professor in the Civil, Environmental, and Architectural Engineering Department at the University of Colorado at Boulder. He received engineering diplomas from prestigious French schools: Ecole Polytechnique and Ecole National des Ponts et Chaussees. He also received his M.S. and Ph.D. degrees in engineering from the Univeeersity of Colorado. He is involved in several research projects related to building energy efficiency. He has published over 100 technical articles and manuscripts related to building energy systems. He is an active member of the American Society for Heating, Refrigerating, and Air-Conditioning Engineers (ASHRAE), and the American Society of Mechanical Engineers (ASME), and the American Society of Engineering Education (ASEE). In the last decade, Dr. Krarti has been consulting engineer in the design and analysis of energy efficient buildings in the US and abroad. In Particular, he has conducted hundreds of building energy audits as well as over a dozen energy management workshops throughout the world. In 1999, Dr Krarti was a visiting professor at the Centre Energetique of the Ecole National des Mines de Paris, France, where he wrote a significant part of this book.

TABLE OF CONTENTS

1

INTRODUCTION TO ENERGY AUDIT

2

ENERGY SOURCES AND UTILITY RATE

STRUCTURES

3

ECONOMIC ANALYSIS

4

ENERGY ANALYSIS TOOLS

5

ELECTRICAL SYSTEMS

6

BUILDING ENVELOPE

7

SECONDARY HVAC SYSTEMS RETROFIT

xii

10   ENERGY MANAGEMENT CONTROL SYSTEMS

11   COMPRESSED AIR SYSTEMS

12 THERMAL ENERGY STORAGE SYSTEMS

13 COGENERATION SYSTEMS

14 HEAT RECOVERY SYSTEMS

15 WATER MANAGEMENT

16 METHODS FOR ESTIMATING ENERGY SAVINGS

PREFACE

In several countries, including the United States, new and/or updated energy efficiency programs are being developed as cost-effective means to reduce national energy consumption, protect the environment, and increase economic competitiveness. Since buildings consume a significant portion of the national energy resources in most countries, they are the primary target of energy efficiency programs. For new buildings, energy efficiency standards and codes have been developed. For existing buildings, energy conservation programs have been implemented with various degrees of success. While in several countries, governmental agencies are still providing financial assistance to improve energy efficiency of existing buildings, private energy service companies in the US are more often providing the needed financing for energy efficiency projects. However, it is a consensus among all countries, that well trained energy auditors are essential to the success of any building energy efficiency program. It is the purpose of this book to provide a training guide for energy auditors and energy managers.

The book presents simplified analysis methods to evaluate energy conservation opportunities in buildings. These simplified methods are based on well-established engineering principles. In addition, several innovative yet proven energy efficiency technologies and strategies are presented. The book is designed to be a self-contained textbook aimed at seniors and/or first-year graduate students. The contents of this book can be covered in a one-semester course in energy management and/or building energy efficiency. However, the book can be used as a reference for practitioners and as a text for continuing education short courses. The users of this book are assumed to have a basic understanding of building energy systems including fundamentals of heat transfer and principles of heating, ventilating, and air-conditioning (HVAC). General concepts of engineering economics, building energy simulation, and building electrical systems are also recommended.

The book is organized in sixteen self-contained chapters. The first three chapters provide basic tools that are typically required to perform energy audits of buildings. Each of the following twelve chapters addresses a specific building subsystem and/or energy efficiency technology. The final chapter provides an overview of basic engineering methods used to verify and measure actual energy savings attributed to implementation of energy efficiency projects. Each chapter includes some worked-out examples that illustrate the use of some simplified analysis methods to evaluate the benefits of energy efficient measures or technologies. Some problems are provided at the end of most chapters to serve as review or homework problems for the users of the book. However, as instructor of an energy management course at the University of Colorado, I found that the best approach for the students to understand and apply the various analysis methods and tools discussed in this book is through group projects. These projects consist of energy audits of real buildings.

When using this book as a textbook, the instructor should start from Chapter 1 and proceed through Chapter 16 in order. However, some of the chapters can be skipped or covered lightly depending on the time constraints and the background of the students. First, general procedures suitable for building energy audits are presented (Chapter 1). Some of the analysis tools and techniques needed to perform building energy audits are then discussed. In particular, analysis methods are briefly provided for utility rate structures (Chapter 2), economic evaluation of energy efficiency projects (Chapter 3), and energy simulation of buildings (Chapter 4). In buildings, electrical systems consume a significant amount of energy. Several energy efficiency strategies and technologies are discussed to reduce energy use from lighting, motors, and appliances (Chapter 5). Various approaches to improve the building envelope are also outlined (Chapter 6). These approaches are particularly suitable for residential buildings characterized by skin-dominated heating/cooling loads. To maintain acceptable comfort levels, heating and cooling systems typically consume the most energy in a building. Several measures are described to improve the energy efficiency of secondary HVAC systems (Chapter 7), central heating and cooling plants (Chapters 8 and 9), and energy management control systems (Chapter 10). In addition, simple strategies are described to reduce the energy used by compressed air systems, especially in industrial facilities (Chapter 11). Selected technologies to reduce energy use and cost in buildings are discussed, including thermal energy storage systems (Chapter 12), cogeneration (Chapter 13), and heat recovery systems (Chapter 14). Cost-effective measures to improve water management inside and outside buildings are presented (Chapter 15). Finally, analysis methods used for the measurement and verification of actual energy savings attributed to energy efficiency projects are briefly summarized (Chapter 16).

A significant part of this book was written during my sabbatical leave at the Centre Energetique at the Ecole National des Mines de Paris (ENMP), France. In particular, some chapters are prepared to be parts of training guides for building energy auditors in Europe. Thus, a special effort has been made to use metric (SI) units throughout the book. However, in several chapters English (IP) units are also used since they are still the standard set of units used in the United States. Conversion tables between the two unit systems (from English to metric and metric to English units) are provided in Appendix A.

I wish to acknowledge the assistance of several people in the conception and the preparation of this book. Special thanks go to Professor Dominique Marchio, who provided valuable technical and financial support during my visit to the Centre Energetique at the Ecole National des Mines de Paris (ENMP). The input of Irene Arditi and Cederic Carretero as well as encouragement of Dr. Frank Kreith, Prof. Jerome Adnot, and Prof. Denis Clodic are also acknowledged. The help of Pirawas Chuangchid and Pyeongchan Ihm in formatting the book is highly appreciated. Finally, I am greatly indebted to my wife Hajer and my children for their continued patience and support throughout the preparation of this book.

Moncef Krarti

October 2000

1
INTRODUCTION TO ENERGY AUDIT

Abstract

This chapter provides an overview of a general energy audit procedure that is suitable for commercial and industrial buildings. Today, energy auditing is commonly performed by energy service companies to improve the energy efficiency of buildings. Indeed, energy auditing has a vital role for the success of performance contracting projects. There are several types of energy audits that are commonly performed by energy service engineers with various degrees of complexity. This chapter describes briefly the key aspects of a detailed energy audit and provides a comprehensive and systematic approach to identify and recommend cost effective energy conservation measures for buildings.

1.1   Introduction

Since the oil embargo of 1973, significant improvements have been made in the energy efficiency of new buildings. However, the vast majority of the existing stock of buildings are more than a decade old and do not meet current energy efficiency construction standards (EIA, 1991). Therefore, energy retrofits of existing buildings will be required for decades to come if the overall energy efficiency of the building stock is to meet the standards.

Investing to improve the energy efficiency of buildings provides an immediate and relatively predictable positive cash flow resulting from lower energy bills. In addition to the conventional financing options available to owners and building operators (such as loans and leases), other methods are available to finance energy retrofit projects for buildings. One of these methods that is becoming increasingly common is performance contracting in which payment for a retrofit project is contingent upon its successful outcome. Typically, an energy services company (ESCO) assumes all the risks for a retrofit project by performing the engineering analysis and obtaining the initial capital to purchase and install equipment needed for energy efficiency improvements. Energy auditing is an important step used by energy service companies to insure the success of their performance contracting projects.

Moreover, several large industrial and commercial buildings have established internal energy management programs based on energy audits to reduce waste in energy use or to comply with the specifications of some regulations and standards. Other building owners and operators take advantage of available financial incentives typically offered by utilities or state agencies to perform energy audits and implement energy conservation measures.

In the 1970s, building energy retrofits consisted of simple measures such as shutting off lights, turning down heating temperatures, turning up air conditioning temperatures, and reducing the hot water temperatures. Today, building energy management includes a comprehensive evaluation of almost all the energy systems within a facility. Therefore, the energy auditor should be aware of key energy issues such as the subtleties of electrical utility rate structures and of the latest building energy-efficiency technologies and their applications.

This chapter describes a general but systematic procedure for energy auditing suitable for both commercial buildings and industrial facilities. Some of the commonly recommended energy conservation measures are briefly discussed. A case study for an office building is presented to illustrate the various tasks involved in an energy audit. Finally, an overview is provided to outline the existing methods for measurement and verification of energy savings incurred by the implementation of energy conservation measures.

1.2   Types of Energy Audits

The term "energy audit" is widely used and may have different meaning depending on the energy service companies. Energy auditing of buildings can range from a short walk-through of the facility to a detailed analysis with hourly computer simulation. Generally, four types of energy audits can be distinguished as briefly described below:

1.2.1   Walk-Through Audit

This audit consists of a short on-site visit of the facility to identify areas where simple and inexpensive actions can provide immediate energy use and/or operating cost savings. Some engineers refer to these types of actions as operating and maintenance (O&M) measures. Examples of O&M measures include setting back heating set point temperatures, replacing broken windows, insulating exposed hot water or steam pipes, and adjusting boiler fuel-air ratio.

1.2.2   Utility Cost Analysis

The main purpose of this type of audit is to carefully analyze the operating costs of the facility. Typically, the utility data over several years are evaluated to identify the patterns of energy use, peak demand, weather effects, and potential for energy savings. To perform this analysis, it is recommended that the energy auditor conducts a walk-through survey to get acquainted with the facility and its energy systems.

It is important that the energy auditor understands clearly the utility rate structure that applies to the facility for several reasons including:

- To check the utility charges and insure that no mistakes were made in calculating the monthly bills. Indeed, the utility rate structures for commercial and industrial facilities can be quite complex with ratchet charges and power factor penalties.
- To determine the most dominant charges in the utility bills. For instance, peak demand charges can be a significant portion of the utility bill especially when ratchet rates are applied. Peak shaving measures can be then recommended to reduce these demand charges.
- To identify whether or not the facility can benefit from using other utility rate structures to purchase cheaper fuel and reduce its operating costs. This analysis can provide a significant reduction in the utility bills especially with implementation of the electrical deregulation and the advent of real time pricing (RTP) rate structures.

Moreover, the energy auditor can determine whether or not the facility is prime for energy retrofit projects by analyzing the utility data. Indeed, the energy use of the facility can be normalized and compared to indices (for instance, the energy use per unit of floor area –for commercial buildings- or per unit of a product –for industrial facilities- as discussed in Chapter 4).

1.2.3   Standard Energy Audit

The standard audit provides a comprehensive energy analysis for the energy systems of the facility. In addition to the activities described for the walk-through audit and for the utility cost analysis described above, the standard energy audit includes the development of a baseline for the energy use of the facility and the evaluation of the energy savings and the cost effectiveness of appropriately selected energy conservation measures. The step by step approach of the standard energy audit is similar to that of the detailed energy audit which is described later on in the following section.

Typically, simplified tools are used in the standard energy audit to develop baseline energy models and to predict the energy savings of energy conservation measures. Among these tools are the degree-day methods, and linear regression models (Fels, 1986). In addition, a simple payback analysis is generally performed to determine the cost-effectiveness of energy conservation measures.

1.2.4   Detailed Energy Audit

This audit is the most comprehensive but also time-consuming energy audit type. Specifically, the detailed energy audit includes the use of instruments to measure energy use for the whole building and/or for some energy systems within the building (for instance by end uses: lighting systems, office equipment, fans, chillers, etc.). In addition, sophisticated computer simulation programs are typically considered for detailed energy audits to evaluate and recommend energy retrofits for the facility.

The techniques available to perform measurements for an energy audit are diverse. During an on-site visit, hand-held and clamp-on instruments can be used to determine the variation of some building parameters such as the indoor air temperature, the luminance level, and the electrical energy use. When long-term measurements are needed, sensors are typically used and connected to a data-acquisition system so measured data can be stored and be remotely accessible. Recently, non-intrusive load monitoring (NILM) techniques have been proposed (Shaw et al., 1988). The NILM technique can determine the real time energy use of the significant electrical loads in a facility using only a single set of sensors at the facility service entrance. The minimal effort associated with using the NILM technique when compared to the traditional sub-metering approach (which requires separate set of sensors to monitor energy consumption for each end-use) makes the NILM a very attractive and inexpensive load monitoring technique for energy service companies and facility owners.

The computer simulation programs used in the detailed energy audit can typically provide the energy use distribution by load type (i.e., energy use for: lighting, fans, chillers, boilers, etc.). They are often based on dynamic thermal performance of the building energy systems and require typically high level of engineering expertise and training. These simulation programs range from those based on the bin method (Knebel, 1983) to those that provide hourly building thermal and electrical loads such as DOE-2 (LBL, 1980). The reader is referred to Chapter 4 for more detailed discussion of the energy analysis tools that can be used to estimate energy and cost savings attributed to energy conservation measures.

In the detailed energy audit, more rigorous economical evaluation of the energy conservation measures are generally performed. Specifically, the cost-effectiveness of energy retrofits may be determined based on the life-cycle cost (LCC) analysis rather than the simple payback period analysis. Life-cycle cost (LCC) analysis takes into account a number of economic parameters such as interest, inflation, and tax rates. Chapter 3 describes some of the basic analysis tools that are often used to evaluate energy efficiency projects.

1.3   General Procedure for a Detailed Energy Audit

To perform an energy audit, several tasks are typically carried out depending on the type of the audit and the size and function of the audited building. Some of the tasks may have to be repeated, reduced in scope, or even eliminated based on the findings of other tasks. Therefore, the execution of an energy audit is often not a linear process and is rather iterative. However, a general procedure can be outlined for most buildings.

Step 1: Building and Utility Data Analysis

The main purpose of this step is to evaluate the characteristics of the energy systems and the patterns of energy use for the building. The building characteristics can be collected from the architectural/mechanical/electrical drawings and/or from discussions with building operators. The energy use patterns can be obtained from a compilation of utility bills over several years. Analysis of the historical variation of the utility bills allows the energy auditor to determine if there are any seasonal and weather effects on the building energy use. Some of the tasks that can be performed in this step are presented below with the key results expected from each task noted:

♦ Collect at least three years of utility data [*to identify a historical energy use pattern*]
♦ Identify the fuel types used (electricity, natural gas, oil, etc.) [*to determine the fuel type that accounts for the largest energy use*]
♦ Determine the patterns of fuel use by fuel type [*to identify the peak demand for energy use by fuel type*]
♦ Understand utility rate structure (energy and demand rates) [*to evaluate if the building is penalized for peak demand and if cheaper fuel can be purchased*]
♦ Analyze the effect of weather on fuel consumption [*to pinpoint any variations of energy use related to extreme weather conditions*]
♦ Perform utility energy use analysis by building type and size (building signature can be determined including energy use per unit area [*to compare against typical indices*].

Step 2: Walk-through Survey

From this step, potential energy savings measures should be identified. The results of this step are important since they determine if the building warrants any further energy auditing work. Some of the tasks involved in this step are:

♦ Identify the customer concerns and needs
♦ Check the current operating and maintenance procedures
♦ Determine the existing operating conditions of major energy use equipment (lighting, HVAC systems, motors, etc.)
♦ Estimate the occupancy, equipment, and lighting (energy use density and hours of operation)

Step 3: Baseline for Building Energy Use

The main purpose of this step is to develop a base-case model that represents the existing energy use and operating conditions for the building. This model is to be used as a reference to estimate the energy savings incurred from appropriately selected energy conservation measures. The major tasks to be performed during this step are:

♦ Obtain and review architectural, mechanical, electrical, and control drawings
♦ Inspect, test, and evaluate building equipment for efficiency, performance, and reliability
♦ Obtain all occupancy and operating schedules for equipment (including lighting and HVAC systems)
♦ Develop a baseline model for building energy use
♦ Calibrate the baseline model using the utility data and/or metered data.

Step 4: Evaluation of Energy Savings Measures

In this step, a list of cost-effective energy conservation measures is determined using both energy savings and economic analysis. To achieve this goal, the following tasks are recommended:

♦ Prepare a comprehensive list of energy conservation measures (using the information collected in the walk-through survey)
♦ Determine the energy savings due to the various energy conservation measures pertinent to the building using the baseline energy use simulation model developed in phase 3
♦ Estimate the initial costs required to implement the energy conservation measures
♦ Evaluate the cost-effectiveness of each energy conservation measure using an economical analysis method (simple payback or life-cycle cost analysis).

Tables 1.1 and 1.2 provide summaries of the energy audit procedure recommended respectively for commercial buildings, and for industrial facilities. Energy audits for thermal and electrical systems are separated since they are typically subject to different utility rates.

Table 1.1: Energy Audit Summary for Residential and Commercial Buildings

PHASE	THERMAL SYSTEMS	ELECTRIC SYSTEMS
UTILITY ANALYSIS	<ul><li>Thermal energy use profile (building signature)</li><li>Thermal energy use per unit area (or per student for schools or per bed for hospitals)</li><li>Thermal energy use distribution (heating, DHW, process, etc.)</li><li>Fuel types used</li><li>Weather effect on thermal energy use</li><li>Utility rate structure</li></ul>	<ul><li>Electrical energy use profile (building signature)</li><li>Electrical energy use per unit area (or per student for schools or per bed for hospitals)</li><li>Electrical energy use distribution (cooling, lighting, equipment, fans, etc.)</li><li>Weather effect on electrical energy use</li><li>Utility Rate structure (energy charges, demand charges, power factor penalty, etc.)</li></ul>
ON-SITE SURVEY	<ul><li>Construction Materials (thermal resistance type and thickness)</li><li>HVAC system type</li><li>DHW system</li><li>Hot water /steam use for heating</li><li>Hot water/steam for cooling</li><li>Hot water/steam for DHW</li><li>Hot water/steam for specific applications (hospitals, swimming pools, etc.)</li></ul>	<ul><li>HVAC system type</li><li>Lighting type and density</li><li>Equipment type and density</li><li>Energy use for heating</li><li>Energy use for cooling</li><li>Energy use for lighting</li><li>Energy use for equipment</li><li>Energy use for air handling</li><li>Energy use for water distribution</li></ul>
ENERGY USE BASELINE	<ul><li>Review architectural, mechanical, and control drawings</li><li>Develop a base-case model (using any baselining method ranging from very simple to more detailed tools)</li><li>Calibrate the base-case model (using utility data or metered data)</li></ul>	<ul><li>Review architectural, mechanical, electrical, and control drawings</li><li>Develop a base-case model (using any baselining method ranging from very simple to more detailed tools)</li><li>Calibrate the base-case model (using utility data or metered data)</li></ul>
ENERGY CONSERVATION MEASURES	<ul><li>Heat recovery system (heat exchangers)</li><li>Efficient heating system (boilers)</li><li>Temperature Setback</li><li>EMCS</li><li>HVAC system retrofit</li><li>DHW use reduction</li><li>Cogeneration</li></ul>	<ul><li>Energy efficient lighting</li><li>Energy efficient equiment (computers)</li><li>Energy efficient motors</li><li>HVAC system retrofit</li><li>EMCS</li><li>Temperature Setup</li><li>Energy efficient cooling system (chiller)</li><li>Peak demand shaving</li><li>Thermal Energy Storage System</li><li>Cogeneration</li><li>Power factor improvement</li><li>Reduction of harmonics</li></ul>

Table 1.2: Energy Audit Summary for Industrial Facilities

PHASE	THERMAL SYSTEMS	ELECTRIC SYSTEMS
UTILITY ANALYSIS	• Thermal energy use profile (building signature) • Thermal energy use per unit of a product • Thermal energy use distribution (heating, process, etc.) • Fuel types used • Analysis of the thermal energy input for specific processes used in the production line (such as drying) • Utility rate structure	• Electrical energy use profile (building signature) • Electrical energy use per unit of a product • Electrical energy use distribution (cooling, lighting, equipment, process, etc.) • Analysis of the electrical energy input for specific processes used in the production line (such as drying) • Utility Rate structure (energy charges, demand charges, power factor penalty, etc.)
ON-SITE SURVEY	• List of equipment that uses thermal energy • Perform heat balance of the thermal energy • Monitor of thermal energy use of all or part of the equipment • Determine the by-products of thermal energy use (such as emissions and solid waste)	• List of equipment that uses electrical energy • Perform heat balance of the electrical energy • Monitor of electrical energy use of all or part of the equipment • Determine the by-products of electrical energy use (such as pollutants)
ENERGY USE BASELINE	• Review mechanical drawings and production flow charts • Develop a base-case model (using any baselining method) • Calibrate the base-case model (using utility data or metered data)	• Review electrical drawings and production flow charts • Develop a base-case model (using any baselining method) • Calibrate the base-case model (using utility data or metered data)
ENERGY CONSERVATION MEASURES	• Heat recovery system • Efficient heating and drying system • EMCS • HVAC system retrofit • Hot water and steam use reduction • Cogeneration (possibly with solid waste from the production line)	• Energy efficient motors • Variable speed drives • Air compressors • Energy efficient lighting • HVAC system retrofit • EMCS • Cogeneration (possibly with solid waste from the production line) • Peak demand shaving • Power factor improvement • Reduction of harmonics

1.4   Common Energy Conservation Measures

In this section some energy conservation measures (ECMs) commonly recommended for commercial and industrial facilities are briefly discussed. It should be noted that the list of ECMs presented below does not pretend to be exhaustive nor comprehensive. It is provided merely to indicate some of the options that the energy auditor can consider when performing an energy analysis of a commercial or an industrial facility. More discussion of energy efficiency measures for various building energy systems is provided in later chapters of this book. However, it is strongly advised that the energy auditor keeps abreast of any new technologies that can improve the building energy efficiency. Moreover, the energy auditor should recommend the ECMs only based on a sound economical analysis for each ECM.

1.4.1   Building Envelope

For some buildings, the envelope (i.e., walls, roofs, floors, windows, and doors) can have an important impact on the energy used to condition the facility. The energy auditor should determine the actual characteristics of the building envelope. During the survey, a descriptive sheet for the building envelope should be established to include information such as materials of construction (for instance, the level of insulation in walls, floors, and roofs) and the area and the number of building envelope assemblies (for instance, the type and the number of panes for the windows). In addition, comments on the repair needs and recent replacement should be noted during the survey.

Some of the commonly recommended energy conservation measures to improve the thermal performance of the building envelope are:

a) Addition of Thermal Insulation. For building surfaces without any thermal insulation, this measure can be cost effective.
b) Replacement of Windows. When windows represent a significant portion of the exposed building surfaces, using more energy efficient windows (high R-value, low-emissivity glazing, air tight, etc.) can be beneficial in both reducing the energy use and improving the indoor comfort level.
c) Reduction of Air Leakage. When infiltration load is significant, leakage area of the building envelope can be reduced by simple and inexpensive weather-stripping techniques.

The energy audit of the envelope is especially important for residential buildings. Indeed, the energy use from residential buildings is dominated by weather since heat gain and/or loss from direct conduction of heat or from air infiltration/exfiltration through building surfaces accounts for a major portion (50 to 80%) of the energy consumption. For commercial buildings, improvements to the building envelope are often not cost-effective due to the fact that modifications to the building envelope (replacing windows, adding thermal insulation in walls) are typically considerably expensive. However, it is recommended to systematically audit the envelope components not only to determine the potential for energy savings but also to insure the integrity of its overall condition. For instance, thermal bridges – if present – can lead to heat transfer increase and to moisture condensation. The moisture condensation is often more damaging and costly than the increase in heat transfer since it can affect the structural integrity of the building envelope.

1.4.2   Electrical Systems

For most commercial buildings and a large number of industrial facilities, the electrical energy cost constitutes the dominant part of the utility bill. Lighting, office equipment, and motors are the electrical systems that consume the major part of energy in commercial and industrial buildings.

a)   Lighting. Lighting for a typical office building represents on average 40% of the total electrical energy use. There are a variety of simple and inexpensive measures to improve the efficiency of lighting systems. These measures include the use of energy efficient lighting lamps and ballasts, the addition of reflective devices, delamping (when the luminance levels are above the levels recommended by the standards), and the use of daylighting controls. Most lighting measures are especially cost-effective for office buildings for which payback periods are less than one year.

b)   Office Equipment. Office equipment constitutes the fastest growing part of electrical loads, especially in commercial buildings. Office equipment includes computers, fax machines, printers, and copiers. Today, there are several manufacturers that provide energy-efficient office equipment (such as those that comply with the US EPA Energy Star specifications). For instance, energy-efficient computers automatically switch to a low-power "sleep" mode or off mode when not in use.

c)   Motors. The energy cost to operate electric motors can be a significant part of the operating budget of any commercial and industrial building. Measures to reduce the energy cost of using motors include reducing operating time (turning off unnecessary equipment), optimizing motor systems, using controls to match motor output with demand, using variable speed drives for air and water distribution, and installing energy-efficient motors. Table 1.3 provides typical efficiencies for several motor sizes.

Table 1.3: Typical efficiencies of motors

MOTOR SIZE (HP)	STANDARD EFFICIENCY	HIGH EFFICIENCY
1	72%	81%
2	76%	84%
3	77%	89%
5	80%	89%
7.5	82%	89%
10	84%	89%
15	86%	90%
20	87%	90%
30	88%	91%
40	89%	92%
50	90%	93%

In addition to the reduction in the total facility electrical energy use, retrofits of the electrical systems decrease space cooling loads and therefore further reduce the electrical energy use in the building. These cooling energy reductions as well as possible increases in thermal energy use (for space heating) should be accounted for when evaluating the cost-effectiveness of improvements in lighting and office equipment.

1.4.3 HVAC Systems

The energy use due to HVAC systems can represent 40% of the total energy consumed by a typical commercial building. The energy auditor should obtain the characteristics of major HVAC equipment to determine the condition of the equipment, their operating schedule, their quality of maintenance, and their control procedures. A large number of measures can be considered to improve the energy performance of both primary and secondary HVAC systems. Some of these measures are listed below:

a) Setting Up/Back Thermostat Temperatures: When appropriate, setback of heating temperatures can be recommended during unoccupied periods. Similarly, setup of cooling temperatures can be considered.
b) Retrofit of Constant Air Volume Systems: For commercial buildings, variable air volume (VAV) systems should be considered when the existing HVAC systems rely on constant volume fans to condition part or the entire building.

c) Installation of Heat Recovery Systems: Heat can be recovered from some HVAC equipment. For instance, heat exchangers can be installed to recover heat from air handling unit (AHU) exhaust air streams and from boiler stacks.

d) Retrofit of Central Heating Plants: The efficiency of a boiler can be drastically improved by adjusting the fuel air ratio for proper combustion. In addition, installation of new energy-efficient boilers can be economically justified when old boilers are to be replaced.

e) Retrofit of Central Cooling Plants: Currently, there are several chillers that are energy-efficient and easy to control and operate and are suitable for retrofit projects.

It should be noted that there is a strong interaction between various components of heating and cooling system. Therefore, a whole-system analysis approach should be followed when retrofitting a building HVAC system. Optimizing the energy use of a central cooling plant (which may include chillers, pumps, and cooling towers) is one example of using a whole-system approach to reduce the energy use for heating and cooling buildings.

1.4.4   Compressed Air Systems

Compressed air has become an indispensable tool for most manufacturing facilities. Its uses range from air-powered hand tools and actuators to sophisticated pneumatic robotics. Unfortunately, staggering amounts of compressed air are currently wasted in a large number of facilities. It is estimated that only 20 to 25% of input electrical energy is delivered as useful compressed air energy. Leaks are reported to account for 10 to 50% of the waste while misapplication accounts for 5 to 40% of loss in compressed air (Howe and Scales, 1998).

To improve the efficiency of compressed air systems, the auditor can consider several issues including whether or not compressed air is the right tool for the job (for instance, electric motors are more energy efficient than air-driven rotary devices), how compressed air is applied (for instance, lower pressures can be used to supply pneumatic tools), how it is delivered and controlled (for instance, the compressed air needs to be turned off when the process is not running), and how a compressed air system is managed (for each machine or process, the cost of compressed air needs to be known to identify energy and cost savings opportunities).

1.4.5   Energy Management Controls

With the constant decrease in the cost of computer technology, automated control of a wide range of energy systems within commercial and industrial buildings is becoming increasingly popular and cost-effective. An energy management and control system (EMCS) can be designed to control and reduce the building energy consumption within a facility by continuously monitoring the energy use of various equipment and making appropriate adjustments. For instance, an EMCS can automatically monitor and adjust indoor ambient temperatures, set fan speeds, open and close air handling unit dampers, and control lighting systems.

If an EMCS is already installed in the building, it is important to recommend a system tune-up to insure that the controls are operating properly. For instance, the sensors should be calibrated regularly in accordance with manufacturers' specifications. Poorly calibrated sensors may cause increase in heating and cooling loads and may reduce occupant comfort.

1.4.6   Indoor Water Management

Water and energy savings can be achieved in buildings by using water-saving fixtures instead of the conventional fixtures for toilets, faucets, showerheads, dishwashers, and clothes washers. Savings can also be achieved by eliminating leaks in pipes and fixtures.

Table 1.4 provides typical water use of conventional and water-efficient fixtures for various end-use. In addition, Table 1.4 indicates the hot water use by each fixture as a fraction of the total water. With water-efficient fixtures, savings of 50% of water use can be achieved for toilets, showers, and faucets.

Table 1.4: Usage characteristics of water-using fixtures

End-Use	Conventional Fixtures	Water-Efficient Fixtures	Usage Pattern	% Hot water
Toilets	3.5 gal/flush	1.6 gal/flush	4 flushes/pers/day	0%
Showers	5.0 gal/min	2.5 gal/min	5 min./shower	60%
Faucets	4.0 gal/min	2.0 gal/min	2.5 min/pers/day	50%
Dishwashers	14.0 gal/load	8.5 gal/load	0.17 loads/pers/day	100%
Clothes Washers	55.0 gal/load	42.0 gal/load	0.3 loads/pers/day	25%
Leaks	10% of total use	2% of total use	N/A	50%

1.4.7   New Technologies

The energy auditor may consider the potential of implementing and integrating new technologies within the facility. It is therefore important that the energy auditor understands these new technologies and knows how to apply them. New technologies that can be considered for commercial and industrial buildings include:

a) Building Envelope technologies: Recently several materials and systems have been proposed to improve the energy efficiency of the building envelope and especially windows including:
  - Spectrally selective glasses which can optimize solar gains and shading effects.
  - Chromogenic glazings which change properties automatically depending on temperature and/or light level conditions (similar to sunglasses that become dark in sunlight).
  - Building integrated photovoltaic panels that can generate electricity while absorbing solar radiation and reducing heat gain through building envelope (typically roofs).

b) Light Pipe technologies: While the use of daylighting is straightforward for perimeter zones that are near windows, it is not usually feasible for interior spaces, particularly those without any skylights. Recent but still emerging technologies "pipe" light from roof or wall-mounted collectors to interior spaces that are not close to windows or skylights.

c) HVAC systems and controls: Several strategies can be considered for energy retrofits including
  - Heat recovery technologies such as rotary heat wheels and heat pipes can recover 50 to 80% of the energy used to heat or cool ventilation air supplied to the building.
  - Dessicant-based cooling systems are now available and can be used in buildings with large dehumidification loads (such as hospitals, swimming pools, and supermarket fresh produce areas) during long periods.
  - Geothermal heat pumps can provide an opportunity to take advantage of the heat stored underground to condition building spaces.
  - Thermal energy storage (TES) systems offer a means of using less expensive off-peak power to produce cooling or heating to condition the building during on-peak periods. Several optimal control strategies have been developed in recent years to maximize the cost savings of using TES systems.

d) Cogeneration: This is not really a new technology. However, recent improvements in its combined thermal and electrical efficiency made cogeneration cost effective in several applications including institutional buildings such hospitals and universities.

1.5   Case Study

To illustrate the energy audit process described above, a case study is presented in this section. The activities performed for each step of the energy audit are briefly described. For more details about the case study, the reader is referred to Kim et al. (1998).

The building analyzed in this case study is a medium size office building located in Seoul, Korea. Figure 1.1 shows the front view of the building.

Figure 1.1: Front view of the audited office building

Step 1: Building and Utility Data Analysis

The first step in the building energy audit process is to collect all available information about the energy systems and the energy use pattern of the building. This information was collected before the field survey. In particular, from the architectural/mechanical/electrical drawings and utility bills, the following information and engineering data were gathered:

Building Characteristics: The building is a 26-story office building with 2-story penthouse and 4-story basement. It is located in downtown Seoul, Korea. The structure of the building consists of modular concrete and steel frame. The building area is 3,920 m^2 and the site area is 6,555 m^2. Single glazed windows are installed throughout the building. Figure 1.2 shows a typical floor plan of the building. Table 1.5 describes the various construction materials used throughout the building.

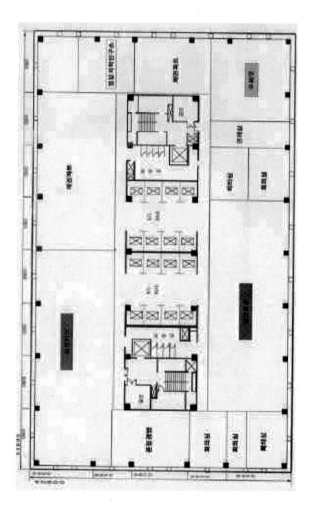

Figure 1.2: Typical floor plan of the audited office building

Energy Use: Figure 1.3 summarizes the monthly electrical energy use of the building for 1993. The monthly average dry-bulb outdoor air temperatures recorded during 1993 are also presented in Figure 1.3. It is clear that the electrical energy use increases during the summer months (June through

October) when the outdoor temperatures are high. During the other months, the electrical energy use is almost constant and can be attributed mostly to lighting and office equipment. Preliminary analysis of the metered building energy use indicated that natural gas consumption is inconsistent from month to month. For example, gas consumption during January is six times higher than during December, even though the weather conditions are similar for both months. Therefore, the recorded data for the natural gas use were considered unreliable and only metered electrical energy use data were analyzed.

Table 1.5: Building construction materials.

Component	Materials
Exterior wall	5 cm tile 16 cm concrete 2.5 cm foam insulation 0.6 cm finishing material
Roof	5 cm light weight concrete 15 cm concrete 2.5 cm foam insulation
Interior wall	2 cm finishing cement mortar 19 cm concrete block 2 cm finishing cement mortar
Glazing	1.2 cm thick single pane glazing
Underground wall	25 cm concrete asphalt shingle air-space 10 cm brick 2 cm finishing cement mortar
Underground floor	15 cm concrete asphalt shingle 12 cm concrete 2 cm finishing cement mortar

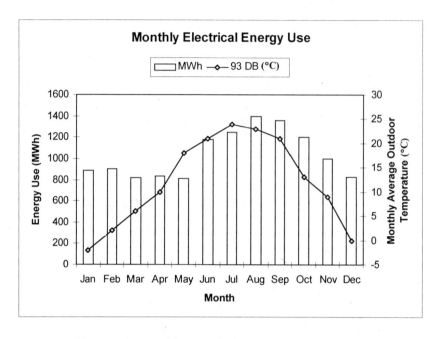

Figure 1.3: Monthly Actual Electrical Energy Consumption.

Step 2: On-site Survey

A one day field survey was conducted with the assistance of the building operator during the summer of 1996. During the survey, several useful and revealing information and engineering data were collected. For instance:

➢ It was found that the building had been retrofitted with energy efficient lighting systems. The measurement of luminance levels throughout the working areas indicated adequate lighting. To determine an estimate of the energy use for lighting, the number and type of lighting fixtures were recorded.

➢ It was observed that the cooling and heating temperature set points were set to be 25.5°C and 24.5 °C, respectively. However, indoor air temperature and relative humidity measurements during the field survey revealed that during the afternoon, the thermal conditions are uncomfortable in several office spaces with average air dry-bulb temperature of 28 °C and relative humidity of 65%. A discussion with the building operator indicated that the chillers are no longer able to meet the cooling loads after the addition of several computers in the building during the last few years. As a solution an ice storage tank was then added to reduce the peak cooling load.

➤ It was discovered during the survey, that the building is heated and cooled simultaneously by two systems: Constant Air Volume (CAV) and Fan Coil Unit (FCU) systems. The CAV system is complemented by the FCU system as necessary. Two air-handling units serve the entire building, and about 58 FCUs are located on each floor.

➤ The heating and cooling plant consists of 3 boilers, 6 chillers, 3 cooling towers, and one ice storage tank. The capacity of the boilers and the chillers is provided below:

> Boilers: 13 MBtuh (2 units) and 3.5 MBtuh (1 unit)
> Chillers: 215 Tons (5 units) and 240 Tons (1 unit)

The thermal energy storage (TES) system consists of a brine ice-on-coil tank. The hours of charging and discharging are 10 and 13, respectively. The TES system is currently controlled using simple and non-predictive storage-priority controls.

The building has relatively high internal heat gains. Some of the building internal heat gain sources are listed in Table 1.6. Operating schedules were based on the discussion with the building operators and on observations during the field survey.

Table 1.6: Internal heat gain level for the office building

Internal Heat Gain	Design Load
Occupancy	17 m²/person;
	Latent heat gain: 45W
	Sensible heat gain: 70W
Lighting	14W/ m²
Equipment	16W/ m²
Ventilation	14.7 CFM/person

Step 3: Energy Use Baseline Model

To model the building using DOE-2, each floor was divided into 2 perimeter and 2 core zones. Figure 1.4 shows the zone configuration used to model the building floors. The main reason for this zoning configuration is the lack of flexibility in the DOE-2 SYSTEMS program. Although the actual building is conditioned by the combination of constant air volume (CAV) and fan coil unit (FCU) systems, the SYSTEM module of DOE-2.1E cannot model two different types of HVAC systems serving one zone. Therefore, a simplification has been made to simulate the actual HVAC system of the building. This simplification consists of the following: the perimeter zone is conditioned by the FCUs, while the core zone is conditioned by CAV. Since all the FCUs are located at the perimeter, this simplification is consistent with the actual HVAC system operation.

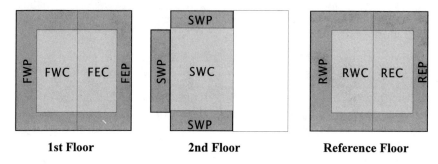

1st Floor   **2nd Floor**   **Reference Floor**

Figure 1.4: Building zoning configuration for DOE-2 computer simulation.

Figure 1.5 shows the monthly electrical energy consumption predicted by the DOE-2 base model and the actual energy use recorded in 1993 for the building. It shows that DOE-2 predicts the actual energy use pattern of the building fairly well, except for the months of September and October. The difference between the annual metered energy use in the building and the annual predicted electricity use by the DOE-2 base-case model is about 762 MWh. DOE-2 predicts that the building consumes 6% more electricity than the actual metered annual energy use. To develop the DOE-2 base-case model, a TRY-type weather file of Seoul was created using the raw weather data collected for 1993. Using the DOE-2 base-case model, a number of ECOs can be evaluated.

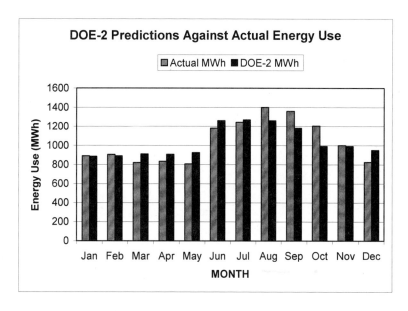

Figure 1.5: Comparison of DOE-2 prediction and actual building electrical

energy use.

Figure 1.6 shows the distribution by end-uses of the building energy use. The electrical energy consumption of the building is dominated by lighting and equipment. The electricity consumption for lighting and office equipment represents about 75% of the total building electricity consumption. As mentioned earlier, a recent lighting retrofit has been performed in the building using electronic ballasts and energy efficient fluorescent fixtures. Therefore, it was decided to not consider a lighting retrofit as an ECO for this study. The electricity consumption for cooling is about 13.1%. The ECOs selected for this building attempt to reduce the cooling loads in order to improve indoor thermal comfort as well as save building energy cost.

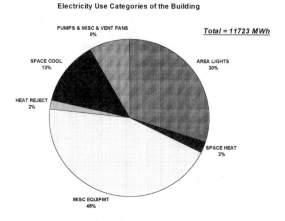

Figure 1.6: Electricity use distribution

Step 4: Evaluation of Energy Conservation Opportunities (ECOs)

Based on the evaluation of the energy use pattern of the building, several energy conservation opportunities (ECOs) for the building were analyzed. Among the ECOs considered in the study, six successfully reduced energy consumption:

1) ECO #1: CAV to VAV conversion: The present AHU fans are all constant speed fans. They supply conditioned air through a constant volume air supply system to the conditioned zones. The system is designed to supply enough air to heat or cool the building under design conditions. Under non-design conditions, more air than needed is supplied. Changing the system to a variable-air-volume (VAV) system would reduce the amount of air supplied by the AHUs and result in less energy to condition the various zones. For this ECO, a constant-volume reheat fan system assigned for the core zones in the building was changed to a VAV system. In particular, VAV boxes – controlled by the space thermostat – are proposed to vary the amount of conditioned air supplied to the building zones to control the indoor temperature. Both labor and equipment costs were included in the estimation of the payback period for this measure.

2) ECO #2: Optimal ice storage control: The current TES system is operated using a non predictive storage-priority control. To improve the benefits of the TES system, a near-optimal controller is suggested. This ECO is analyzed to determine if the cost of electrical energy consumption for the building is reduced when a near-optimal control strategy is used. To

determine the savings of this option, the simulation environment developed by Henze et al. (1997a) is used. This simulation environment is based on dynamic programming technique and determines the best operating controls for the TES system given the cooling and non-cooling load profiles and electrical rate structure. No DOE-2 simulation is performed for this ECO. The results of the dynamic programming simulation indicate that an energy use reduction of 5% can be achieved using a near-optimal control in lieu of the storage-priority control. To implement this near-optimal control, a predictor is required to determine future building cooling/non-cooling loads. An example of such a predictor could be based on neural networks (Kreider et al. 1995). The labor cost and the initial cost of adding some sensors and a computer was considered to determine the payback period for this measure.

3) ECO #3: Glazing retrofit: For this building, low-e glazing systems are considered to reduce the internal heat gain due to solar radiation. Thus, the cooling load is reduced. In addition, the increased U-value of the glazing reduces the heating load. For this ECO, the existing single pane windows with the glass conductance of 6.17 (W/m^2-K) and the shading coefficient of 0.69 were changed to the double-pane windows. These double-pane windows reduce both the glass conductance and the shading coefficient to 1.33 (W/m^2-K) and 0.15, respectively.

4) ECO #4: Indoor temperature set back/up: In this ECO, the impact of the indoor temperature setting on the building energy use is analyzed using DOE-2 simulation program; the heating temperature was set from 24.5°C to 22.5 °C and the cooling temperature was set from 25.5 °C to 27.5 °C. There is only labor cost associated with this measure.

5) ECO #5: Motor replacement: Increasing the efficiency of the motors for fans and pumps can reduce the total electric energy consumption in the building. In this ECO, the existing efficiencies for the motors were assumed to range from 0.85 (for 10 HP motors) to 0.90 (for 50 HP motors). Energy efficient motors have efficiencies that range from 0.91 (for 10 HP motors) to 0.95 (for 50 HP motors). Only the differential cost was considered in the economic analysis.

6) ECO #6: Daylighting control plan: A continuous dimming control would regulate the light level so that the luminance level inside the zones remains constant. The electricity consumption of the building can be significantly reduced, while the gas consumption can be slightly increased because of the reduced space heat gain from the lighting system. For this ECO, a daylighting system with continuous dimming control was considered for perimeter office zones.

The impact of the selected ECOs on the electricity use in the building as predicted by DOE-2 simulations is shown in Table 1.7. Based on these results, "converting CAV to VAV" and "implementing daylighting control system with dimming control" reduces the total electricity consumption of the building 5.2% and 7.3%, respectively. These savings are significant considering that the electricity consumption for the cooling plant alone is about 13.1% of the total electricity consumption of the building.

Table 1.7: Economic analysis of the ECOs

	Electricity Cost (Mwon)	LNG Cost (Mwon)	Total Cost (MWon)	Capital Cost (MWon)	Saving (%)	Savings (Mwon)	Payback Period (years)
Base Case	984.4	139.1	1123.5	-	-	-	-
ECO #1	940.8	49.8	990.5	465.5	11.8	133.0	3.5
ECO #2	979.1	139.1	1118.2	42.4	0.6	5.3	8.0
ECO #3	977.9	126.9	1104.8	280.5	1.7	18.7	15.0
ECO #4	983.7	106.4	1090.1	16.7	3.0	33.4	0.5
ECO #5	972.6	138.7	1111.4	60.5	1.1	12.1	5.0
ECO #6	911.6	144.8	1056.4	268.4	6.0	67.1	4.0

The economic analysis was performed using the utility rate of Seoul, Korea. Table 7 presents the energy cost savings of ECOs in Korean currency (1000 Won = 1 $). In addition to the electricity cost, the natural gas cost was also included in the economic analysis. The natural gas is used only for heating the building. The economic analysis shows that the VAV conversion reduces the total building energy cost by more than 10%, and the daylighting control saves about 6% of the total energy costs.

Step 5: Recommendations

From the results for economic analysis, the VAV conversion (ECO #1), adjustment of temperature set-point (ECO #4), and the daylighting control (ECO #6) are the recommended energy saving opportunities to be implemented for the audited office building.

1.6   Verification Methods of Energy Savings

Energy conservation retrofits are deemed cost-effective based on predictions of energy and cost savings. However, several studies have found that large discrepancies exist between actual and predicted energy savings. Due to the significant increase in the activities of Energy Service Companies (ESCOs), the need became evident for standardized methods for measurement and verification of energy savings. This interest has led to the development of the *North American Energy Measurement and Verification Protocol* published in 1996 and later expanded and revised under the *International Performance Measurement and Verification Protocol.*

In principle, the measurement of the retrofit energy savings can be obtained by simply comparing the energy use during pre- and post-retrofit periods. Unfortunately, the change in energy use between the pre- and post-retrofit periods is not only due to the retrofit itself but also to other factors such as changes in weather conditions, levels of occupancy, and HVAC operating procedures. It is important to account for all these changes to determine accurately the retrofit energy savings.

Several methods have been proposed to measure and verify energy savings of implemented energy conservation measures in commercial and industrial buildings. Chapter 16 describes a number of methods suitable for measurement and verification of energy savings. Some of these techniques are briefly described below:

I. <u>Regression Models:</u> The early regression models used to measure savings adapted the Variable base Degree Day (VBDD) method. Among these early regression models is the PRInceton Scorekeeping Method (PRISM) which uses measured monthly energy consumption data and daily average temperatures to calibrate a linear regression model and determine the best values for non weather dependent consumption, the temperature at which the energy consumption began to increase due to heating or cooling (the change-point or base temperature), and the rate at which the energy consumption increased. Several studies have indicated that the simple linear regression model is suitable for estimating energy savings for residential buildings. However, subsequent work has shown that the PRISM model does not provide accurate estimates for energy savings for most commercial buildings (Ruch and Claridge, 1992). Single-variable (temperature) regression models require the use of at least four-parameter segmented linear or change-point regressions to be suitable for commercial buildings. Katipamula et al. (1994) proposed multiple linear regression models to include as independent variables internal gain, solar radiation, wind, and humidity ratio, in addition to the outdoor temperature. For the buildings considered in their analysis, Katipamula et al. found that wind and solar radiation have small effects on the energy consumption. They also found that internal gains have generally modest impact on energy consumption. Katipamula et al. (1998) discuss in more detail the advantages and the limitations of multivariate regression modeling.

II. <u>Time Variant Models:</u> There are several techniques that are proposed to include the effect of time variation of several independent variables on estimating the energy savings due to retrofits of building energy systems. Among these techniques are the artificial neural networks (Krarti et al.,

1998), Fourier series (Dhar et al., 1998), and non-intrusive load monitoring (Shaw et al., 1998). These techniques are typically involved and require high level of expertise and training.

1.7   Summary

Energy audit of commercial and industrial buildings encompasses a wide variety of tasks and requires expertise in a number of areas to determine the best energy conservation measures suitable for an existing facility. This chapter provided a description of a general but systematic approach to perform energy audits. If followed carefully, the approach helps facilitate the process of analyzing a seemingly endless array of alternatives and complex interrelationships between building and energy system components.

2
ENERGY SOURCES AND UTILITY RATE STRUCTURES

Abstract

In this chapter, an overview of the primary energy sources commonly used in the US is presented. In particular, the evolution over the last three decades of both the consumption levels and the end-use prices for primary energy sources is discussed. Moreover, the chapter outlines the most important features of utility rate schedules used throughout the US for various energy sources with special attention to electric utility rate structures. Several examples are presented to illustrate how the utility bills are typically calculated.

2.1   Introduction

The energy cost is an important part of the economic viability of several energy conservation measures. Therefore, it is crucial that an energy auditor and/or building manager understands how energy costs are determined. Generally, a considerable number of utility rate structures do exist within the same geographical location. Each utility rate structure may include several clauses and charges that sometimes make following the energy billing procedure a complicated task. The complexity of utility rate structures is becoming even more acute with the deregulation of electric industry. However, with new electric utility rate structures (such as real-time-pricing rates), there can be more opportunities to reduce energy cost in buildings.

At the beginning of this chapter, the primary energy sources consumed in the US are described. The presentation emphasizes the energy use and price by end-use sectors including residential, commercial, and industrial applications. In the US, buildings and industrial facilities are responsible respectively for 36% and 38% of the total energy consumption. The transportation sector, which accounts for the remaining 26% of the total US energy consumption, uses mostly fuel products. However, buildings and industries consume predominantly electricity and natural gas. Coal is primarily used as an energy source for electricity generation due its low price.

At the end of this chapter, the various features of utility rate structures available in the US are outlined. More emphasis is given to the electrical rate structures since a significant part of total energy cost in a typical facility is attributed to electricity. However, the price rate structures of other energy sources are discussed. The information provided in this chapter is based on recent surveys of existing utility rate structures. However, the auditor should be aware that most utilities revise their rates on a regular basis. If detailed information is required on the rates available from a specific utility, the auditor should contact the utility directly.

2.2   Energy Resources

The sources of energy used in the US include: coal, natural gas, petroleum products, and electricity. The electricity can be generated from either power plants fueled from primary energy sources (i.e., coal, natural gas, or fuel oil) or from nuclear power plants or renewable energy sources (such as hydroelctric, geothermal, biomass, wind, photovoltaic, and solar thermal sources).

In the US, the energy consumption has fluctuated in response to significant changes in oil prices, economic growth rates, and environmental concerns especially since the oil crisis of early 1970's. For instance, the US energy consumption increased from 66 quadrillion British thermal units (Btu) in 1970 to 94 quadrillion Btu in 1998 (EIA, 1998). Table 3.1 summarizes the changes in the US energy consumption by source from 1972 to 1998.

Table 2.1: Annual US Energy Consumption by Primary Energy Sources in Quadrillion Btu *(Source:* EIA, 1998*)*

Primary Energy Source	1972	1982	1992	1998*
Coal	12.077	15.322	19.158	21.620
Natural Gas	22.469	18.505	20.131	21.840
Petroleum Products	32.947	30.232	33.527	36.537
Nuclear Power	0.584	3.131	6.607	7.157
Renewable Energy	4.478	6.293	6.308	7.073
Total	72.758	73.442	85.495	94.231

* The data for 1998 are preliminary data that may be revised.

It is clear from the data summarized in Table 2.1 that the consumption of coal has increased significantly from 12 quadrillion Btu in 1972 to 21.6 quadrillion Btu in 1998. However, the US consumption of natural gas actually declined from 22.5 quadrillion Btu in 1972 to 18.5 quadrillion in 1982 before increasingly slightly to 21.8 quadrillion Btu in 1998. This decline in natural gas consumption is due to uncertainties about supply and regulatory restrictions especially in the 1980s. Between 1972 and 1998, consumption of other energy sources have generally increased. The increase is from 33.0 quadrillion Btu to 33.5 quadrillion Btu for petroleum products, from 0.6 quadrillion Btu to 7.2 quadrillion Btu for nuclear power, and from 4.5 quadrillion Btu to 7.1 quadrillion Btu for renewable energy which consists almost exclusively of hydroelectric power.

Table 2.2 provides the average nominal energy prices for each primary fuel type. Over the years, coal remains the cheapest energy source. The cost of electricity is still high relative to the other fuel types. As illustrated in Table 2.2, the prices of all energy sources have increased significantly after the energy crisis of 1973. However, over the last decade energy cost of all energy sources has actually declined when inflation is taken into account.

Table 2.2: Consumer Price Estimates for Energy in Nominal Values in US $/Million Btu *(Source:* EIA, 1998*)*

Primary Energy Source	1972	1982	1992	1998*
Coal	0.45	1.73	1.46	1.37
Natural Gas	-	4.23	3.89	3.81
Petroleum Products	1.78	8.35	7.04	7.23
Electricity	5.54	18.16	20.07	20.30

* The data for 1998 are preliminary data that may be revised.

2.2.1   Electricity

Overall Consumption and Price

In the US, coal is the fuel of choice for most existing electrical power plants as shown in Table 2.3. However, gas-fired power plants are expected to be more common in the future due to more efficient and reliable combustion turbines.

Table 2.3: Annual US Electric Energy Generated by Utilities by Primary Energy Sources in Billion kWh *(Source:* EIA, 1998*)*

Primary Energy Source	1972	1982	1992	1998*
Coal	771	1,192	1,576	1,807
Natural Gas	376	305	264	309
Petroleum Products	274	147	89	110
Nuclear Power	54	283	619	674
Renewable Energy	274	314	254	316
Total	1749	2,241	2,797	3,212

* The data for 1998 are preliminary data that may be revised.

The electricity sold by US utilities has increased steadily for all end-use sectors as indicated by the data summarized in Table 2.4. The increase in electricity consumption could be even higher without the various energy conservation programs implemented by the Federal or State governments and utilities. For instance, it is estimated that the demand-side-management (DSM) programs provided by utilities have saved about 35 billion kWh in electrical energy use during 1992 and over 56 billion kWh in 1997 (EIA, 1998).

Table 2.4: Annual US Electric Energy Sold by Utilities by Sector in Billion kWh *(Source:* EIA, 1998)

End-Use Sector	1972	1982	1992	1998*
Residential	539	730	936	1,124
Commercial	359	526	761	949
Industrial	641	745	973	1,047

* The data for 1998 are preliminary data that may be revised.

The prices of electricity for all end-use sectors have actually decreased since 1982 after a recovery period from the 1973 energy crisis as illustrated in Table 2.5. As expected, industrial customers enjoyed the lowest electricity price over the years. Meanwhile, the cost of electricity for residential customers remained the highest.

Table 2.5: Average Retail Prices of Electric Energy Sold by US Utilities by Sector in 1992 cents per kWh *(Source:* EIA, 1998*)*

End-Use Sector	1972	1982	1992	1998*
Residential	7.2	9.8	8.2	7.4
Commercial	6.9	9.8	7.7	6.6
Industrial	3.6	7.1	4.8	4.0

* The data for 1998 are preliminary data that may be revised.

Future of US Electricity Generation

Currently, the electricity market is in the midst of a restructuring period and is becoming increasingly competitive. Several innovative technologies are being considered and tested to generate electricity. A relatively recent approach to produce electricity using small and modular generators is the distributed generation concept. The small generators with capacities in the range of 1 kW to 10 MW can be assembled and relocated in strategic locations (typically near customer sites) to improve power quality and reliability, and offer the flexibility to meet a wide range of customer and distribution system needs.

A number of technologies have emerged in the last decade that allow generation of electricity with reduced waste, cost, and environmental impact. It is expected that these emerging technologies will improve the viability of distribution generation in a competitive deregulated market. Among these technologies are fuel cells, micro-turbines, combustion turbines, gas engines, and diesel engines. Chapter 13 discusses some of the emerging technologies in electricity generation.

Utility Deregulation Impact

Following the acceptance of the Energy Policy Act of 1992, which requires open access to utility transmission lines, the US Federal Energy Regulatory Commission (FERC) issued orders to allow the establishment of wholesale power market with independent system operators. As the result of significant increase in the quantities of bulk power sale, delivery of energy to users has become increasingly difficult especially through the existing transmission and distribution networks. The frequent power outages, experienced in the last few years especially in western US, illustrate the precarious stability of the transmission system.

Moreover, several States have started to implement retail access, which allows customers to choose among several electric service providers based on a competitive market that may offer a variety of customized services such as a premium power quality. Unfortunately, existing distribution networks are not designed to support multiple suppliers or to channel value-added services. Ironically, utilities have built fewer than half the transmission capacity between 1990 and 1995 they built in the previous five years (1985-1990). This reduction of investment in the transmission grid is largely due to the uncertainty about on-going electric utility deregulation and restructuring (EPRI, 1999).

In addition to adding new transmission capacity, it is believed that existing transmission and distribution networks and their control have to be upgraded using advanced and new technologies to ensure high reliability and safety of the power delivery system. Among the technologies that are being considered to upgrade the power delivery system are:

- Super-conducting transmission cables: the discovery of high-temperature superconducting (HTS) materials using ceramic oxides in 1986 has lowered the cost of superconducting transmission cables to a reasonable level. It is estimated that an HTS cable could carry 500 MW of electric power at voltages as low as 50 kV.

- High-voltage electronic flexible AC transmission system (FACTS) controllers are now used by several utilities to increase the capacity of transmission lines and improve overall delivery system reliability. Unlike conventional electromechanical controllers, FACTS controllers are sufficiently fast to reduce bottlenecks and transient disturbances in power flow and thus reduce transmission system congestion and improve overall delivery reliability.

- Cost-effective distributed generation and storage technologies that offer flexibility to meet a wide variety of customer needs. Among distribution generation systems under development and testing are microturbines with capacities ranging from 10 to 250 kW, and fuel cells which offer clean, efficient, compact, and modular generation units.

- Diversified and integrated utility services to meet the divergent needs of various market segments. For instance, innovative rate structures such real-time-pricing rates are being offered to customers that are demanding lower rates. Moreover, some utilities are integrating electricity with other services such as internet access, telecommunications, and cable television using fiber-optic networks. However, the move to integrate utility functions requires new hardware and software technologies. For instance, low-cost electronic meters with two-way communications are needed to provide real-time pricing and billing options for multiple utility services.

2.2.2   Natural Gas

As indicated in Table 2.6, the total US consumption of natural gas has actually declined between 1972 and 1992. The industrial sector experienced the highest reduction in natural gas use especially in the 1980's. The main reason for the decline in natural gas use is attributed to the restructuring and the deregulation of several segments of the gas industry during most of the 1970's. Indeed, the regulation of natural gas markets had the effect of reducing the availability of natural gas. As indicated in Table 2.7, the prices of natural gas increased between 1972 and 1982 because of its limited availability. However, starting from early 1990s, the gas supplies became more certain and some of the regulations were removed. As a consequence, the prices of natural gas have actually decreased between 1982 and 1992.

Table 2.6: Annual US Consumption of Natural Gas by Sector in Trillion cubic feet   (*Source:* EIA, 1998)

End-Use Sector	1972	1982	1992	1998*
Residential	5.13	4.63	4.69	4.51
Commercial	2.61	2.61	2.80	3.09
Industrial	9.62	6.94	8.70	9.71

* The data for 1998 are preliminary data that may be revised.

In the future, it is expected that the natural gas market will continue to expand and its price to be competitive. In particular, the future for natural gas as a primary energy source for electricity generation is considered to be promising. Indeed, gas-fired power plants are competitive because of their high efficiencies (approaching 50%) and are environmentally attractive since they produce significantly lower carbon and sulfur emissions than plants powered by coal or oil.

Table 2.7: Average Retail Prices of Natural Gas by Sector in 1992 dollars per 1000 cubic feet *(Source:* EIA, 1998*)*

End-Use Sector	1972	1982	1992	1998*
Residential	3.62	7.36	5.89	6.05
Commercial	2.63	6.87	4.88	4.84
Industrial	1.35	5.51	2.84	2.72

* The data for 1998 are preliminary data that may be revised.

2.2.3   Petroleum Products

Overall, the US consumption of fuel oil and other petroleum products has remained almost the same between 1972 and 1992 as indicated in Table 2.8. However, the oil price fluctuated significantly over the last three decades after the 1973 energy crisis. Table 2.8 clearly indicates that after a drastic increase (almost four fold) between 1972 and 1982, the crude oil prices have decreased in 1998 to levels even lower than those experienced in 1972.

Table 2.8: US Consumption of Petroleum Products by Sector in Million barrels per day *(Source:* EIA, 1998)

End-Use Sector	1972	1982	1992	1998*
Residential/Commercial	2.25	1.24	1.12	1.13
Industrial	4.19	4.06	4.55	4.81
Transportation	8.57	9.31	10.95	12.22
Electric Utilities	1.36	0.69	0.42	0.52
Total	16.37	15.30	17.03	18.68

* The data for 1998 are preliminary data that may be revised.

In the building sector (i.e., residential and commercial applications), the US consumption level of petroleum products has decreased over the years as indicated in Table 2.9. However, the use of petroleum fuels has steadily increased in the transportation sector.

Table 2.9: Average Crude Oil Price in the US in 1992 dollars per barrel (*Source:* EIA, 1998)

Year	1972	1982	1992	1998*
Price	10.56	40.63	15.99	9.65

* The data for 1998 are preliminary data that may be revised.

2.2.4 Coal

In the US, coal is primarily used as an energy source for power generation by electric utilities as shown in Table 2.10. Indeed, the total US consumption of coal has actually increased between 1972 and 1998 due primarily to the growth in coal use by electric utilities. In all other sectors (i.e., residential, commercial, and industrial), the coal consumption has generally decreased. These consumption trends are expected to be maintained in the near future for all sectors. However, the share of electricity generation attributed to coal will be reduced due to more reliance in the future on other generation technologies as discussed in section 2.2.1.

Table 2.10: Annual US Consumption of Coal by Sector in Million of short tons (*Source:* EIA, 1998)

End-Use Sector	1972	1982	1992	1998*
Residential/Commercial	11.7	8.2	6.2	6.3
Industrial	160.1	103.0	106.4	97.9
Electric Utilities	351.8	593.7	795.1	933.4
Total	524.3	706.9	907.7	1,037.1

* The data for 1998 are preliminary data that may be revised.

The abundant coal reserve base and the lingering excess production capacity have helped maintain low coal prices especially during the last decade as indicated by Table 2.11. In the future, however, the coal price is expected to rise slowly due to reserve depletion and slow growth in labor productivity. The higher coal prices coupled with the environmental concerns may cause a future decline of coal consumption in the US.

Table 2.11: Average Coal Price in the US in 1992 dollars per short ton (*Source: EIA, 1998*)

Year	1972	1982	1992	1998*
Price	23.11	38.82	21.03	15.60

*   The data for 1998 are preliminary data that may be revised.

2.3   Electricity Rates

To generate electricity, utilities have to consider several operating costs to determine their rates. Typically, an electrical utility is faced with the following cost items:

- Generation Plant: The cost of operating the power plant to generate electricity represents typically the highest cost category. Indeed, the power generation plants have to meet several regulations and safety requirements. For instance, several plants are required to meet strict pollution standards especially if they operate in highly populated areas. Meeting these regulations can significantly increase the generating costs.

- Transmission/Distribution Systems: To deliver the electricity from the generation plant where it is produced to areas where it is utilized, transmissions lines, substations, and distribution networks have to be used. The cost of the transmission/distribution systems depends on the distances to be covered as well as transformers, capacitors, and meters to be used. Moreover, the delivery energy losses can be a significant part of the transmission/distribution costs.

- Fuel Costs: Electricity is generated using a primary fuel source. The fuel cost can be small as in case of hydroelectric plants or significant as in conventional fuel oil or coal power plants. The cost of fuel may fluctuate depending on the world markets.

- Administrative Costs: The salaries of management, technical, and office staff as well as insurance and maintenance costs for the power plant equipment are part of the administrative costs. These administrative costs can be large, especially for nuclear power plants.

Other factors that affect the cost of electricity include the generating capacity of the utility, and the demand/supply conditions at a given time (i.e., on-peak and off-peak periods).

Utilities allocate the cost of electricity differently depending on the type of customers, by offering various rate schedules.   Three customer types are generally considered by utilities: residential, commercial, and industrial customers. Each utility may offer several rate structures for each customer type. It is therefore important that the auditor know the various rate structures that can be offered to the audited facility.

Utilities can tailor their rates to the customer needs for electricity using several methods. Some of the common rate structures used by US utilities are summarized below:

- Block Pricing Rates
- Seasonal Pricing Rates
- Innovative Rates

In addition to these rates, utilities provide some riders and discounts to their customers. A rider may modify the structure of a rate based on specific qualifications of the customer. For instance, utilities can change the summer energy charge of their residential customers using an air-conditioning rider. Moreover, utilities can reduce energy and/or demand charges using discount provisions. For instance, several utilities offer a voltage discount when the customer is willing to receive voltages higher than the standard voltage level. To benefit from this discount, the customer may have to install and maintain a properly sized transformer.

More detailed discussion of each of the commonly available rates is provided in the following sections with some examples to illustrate how an energy auditor can apply the utility rate provisions to check the calculation of utility bills. First, common features of all electric utility rates are discussed in detail.

2.3.1   Common Features of Utility Rates

There are several utility rate features and concepts that the auditor should be familiar with to be able to interpret and correctly analyze the utility billing procedure. Some of these concepts are described in the following sections.

Billing Demand:

The demand that is billed by the utility is referred to as the billing demand. The billing demand is often determined from the peak demand obtained for one month (or any billing cycle). The peak demand, also known as actual demand, is defined as the maximum demand or maximum average measured demand in any 15-minute period in the billing cycle. To better understand the concept of billing demand, consider two different monthly load profiles: a rugged profile A and a flat profile B as illustrated in Figure 2.1. Assume that the average demand for profile A coincides with that of profile B (and thus the total energy use in kWh for both profiles A and B is the same). It is not equitable for the utility to bill the same charges for the two profiles. Indeed, profile A requires that the utility supply higher demand and thus increases its generation capacity for only a short period of time. Meanwhile, profile B is ideal for the utility since it does not change over time. Therefore, some utilities charge their customers for the peak demand incurred during the billing period. This demand charge may serve as an incentive to the customers to reduce or "shave" their peak demand.

In some rates, the billing demand is determined based on the utility specifications and may be different from the maximum demand actually measured during a billing period. For instance, ratchet and power factor clauses can change the determination of the billing demand. Moreover, a minimum billing demand can be specified in a contractual agreement between the utility and the customer. This minimum demand is often called the contract demand.

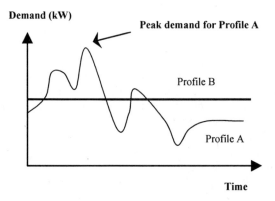

Figure 2.1: Peak demand for two different electric load profiles

Power Factor Clause:

The power factor is defined as the ratio of actual power used by the consumer (expressed in kW) to the total power supplied by the utility (expressed in kVA). Figure 2.2 shows the power triangle to illustrate the concept of power factor. The reader is referred to Chapter 5 for more details on the concept of power factor.

For the same actual power consumed by two customers but with different power factor values, the utility has to supply higher total power to the customer with the lower power factor. To penalize for low power factors (generally lower than 0.85), some utilities use a power factor clause to change the billing demand or to impose new charges (on total power demand or reactive power demand) according to the value of the power factor. For instance, the billing demand can be increased whenever the measured power factor of the customer is below a reference value or base power factor:

$$Billed - Demand = Actual - Demand * \left(\frac{pf_{base}}{pf_{actual}} \right)$$ (2.1)

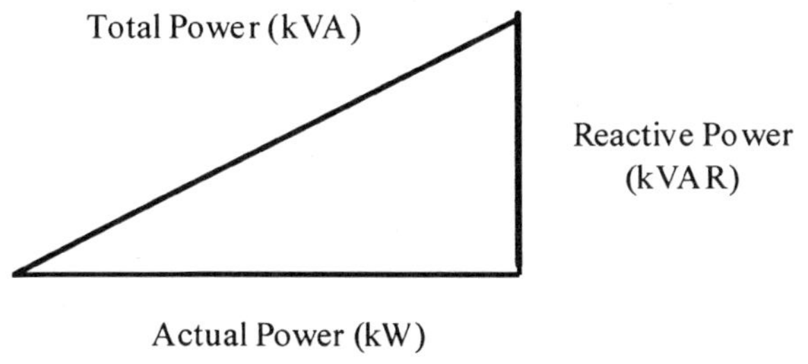

Total Power (kVA)

Reactive Power
(kVAR)

Actual Power (kW)

$$PowerFactor(pf) = \frac{ActualPower}{TotalPower}$$

Figure 2.2: Power Triangle Diagram

Example 2.1 illustrates how the power factor clause may affect the utility bill.

Example 2.1: *A utility has the following monthly billing structure for their industrial customers:*

- *Customer charge: $450/month*
- *Demand charge: $20/kW*
- *Energy charge: $0.03/kWh*

A power factor clause states that the billing demand, upon which the demand charge is based, shall be the actual demand for the month corrected for the power factor by multiplying the measured actual demand by 80 and dividing the product by the actual average power factor expressed as a percent. The power factor clause applies whenever the average monthly power factor is less than 80%.

(i) Calculate the utility bill for an industrial facility with the following energy use characteristics during a specific month:

- *Actual demand: 300 kW*
- *Energy Consumption: 50,000 kWh*
- *Average Monthly power factor: 60%*

(ii) Determine the cost savings achieved for the month if the facility power factor is improved to be always above 80%.

Solution:

(i) First, the billing demand for the month is determined using the power factor clause since the average power factor is below 80%:

Billed demand = Actual demand * 80 / pf = 300 * 80 / 60 = 400 kW

Then, the monthly bill can be calculated using the utility rate structure:

(a) Customer charge = $ 450.00
(b) Demand charge = 400 kW * $20/kW = $ 8,000.00
(c) Energy charge = 50,000 kWh * $0.03/Wh = $ 1,500.00

Total monthly charges = (a) + (b) + (c) = $ 9,950.00

The actual bill may include other components such as fuel adjustment cost and sales taxes.

(ii) The cost savings from an improved power factor are due to a reduction in the billing demand. Indeed, the billing demand for the considered month will be the actual demand (i.e., 300 kW). Thus, a reduction of 100 kW in billing demand which results in a reduction of demand charges of:

Cost Savings = 100 kW * $20/kW = $ 2,000/ month

The reader is referred to Chapter 5 to determine the cost-effectiveness for power factor improvement measures.

Ratchet Clause:
Typically, the utility charges are billed monthly. In particular, the demand charges are based on monthly peak demand. However, when the peak demand for one month is significantly higher than for the other months (such as the case for buildings with high cooling loads in the summer months), the utility has to supply the required peak demand and thus operate additional generators for only one or two months. For the rest of the year, the utility will have to maintain these additional generators. To recover some of this maintenance cost and to encourage demand shaving, some utilities use a ratchet clause in the determination of the billed demand. For instance, the billed demand for any given month is a fraction of the highest maximum demand of the previous 6 months (or 12 months) or the actual demand incurred in the month. Example 2.2 illustrates the calculation procedure of the utility bill with a ratchet clause.

Example 2.2: *The utility has added a ratchet clause to the rate structure described of Example 2.1. Specifically, the ratchet clause states that no billing demand shall be considered as less than the 70% of the highest on-peak season maximum demand corrected for the power factor previously determined during the 12 months ending with the current month.*

Calculate the utility bill for the industrial facility considered in Example 2.1 taking into account the ratchet clause. Assume that the previous highest demand (during the last 12 months) is 700 kW.

Solution:
First, the minimum billing demand determined by the ratchet clause is calculated:

Minimum billing demand (ratchet clause) = 700 kW * 0.70 = 490 kW

Since the billing demand based on the power factor clause was found to be 400 kW (see Example 2.1), the billing demand for the month is:

Billed demand = max (490 kW, 400 kW) = 490 kW
Then, the monthly bill can be calculated using the utility rate structure:

(a) Customer charge = $ 450.00
(b) Demand charge = 490 kW * $20/kW= $ 9,800.00
(c) Energy charge = 50,000 kWh * $0.03/Wh = $ 1,500.00

Total monthly charges = (a) + (b) + (c) = $ 11,750.00

Thus, the ratchet clause increases the utility bill for the month.

Fuel Cost Adjustment:

Most utilities have to purchase primary energy sources (fuel oil, natural gas, and coal) to generate electricity. Since the cost of these commodities changes over time, the utilities impose an adjustment to their energy charges to account for any cost variation of their primary energy sources. Generally, utilities provide in the description of their rate structure a formula that they use to calculate the fuel cost adjustment. In addition to fuel cost adjustment, utilities may levy taxes and surcharges to recover imposts required from them by federal or state governments or agencies. In some cases, the fuel cost adjustment can be a significant proportion of the utility bill. Example 2.3 illustrates the effect of both the fuel cost adjustment and sales tax on a monthly utility bill.

Example 2.3: *The utility has added a ratchet clause to the rate structure described in Example 2.2. Specifically, the ratchet clause states that no billing demand shall be considered as less than 70% of the highest on-peak season maximum demand corrected for the power factor previously determined during the 12 months ending with the current month.*

Calculate the utility bill for the industrial facility considered in Example 2.2 taking into account both a fuel cost adjustment of 0.015/kWh and a sales tax of 7%.

Solution:

The fuel cost adjustment should be applied to the energy use, while the sales tax should be applied to the total cost. Thus, the monthly bill for the industrial facility considered in Example 2.2 becomes:

(a) Customer charge   = $ 450.00
(b) Demand charge = 490 kW * $20/kW   = $ 9,800.00
(c) Energy charge = 50,000 kWh * [$0.03/kWh +$0.015/kW]= $ 2,250.00

Total monthly charges (before sales tax) = (a) + (b) + (c) = $12,500.00

Total monthly charges (after sales tax) = $12,500.00 * 1.07 = $ 13,375.00

The cumulative effect of both fuel cost adjustment and sales tax increased the utility bill for the month by [$13,375 - $11,750 = $1,625].

Service Level:

Utilities typically offer several rate structures for a given customer depending on the type of service. For instance, utilities may have different rates depending on the voltage level provided to the customers. The higher the delivery voltage level, the cheaper is the energy rate. In particular, utilities offer reduced rate for demand and/or energy charges to customers that own their service transformers This type of rate is generally described under various clauses of a utility bill including customer-owned transformer, voltage level, or service type. The owner has to determine the cost-effectiveness of owning a transformer, and thus receive higher voltage level, by comparing the utility bill savings against the cost of purchasing, leasing, and maintaining the transformer. Typically, transformer energy losses are charged to the owner. In addition, the owner should have a standby transformer or make another arrangement (such as standby utility service) in case of breakdown of the service transformer. Example 2.4 presents an estimation of savings in utility bills due to owning a service transformer.

Example 2.4: *The industrial facility of Example 2.1 has an option of owning and operating its own service transformer with the advantage of reduced rate structure as described below:*

- *Customer charge: $650/month*
- *Demand charge: $15/kW*
- *Energy charge: $0.025/kWh*

    Calculate the reduction in the utility bill during the month for which the energy use characteristics are given in example 2.1. In the savings calculation, do not consider the power factor clause, ratchet clause, sales tax, and fuel cost adjustment.

Solution:

    First, the utility bill is determined in the case the facility owns its service transformer:

         (a) Customer charge   = $ 650.00
         (b) Demand charge = 400 kW * $15/kW   = $ 6,000.00
         (c) Energy charge = 50,000 kWh * $0.025/Wh   = $ 1,250.00

         Total monthly charges = (a) + (b) + (c) = $ 7,900.00

    Thus, the savings in the utility bill for the month is:

         Utility bill savings = $9,950 – $7,900 = $2,050

    The reader is referred to Chapter 5 to determine the cost-effectiveness of replacing or owning transformers.

2.3.2   Block Pricing Rates

In these rates, the energy price depends on the rate of electricity consumption using either inverted blocks or descending blocks. An inverted block pricing rate structure increases the energy price as the consumption increases. On the other hand, a descending (also referred to as a declining) block rate structure reduces the price as the energy consumption increases. Typically, the rate is referred to as a "flat" rate when the energy price does not vary with the consumption level. Table 2.12 and Table 2.13 show examples of block pricing rates for energy and demand charges. Examples 2.5 and 2.6 illustrate the calculation details of utility bills.

Based on a comprehensive survey of existing rate structures offered by US utilities (GRI, 1993), block pricing rate structures are commonly used. Indeed, almost 60% of the surveyed electric utilities offer descending or inverted block pricing structure for the energy charges. For the residential customers, a combination of descending, flat, and inverted rate structures is used throughout the US. However, descending energy rate structures are used almost exclusively for commercial and industrial customers. Some electric utilities offer block pricing rates for demand as well as energy charges.

Table 2.12: An electric utility rate with no-demand charges for residential general service (utility rate A)

Billing Item	Winter (Nov-Apr)	Summer (May-Oct)
Customer Charge	$7.50	$7.50
Minimum Charge	$7.50	$7.50
Fuel Cost Adjustment	0.00	0.00
Tax Rate	6.544%	6.544%
No. of Energy Blocks	1	3
Block 1 Energy Size (kWh)	0	400
Block 2 Energy Size (kWh)	0	400
Block 3 Energy Size (kWh)	0	800
Block 1 Energy Charge ($/kWh)	0.088	0.0874
Block 2 Energy Charge ($/kWh)	0.000	0.1209
Block 3 Energy Charge ($/kWh)	0.000	0.1406

Example 2.5: *Using utility rate structure A, calculate the utility bill for a residence during a summer month when the energy use is 800 kWh.*

Solution:

Considering all the charges imposed by utility rate A (see Table 2.12), the monthly bill can be calculated as follows:

(a) Customer charge   = $ 7.50
(b) Energy charge = 400kWh*$0.0874/kWh+400kWh*$0.1209/kWh = $ 83.32
(c) Fuel cost adjustment   = $ 0.00
(d) Taxes = 6.544% [(a)+(b)+(c)]   = $ 5.94

   Total monthly charges = (a) + (b) + (c) + (d) = $ 96.76

Thus, the average cost of electricity for the month is $96.76/800 kWh or $0.12095/kWh.

2.3.3   Seasonal Pricing Rates

Some electric utilities offer seasonal rate structures to reflect the monthly variations in their generation capacity and energy cost differences. Generally, the utilities that provide seasonal rate structures use different energy and/or demand charges during winter and summer months. The summer charges are typically higher than winter charges for most electric utilities due to higher energy consumption attributed to cooling of buildings. Table 2.14 and Table 2.15 show examples of seasonal pricing rates while examples 2.5 an 2.6 illustrate the calculation procedure for monthly utility bills.

Based on a survey conducted by the GRI (1993), over than 55% of US electric utilities offer seasonal pricing rates for residential customers. Only 5% of electric utilities had residential rates where the winter rate is actually higher than the summer rate. These utilities are located in the Northeast and the West regions of the US. The same survey reveals that over 42% of US utilities use seasonal pricing rates for commercial and industrial customers. Only 7% of the utilities surveyed offer rates with higher winter prices for their commercial and industrial customers.

Table 2.13: An electric utility rate with demand charges for commercial general service (utility rate B)

Billing Item	Winter (Oct-May)	Summer (Jun-Sep)
Customer Charge	0.00	0.00
Minimum Charge	$25.00	$25.0
Fuel Cost Adjustment	0.01605	0.01377
Tax Rate	0.00	0.00
No. of Energy Blocks	3	3
Block 1 Energy Size (kWh)	40000	40000
Block 2 Energy Size (kWh)	60000	60000
Block 3 Energy Size (kWh)	>100000	>100000
Block 1 Energy Charge ($/kWh)	0.059	0.065
Block 2 Energy Charge ($/kWh)	0.042	0.047
Block 3 Energy charge ($/kWh)	0.039	0.042
No. of Demand Blocks	2	2
Block 1 Demand Size (kW)	50	50
Block 2 Demand Size (kW)	>50	>50
Block 1 Demand Charge ($/kW)	12.39	13.71
Block 2 Demand Charge ($/kW)	11.29	12.52
Reactive Demand Charge ($/kVAR)	0.20	0.20

Example 2.6: *Using utility rate structure B, calculate the utility bill for a commercial facility during a winter month when the energy use is 70,000 kWh, the billing demand is 400 kW, and the average reactive demand is 150 kVAR.*

Solution:

Accounting for all the charges considered by utility rate B (see Table 2.13), the electric energy bill for the winter month can be calculated as follows:

(a) Customer charge	= $ 0.00
(b) Energy charge =40,000kWh*$0.059/kWh	
+30,000kWh *$0.042/kWh	= $ 3,620.00
(c) Demand charge = 50kW * $12.39/kW	
+ 350kW*$11.29/kW	= $ 4,571.00
(d) Fuel cost adjustment = 70,000kWh*$0.01605	= $ 1,123.50
(e) Reactive demand charge = 150kVAR*$0.20/kVAR	= $ 30.00
(f) Taxes	= $ 0.00

Total monthly charges = (a) + (b) + (c) + (d) + (e) + (f) = $ 9,344.50

The average cost of electricity for the month is then $9,344.50/70,000 kWh or $0.1335/kWh.

2.3.4   Innovative Rates

Due to the increased focus on integrated resource planning, demand-side management, and the competitive energy market, several innovative rates have been implemented by utilities. These innovative rates have the main objective to profitably meet the customer needs. Moreover, some utilities foster new technologies through the use of innovative rates to retain their customers.

There are several categories of rates that can be considered to be innovative rates. In the US, innovative rates can be classified into seven categories (GRI, 1999):

(i) Time-of-Use (TOU) Rates

The time-of-use (TOU) rates are time-differentiated rates with the cost of electricity varying during specific times of the day and/or the year. The TOU rates which appeared first in 1940's set "on -peak" and "off-peak" periods with different energy and/or demand charges. Generally, the on-peak periods occur during daytime hours and have higher costs of energy and demand than the off-peak periods which occur during the night-time. Table 2.14 presents a time-of-use rate structure for residential customers. Example 2.7 illustrates the calculation approach of a utility bill for the TOU rate presented in Table 2.14.

Table 2.14: An electric time-of-use rate for residential service (utility rate C)

Billing Item	Winter (Oct-May)	Summer (Jun-Sep)
Customer Charge	$9.85	$9.85
Minimum Charge	$9.85	$9.85
Tax Rate	3.00%	3.00%
Energy Charge ($/kWh)		
On-Peak Hours:	0.1412	0.1500
Off-Peak Hours	0.0335	0.0335

Notes:
On-Peak Hours are defined as:
  6:00a.m. - 1:00p.m. and 4:00p.m. - 9:00p.m. (M-F) during Oct-Mar
  10:00 a.m. – 9:00 p.m. (M-F) during Apr-Sep
Off-peak Hours are all the remaining hours including holidays.

Example 2.7: *Using utility rate structure C, calculate the utility bill for a residence during a summer month when the energy use is 2,000 kWh with 55% of the energy consumption occurring during the on-peak period.*

Solution:

With all the charges considered by utility rate C (see Table 2.14), the electric energy bill for the winter month can be calculated as follows:

$$
\begin{array}{ll}
\text{(a) Customer charge} = & = \$\,9.85 \\
\text{(b) Energy charge} = 2{,}000\text{kWh}*[0.55*\$0.1500/\text{kWh} & \\
\qquad\qquad +0.45*\$0.0335/\text{kWh}] & = \$\,195.15 \\
\text{(c) Taxes} = 3\% * [(a)+(b)] & = \$\,6.15 \\[6pt]
\text{Total monthly charges} = (a)+(b)+(c) = & = \$\,211.15
\end{array}
$$

The average cost of electricity for the month is then \$211.15/2,000 kWh or \$0.1056/kWh.

(ii) The Real-Time-Pricing (RTP) Rates

The real-time-pricing (RTP) rates were first offered by Pacific Gas & Electric in 1985 (Mont and Turner, 1999). Like TOU rates, RTP rates are time-differentiated rates but the cost of electricity varies on an hourly basis. Typically, the utilities inform their customers of the hourly electricity prices only few hours before they take effect. A more detailed description of the RTP rate structures available in the US is provided later in section 2.3.4.

(iii) The End-Use Rates

To encourage customers to install and operate specific energy-consuming equipment, some US utilities offer end-use rates. With these rates, the utilities can impose operation periods and/or efficiency standards for selected and predefined equipment. For instance, the air conditioning rate allows electric utilities to interrupt service or cycle off the air conditioning equipment during specific times. Typically, the end-use rates require separate metering of the equipment.

(iv) Specialty Rates

The specialty rates are provided by utilities for specific purposes such as energy conservation and dispatchable customer generation. Energy conservation rates are offered by a limited number of US utilities to foster the use of energy efficient equipment and/or high standards of building materials. Dispatchable customer generation rates are provided to customers that have standby generators on their premises. In exchange for a reduced rate or a credit, the customers are requested to operate the generators whenever the utility needs additional generating capacity.

(v) Financial Incentive Rates

Financial incentive rates encompass economic development rates, displacement rates, and surplus power rates. The economic development rates are typically offered to encourage new customers to locate (or existing customers to expand) in specific areas that need to be economically revitalized. The displacement rates are offered to customers that are capable of generating electricity to entice them to use utility-provided electricity. Finally, the surplus power rates are highly reduced energy rates that are offered to large commercial and industrial customers when the utility has an excess of electric capacity.

(vi) Non-Firm Rates

The non-firm rates include interruptible rates, stand-by rates, and load management rates. Interruptible rates are offered to customers that can reduce or even eliminate (interrupt) their electricity needs from the utility. The electricity pricing rates depend on several factors such as the capacity that can be interrupted, the length of interruption, and the notification before interruption.

Stand-by rates are intended for customers that require utility-provided electricity on an intermittent basis since they are capable of generating most of their electricity needs (by using, for instance, cogeneration systems as discussed in Chapter 13). The stand-by rates can be offered under three rates:

(a) Maintenance Rates when the customers need electricity from the utility during predefined down-time periods of the generating equipment to perform routine maintenance.
(b) Supplementary Rates for customers that regularly use more electricity than they are capable of generating.
(c) Back-up Rates intended for customers that prefer to have back-up power from the utility in case of unexpected events (i.e., outages in the customers' generators).

Load-management rates are offered by utilities to control the usage of specific equipment such as space conditioning systems during peak periods. Generally, the load management rates are offered in combination with TOU rates.

(vii) Energy Purchase Rates

The energy purchase rates, also known as buy-back rates, are offered by utilities that want to purchase specific levels of energy or generating capacity from customers. The customers are non-utility electricity generators that qualify under the requirements of Public Utility Regulatory Policies Act (PURPA) such as cogeneration facilities, and independent power producers.

2.3.5   Real Time Pricing Rates

Based on a survey performed recently by Mont and Turner (1999), most of the existing RTP rate structures in the US are rather experimental and are restricted to selected number of customers with large power demands (varying from 250 kW and 10 MW depending on the utility). The available RTP rate structures can be classified into four categories in the survey as described briefly below:

Category 1: Base Bill and Incremental Energy Charge Rates

Under these rates, the customers receive firm hourly electricity prices for the next day before a warning time. Some utilities provide updates of their prices daily (day-ahead-pricing) while other utilities provide hourly prices only once a week (week-ahead-pricing). Typically, the category 1 RTP rates include the following charges:

- Base Bill Charge so that the utility recovers its revenue requirements. This charge depends on the customer baseline load (CBL) which defines the customer's typical energy use (kWh) for each hour of the year an d demand values (kW) for each month.

- Incremental Energy Charge (or credit) reflects the cost of the energy used by the customer above (or below) its CBL profile.

In this category of RTP, an adequate estimation of the CBL profile is an important factor to determine the final energy bill. Therefore, the customer should negotiate for the most favorable CBL values. In general, the customer would benefit from smaller CBL values since the hourly prices of the RTP rate typically fall below those of the standard rate. Moreover, flexible customers can change their electrical load profile to avoid high costs and even obtain a credit for usage below their CBL.

Category 2: Total Energy Charge Rates

The hourly electricity prices for this category are applied to the total energy consumption. In addition, the utility imposes an additional charge to recover its revenue requirements. Some utilities set a monthly fixed charge based on the customer's average apparent demand (kVA) and actual demand (kW) determined using the last 12 months load profile. Other utilities charge for actual demand, usually ratcheted for the last 12 –month period.

The category 2 RTP rates are not suitable for inflexible customers that cannot change their load profile since they cannot protect themselves against higher electricity prices.

Category 3: Day-Type Rates

These rates are similar to the category 2 RTP rates but with several predefined day types (such as off-peak day, normal day, etc.) associated with different hourly firm electricity prices. These hourly prices are based on the total energy consumption with possibly a charge for actual demand. The utility informs the customer of the applicable day-type rate one day in advance. Again these rates are only suitable for flexible customers that can vary their energy use profile to accommodate the changes in the electricity price.

Category 4: Index-Type Rates

The electricity prices for these rates are determined based on the financial market electricity indexes (in Dow Jones) or the trading prices for electricity future contracts (in New York Mercantile Exchange). Generally, the customers have to forecast the electricity prices (using various tools and sources) since the utility does not provide the prices in advance.

Case Study of RTP rates

In this section, a detailed description is presented to calculate a monthly utility bill using RTP rate structures. The RTP rates used in this section are based on the day-ahead-pricing (DAP) pilot rates proposed by Oklahoma Gas and Electric (OG&E) utility discussed by Mont and Turner (1999).

In the DAP rates offered by OG&E, there are two types of rates offered under either curtailment or non-curtailment services. Customers with curtailment service are allowed to buy electricity (referred to as " buy-through power" or "emergency power") above the contracted customer base load (CBL) profile during selected hours (or a curtailment period). The total cost of the buy-through power during a curtailment period is known as "buy-through period. Typically, the utility notifies in advance regarding the availability and the hourly prices of emergency power. Table 2.15 shows the hourly prices for DAP rates offered by OG&E for both non-curtailment and curtailment days (Mont and

Turner, 1999). The curtailment period is set between 3:00 pm and 8:00 pm with a contracted demand of 500 kW. To illustrate the calculation procedure of utility bills using the RTP rates compared to those based on the conventional rate structures, Example 2.8 is presented using the OG&E DAP rates.

Cost savings in the utility bills based on RTP rate structures can be achieved by shifting the load during the curtailment hours to non-curtailment hours. Example 3.9 illustrates the magnitude of savings achieved by a customer under OG&E DAP rates.

Table 2.15: Day-ahead-Price rates from OG&E utility for typical curtailment and non-curtailment days

Hour	Non-curtailment rate price ($/kWh)	Curtailment rate price ($/kWh)
0:00 – 1:00	0.01673	0.01673
1:00 – 2:00	0.01683	0.01683
2:00 – 3:00	0.01730	0.01730
3:00 – 4:00	0.01780	0.01780
4:00 – 5:00	0.01900	0.01900
5:00 – 6:00	0.02130	0.02130
6:00 – 7:00	0.02730	0.02730
7:00 – 8:00	0.02890	0.02890
8:00 – 9:00	0.02932	0.02932
9:00 – 10:00	0.03213	0.03213
10:00 –11:00	0.03575	0.03575
11:00 – 12:00	0.03423	0.03423
12:00 – 13:00	0.03456	0.03456
13:00 – 14:00	0.03677	0.03677
14:00 – 15:00	0.03977	0.03977
15:00 – 16:00	0.04254	1.06350
16:00 – 17:00	0.04354	1.08950
17:00 – 18:00	0.03595	0.89875
18:00 – 19:00	0.02587	0.64675
19:00 – 20:00	0.01950	0.48750
20:00 – 21:00	0.01523	0.01523
21:00 – 22:00	0.01532	0.01532
22:00 – 23:00	0.01723	0.01723
23:00 – 24:00	0.01722	0.01722

Example 2.8: *Calculate the utility bills for an industrial customer serviced by OG&E under two rates:*

(i) *Conventional rate with the following charges:*
✓ *Customer charge: $150.00*
✓ *Energy charge: $0.0264/kWh*
✓ *Demand charge:$13.1/kW*

(ii) *RTP rate structure with the hourly prices given in Table 2.15. In addition, a customer charge (referred to as an administrative charge) of $300 is incurred monthly by the customer under the RTP rate.*

The hourly energy use and the contracted CBL profiles (for the RTP rate) are summarized in Table 2.16 for typical week and weekend days (data provided by Mont and Turner, 1999).

Assume that the month has 21 weekdays (including two curtailment days) and 9 weekend days. In the utility bill calculations, do not account for any other charges such as taxes and credits.

Solution:

(1) First, the utility bill is determined using the conventional rate structure based on the actual energy use (see Table 2.16):

(a) Customer charge	= $ 150.00
(b) Energy charge = $0.0264/kWh*[21*42,461 kWh + 9*10,510 kWh]	=$26,037.55
(c) Demand charge = $13.1/kW*3,015 kW	= $39,496.50

Total monthly charges = (a) + (b) + (c) = $65,684.05

The average cost of electricity for the month based on the conventional rate structure is then $65,684.05/ [21*42,461 kWh + 9*10,510 kWh] or $0.06660/kWh.

(2) The utility bill based on the RTP rate is described below:

(a) Customer charge	= $ 300.00
(b) Energy charge for week days (non-curtailment days) [see Table 2.17]	= $19*55.99
(c) Energy charge for week days (curtailment days) [see Table 2.18]	=$2*8,196.79
(d) Standard charge based on the CBL profile [see note below]	
i. Customer charge =	$ 150.00
ii. Energy charge = $0.0264/kWh*958,960 (see note)	$26,037.55
iii. Demand charge = $13.1/kW*2,765kW	$36,221.50

Total monthly charge = i. + ii. + iii.   =$62,409.05

Total monthly charge = (a) + (b) + (c) + (d) = $80,166.44

The average cost of electricity for the month based on the RTP rate structure is then $80,166.44/[21*42,461 kWh + 9*10,510 kWh] = $0.08128/kWh.

Therefore, the conventional rate structure is more cost-effective for the industrial facility considered in this example (at least during the month for which the utility bills were calculated as shown above).

Note: The total monthly energy use based on the CBL profiles is calculated as detailed below:

kWh(CBL) = 19 days * 42,040 kWh/day + 2 days * 32,805 kWh/day + 9 days * 10,510 kWh/day

Or,

kWh (CBL) = 968,960 kWh

Table 2.16: Actual and CBL hourly energy use profiles for the industrial facility assumed in Example 2.8

Hour	Typical Weekday Actual Energy Use (kWh)	Typical Weekend day Actual Energy Use (kWh)	Non-Curtailment Typical Day CBL Energy Use (kWh)	Curtailment Typical Day CBL Energy Use (kWh)
0:00 – 1:00	345	250	220	220
1:00 – 2:00	455	270	230	230
2:00 – 3:00	676	305	325	325
3:00 – 4:00	785	320	360	360
4:00 – 5:00	980	360	455	455
5:00 – 6:00	1,200	405	645	645
6:00 – 7:00	1,800	460	1,715	1,715
7:00 – 8:00	2,905	510	2,450	2,450
8:00 – 9:00	2,825	525	2,545	2,545
9:00 – 10:00	2,800	520	2,480	2,480
10:00 –11:00	2,845	535	2,560	2,560
11:00 – 12:00	3,015	550	2,765	2,765
12:00 – 13:00	2,500	565	2,745	2,745
13:00 – 14:00	2,765	560	2,730	2,730
14:00 – 15:00	2,890	535	2,650	2,650
15:00 – 16:00	2,815	525	2,590	500
16:00 – 17:00	2,825	530	2,520	500
17:00 – 18:00	2,780	510	2,370	500
18:00 – 19:00	1,755	495	2,175	500
19:00 – 20:00	1,100	480	2,080	500
20:00 – 21:00	845	395	1,945	1,945
21:00 – 22:00	635	345	1,680	1,680
22:00 – 23:00	535	290	1,230	1,230
23:00 – 24:00	385	270	575	575
Total	42,461	10,510	42,040	32,805

Table 2.17: RTP hourly energy cost during a typical non-curtailment day for the industrial facility of Example 2.8

Hour	Typical Weekday Actual Energy Use (kWh)	Non-Curtailment Typical Day CBL Energy Use (kWh)	Non-Curtailment Typical Day Hourly Energy Price ($/kWh)	Non-Curtailment Typical Day Hourly Energy Cost ($)
0:00 – 1:00	345	220	0.01673	2.09
1:00 – 2:00	455	230	0.01683	3.79
2:00 – 3:00	676	325	0.01730	6.07
3:00 – 4:00	785	360	0.01780	7.57
4:00 – 5:00	980	455	0.01900	9.98
5:00 – 6:00	1,200	645	0.02130	11.82
6:00 – 7:00	1,800	1,715	0.02730	2.32
7:00 – 8:00	2,905	2,450	0.02890	13.15
8:00 – 9:00	2,825	2,545	0.02932	8.21
9:00 – 10:00	2,800	2,480	0.03213	10.28
10:00 –11:00	2,845	2,560	0.03575	10.19
11:00 – 12:00	3,015	2,765	0.03423	8.56
12:00 – 13:00	2,500	2,745	0.03456	−8.47
13:00 – 14:00	2,765	2,730	0.03677	1.29
14:00 – 15:00	2,890	2,650	0.03977	9.54
15:00 – 16:00	2,815	2,590	0.04254	9.57
16:00 – 17:00	2,825	2,520	0.04354	13.28
17:00 – 18:00	2,780	2,370	0.03595	14.74
18:00 – 19:00	1,755	2,175	0.02587	−10.87
19:00 – 20:00	1,100	2,080	0.01950	−19.11
20:00 – 21:00	845	1,945	0.01523	−16.75
21:00 – 22:00	635	1,680	0.01532	−16.01
22:00 – 23:00	535	1,230	0.01723	−11.97
23:00 – 24:00	385	575	0.01722	−3.27
Total	42,461	42,040		55.99

Table 2.18: RTP hourly energy cost during a typical curtailment day for the industrial facility of Example 2.8.

Hour	Typical Weekday Actual Energy Use (kWh)	Curtailment Typical Day CBL Energy Use (kWh)	Curtailment Typical Day Hourly Energy Price ($/kWh)	Curtailment Typical Day Hourly Energy Cost ($)
0:00 – 1:00	345	220	0.01673	2.09
1:00 – 2:00	455	230	0.01683	3.79
2:00 – 3:00	676	325	0.01730	6.07
3:00 – 4:00	785	360	0.01780	7.57
4:00 – 5:00	980	455	0.01900	9.98
5:00 – 6:00	1,200	645	0.02130	11.82
6:00 – 7:00	1,800	1,715	0.02730	2.32
7:00 – 8:00	2,905	2,450	0.02890	13.15
8:00 – 9:00	2,825	2,545	0.02932	8.21
9:00 – 10:00	2,800	2,480	0.03213	10.28
10:00 –11:00	2,845	2,560	0.03575	10.19
11:00 – 12:00	3,015	2,765	0.03423	8.56
12:00 – 13:00	2,500	2,745	0.03456	–8.47
13:00 – 14:00	2,765	2,730	0.03677	1.29
14:00 – 15:00	2,890	2,650	0.03977	9.54
15:00 – 16:00	2,815	500	1.06350	2,462.00
16:00 – 17:00	2,825	500	1.08950	2,533.09
17:00 – 18:00	2,780	500	0.89875	2,049.15
18:00 – 19:00	1,755	500	0.64675	811.67
19:00 – 20:00	1,100	500	0.48750	292.50
20:00 – 21:00	845	1,945	0.01523	–16.75
21:00 – 22:00	635	1,680	0.01532	–16.01
22:00 – 23:00	535	1,230	0.01723	–11.97
23:00 – 24:00	385	575	0.01722	–3.27
Total	42,461	32,805		8,196.79

Example 2.9: *Calculate the cost savings in electric utility bills for an industrial customer serviced by OG&E RTP rate used in Example 2.8 if the excess electrical load experienced during the curtailment hours (between 3 p.m. and 8 p.m.) is shifted uniformly to the next 4 hours (between 8 p.m. and midnight). With this load shifting measure, the demand during the curtailment hours would be 500 kW and the total daily energy use for typical days would remain the same as shown in Table 2.18.*

Solution:

(a) The excess load during the curtailment hours is first determined as follows:

Excess Load = 2,815 kWh + 2,825 kWh + 1,755 kWh + 1,100 kWh – 4 * 500 kWh = 6,495 kWh

(b) For each hour during the next 4 hours (i.e., between 8 p.m. and midnight), the additional load will be 6,495 kWh/4 = 1623.75 kWh. The additional energy cost during the non-curtailment hours associated to this shift is then:

Energy Cost Increase = 1623.75 kWh *[(0.01523+0.01532+0.01723+0.01722) $/kWh]= $105.54

(c) The shift of electrical load during the curtailment hours has the effect of eliminating any hourly energy cost during the curtailment hours from 3 p.m. to 7 p.m. as shown in Table 2.18.

(That is $2,533.09+$2,049.15+$811.67+$292.50 = $5,686.41).

(d) Since all the other charges remain unchanged, the reduction in cost savings is simply the energy charge avoided during the curtailment hours minus any energy cost increase during non-curtailment hours, or:

Utility cost savings = [$5,686.41/day –105.54/day]* 2 days = $11,161.74 for the month.

2.4 Natural Gas Rates

The rate structures for natural gas are similar to those described for electricity. However, the rates are generally easy to understand and apply. For instance, natural gas utilities rarely charge for peak demands. However, energy chares using block rates or seasonal rates are commonly offered. Examples of natural gas utility rates are illustrated in Table 2.19 for energy block pricing rates and Table 2.20 for energy and demand block pricing rates (GRI, 1999). Calculation details of monthly utility bills for natural gas are shown in Examples 2.10 and 2.11.

Table 2.19: A gas utility rate for residential general service (utility rate D)

Billing Item	Any Month (Jan-Dec)
Customer Charge	$4.50
Minimum Charge	$4.50
Fuel Cost Adjustment	0.00
Tax Rate	4.00%
No. of Energy Blocks	2
Block 1 Energy Size (MMBtu)	2.5
Block 2 Energy Size (MMBtu)	> 2.5
Block 1 Energy Charge ($/MMBtu)	5.145
Block 2 Energy Charge ($/MMBtu)	4.033

Example 2.10: *Using utility rate structure D, calculate the gas utility bill for a residence during a month when the energy use is 10.4 MMBtu.*

Solution:

Considering all the charges imposed by utility rate D (see Table 2.19), the monthly gas bill can be calculated as shown below:

(a) Customer charge	= $ 4.50
(b) Energy charge = 2.5MMBtu*$5.145/MMBtu	
+ 7.9MMBtu *$4.033/MMBtu	= $44.72
(c) Fuel cost adjustment	= $ 0.00
(d) Taxes = 4.00% [(a)+(b)+(c)] =	= $ 8.05
Total monthly charges = (a) + (b) + (c) + (d)	= $57.27

Thus, the average gas cost for the month is $57.27/10.4 MMBtu or
5.51/MMBtu.

Table 2.20: A gas utility rate with demand charges for commercial general
service (utility rate E)

Billing Item	Any Month (Jan-Dec)
Customer Charge	0.00
Minimum Charge	$300.00
Fuel Cost Adjustment	0.00
Tax Rate	4.00%
No. of Energy Blocks	3
Block 1 Energy Size (MMBtu)	10000
Block 2 Energy Size (MMBtu)	10000
Block 3 Energy Size (MMBtu)	> 20000
Block 1 Energy Charge ($/MMBtu)	3.482
Block 2 Energy Charge ($/MMBtu)	3.412
Block 3 Energy charge ($/MMBtu)	3.385
No. of Demand Blocks	2
Block 1 Demand Size (MMBtu/day)	10
Block 2 Demand Size (MMBtu/day)	> 10
Block 1 Demand Charge ($/MMBtu/day)	5.50
Block 2 Demand Charge ($/MMBtu/day)	4.50

Example 2.22: *Using utility rate structure E, calculate the gas utility bill for a commercial facility during a month when the energy use is 15,000 MMBtu and the billing demand is 500 MMBtu/day.*

Solution:

Accounting for all the charges considered by utility rate E (see Table 2.20), the electric energy bill for the winter month can be calculated as follows:

(a) Customer charge	=$ 0.00
(b) Energy charge = 10,000MMBtu*$3.482/MMBt + 5,000MMBtu *$3.412/MMBt	=$ 51,880.00
(c) Demand charge = 10* $5.50 + 490 * $4.50	=$ 2,260.00
(d) Taxes = 4.00%*[(a)+(b)+(c)]=	=$ 2,165.60
Total monthly charges = (a) + (b) + (c) + (d)	=$ 56,305.60

> The average cost of electricity for the month is then $56,305.60/15,000 MMBtu or $3.75/MMBtu.

In addition to energy and demand charges, the price of natural gas is determined based on the interruptible priority class selected by the customer. A customer with a low priority has a cheaper rate but can be curtailed whenever a shortage in the gas supply is experienced by the utility. However, some small quantities of gas are generally supplied to prevent the pipes from freezing and to keep the pilot lights burning.

2.5 Utility Rates for Other Energy Sources

The utility rate structures for energy sources other than electricity and natural gas are generally based on a flat rate. For instance, the crude oil is typically charged per gallon while coal is priced on a per ton basis. The prices of oil products and coal are set by the market conditions but may vary within a geographical area depending on local surcharge and tax rates. Moreover, fuel oil or coal can be classified in a number of grades. The grades of fuel oil depend on the distillation process. For instance, No. 1 oil is a distillate and is used as domestic heating oil. In the other hand, No. 6 oil has a very heavy residue left after the other oils have been refined. The grades of coal depend on the sulfur content and percentage of moisture. Chapter 8 describes in more details some of the properties of fuel oils.

In some applications, it may be possible and desirable to purchase steam or chilled water to condition buildings rather than using primary fuel to operate boilers and chillers. Steam can be available from large cogeneration plants. Chilled water and steam may also be produced based on the economics of scale in district heating/cooling systems. Generally, steam and chilled water are both charged based on either a flat rate or a block rate structure for both energy and demand. The steam is charged based on pound per hour (for demand charges) or thousand of pounds (for energy charges). Meanwhile, the chilled water is charged on the basis of tons (for demand charges) or ton-hours (for energy charges).

2.6 Summary

In this chapter, various energy sources used to operate buildings and industrial facilities in the US are discussed. In addition, the energy price rate structures proposed by US utilities are outlined with a special emphasis on electricity pricing features. Several calculation examples are presented to illustrate the effects of various components of rate structures on the monthly utility bills.

The main goal of this chapter is to help the energy auditor understand the complexities of various utility rate structures. Significant energy cost savings can be achieved by selecting the energy pricing rate best suited for the audited facility.

Problems

2.1 (i) Calculate the utility bill for a facility during one month characterized by the following parameters:
- ✓ Actual demand: 450 kW
- ✓ Energy consumption: 85,000 kWh
- ✓ Average power factor: 70%

The facility is subject to the following rate structure:
- Customer charge: $150/month
- Billed demand charge: $10/kW
- Energy charge: $0.025/kWh

The rate structure includes a power factor clause that states that the billing demand is determined as the actual demand multiplied by the ratio of 85 and the actual average power factor expressed in percent.

(ii) Determine the cost savings achieved for the month if the facility power factor is improved to be at least 85%.

2.2 A Company consumed about 355,000 kWh of electrical energy during the month of April. The billed demand is estimated to be 600 kW for April. If the Sales tax is 7%, calculate the electric utility bill for the company.

Estimate the reduction in April electric utility bill for the company during following a peak shaving measure that lowered April billed peak demand to 450 kW.

The company is subject to the following electric rate structure:
Rate-1:
- (i)   Customer charge: $145.00/month
- (ii)   Energy charge: $0.0325/kWh/month
- (iii)   Demand charge applicable to the billing demand:
    On-peak season: $500 for the first 100 kW
                    plus $5.50/kW for any additional kW
    Off-peak season: $300 for the first 100 kW
                    plus $3.50/kW for any additional kW

The on-peak season extends from May to October and the off-peak season spans from November to April.

2.3 A manufacturing company consumes on average 250,000 kWh per month and has a billed demand of 1200 kW during each month. The company can select between two rates: Rate-1 (described in Problem 2.2) and Rate-2 (described below). Determined the best rate suitable for the company.

Rate-2:
(iv)   Customer charge: $250.00/month
(v)   Energy charge: $0.0225/kWh/month
(vi)   Demand charge applicable to the billing demand:
         On-peak season: $400 for the first 100 kW
                         plus $6.00/kW for any additional kW
         Off-peak season: $250 for the first 100 kW
                         plus $4.00/kW for any additional kW

The on-peak season extends from June to September and the off-peak season spans from October to May.

2.4 The monthly energy consumption and peak demand history of a commercial building is shown below for one year.

Month	Energy Use (kWh)	Peak Demand (kW)
Jan	125,000	455
Feb	137,500	505
Mar	155,500	610
Apr	176,000	675
May	185,500	755
Jun	201,000	920
Jul	235,500	1,150
Aug	240,000	1,200
Sep	197,500	895
Oct	184,500	650
Nov	164,000	605
Dec	141,000	550

The commercial building is on rate structure defined in Table 2.13 (utility rate B).

(i)   Estimate the annual electric utility bill for the commercial building assuming that the average power factor is 70% during each month.

(ii)   Determine the reduction in the annual electric utility bill if the power factor is improved to 90%.

3

ECONOMIC ANALYSIS

Abstract

This chapter provides an overview of the basic principles of economic analysis that are essential to determine the cost-effectiveness of various energy conservation measures suitable for commercial buildings and industrial facilities. The purpose of this chapter is to describe various evaluation methods, including the life-cycle cost (LCC) analysis technique, used to make decisions about retrofitting energy systems. The simple payback period method is also presented since it is widely used in the preliminary stages of energy audits.

3.1   Introduction

In most applications, initial investments are required to implement energy conservation measures. These initial costs must be generally justified in terms of a reduction in the operating costs (due to energy cost savings). Therefore, most improvements in the efficiency of energy systems have a delayed reward, that is, expenses come at the beginning of a retrofit project while the benefits are incurred later. For an energy retrofit project to be economically worthwhile, the initial expenses have to be lower than the sum of savings obtained by the reduction in the operating costs over the lifetime of the project.

The lifetime of an energy system retrofit project often spans several years. Therefore, it is important to properly compare savings and expenditures of various amounts of money over the lifetime of a project. Indeed, an amount of money at the beginning of a year is worth less at the end of the year and has even less buying power at the end of the second year. Consequently, the amounts of money due to expenditures or savings incurred at different times of a retrofit project cannot be simply added.

In engineering economics, savings and expenditures of amounts of money during a project are typically called *cash flows*. To compare the various cash flows over the lifetime of a project, a life-cycle cost analysis is typically used. In this chapter, the basic concepts of engineering economics are described. First, some of common economic parameters are defined. In addition, data are provided to help the reader estimate relevant economic parameters. Then, the general procedure of an economic evaluation of a retrofit project is described.

Finally, some of the advantages and disadvantages of the various economic analysis methods are discussed.

3.2   Basic Concepts

There are several economic parameters that affect a decision between various investment alternatives. To perform a sound economic analysis for energy retrofits, it is important that the auditor be (i) familiar with the most important economic parameters, and (ii) aware of the basic economic concepts. The parameters and the concepts that significantly affect the economic decision making include:

- The time value of money and interest rates including simple and compounded interest.
- Inflation rate and composite interest rate
- Taxes including sales, local, state, and federal tax charges.
- Depreciation rate and salvage value

In the following sections, the above listed parameters are described in detail to help the reader better understand the Life-cycle cost analysis procedure to be discussed later on in this chapter.

3.2.1   Interest Rate

When money is borrowed to cover part or all the initial cost of a retrofit project, a fee is charged for the use of this borrowed money. This fee is called *interest (I)* and the amount of money borrowed is called *principal (P)*. The amount of the fee depends on the value of the principal and the length of time over which the money is borrowed.

The interest charges are typically normalized to be expressed as a percentage of the total amount of money borrowed. This percentage is called *the interest rate (i)*:

$$i = \frac{I}{P} \tag{3.1}$$

It is clear that an economy with low interest rates encourages money borrowing (for investment in projects or simply for buying goods) while an economy with high interest rates encourages money saving. Therefore, if money has to be borrowed for a retrofit project, the interest rate is a good indicator of whether or not the project can be cost-effective.

To calculate the total interest charges over the lifetime of a project, two alternatives are typically considered:

Simple Interest charges: The interest fee, I, to be paid at the end of the life of the loan is proportional to both the interest rate, i, and the lifetime, N:

$$I = NiP \tag{3.2}$$

The same unit of time should be used for both N and i (i.e., one year, one month, etc.). The total amount of payment, F, due at the end of the loan period includes both the principal and the interest charges:

$$F = P + I = P(1 + Ni) \tag{3.3}$$

$P + Ni P$
$= P(1 + N_i)$

Compounded Interest Charges: In this case, the lifetime, N, is divided into smaller periods, typically called interest periods (such as one month or often one year). The interest fee is charged at the end of each interest period and is allowed to accumulate from one interest period to the next. The interest charges, I_k, for the total mount, F_k, accumulated after k periods is

$$I_k = iF_k \tag{3.4}$$

Therefore, the total payment, F_{k+1}, at the end of period k (and thus at the beginning of period k+1) is

$$F_{k+1} = F_k + I_k = (1 + i)F_k \tag{3.5}$$

$F_k + i F_k$
$\approx F_k(1 + i)$

If the principal is P which is the total payment at the beginning of first period (i.e., $F_1 = P$), the total amount due at the end of life time N is

$$F = F_{N+1} = (1 + i)^N P \tag{3.6}$$

It is clear that the total amount of money, F, increases exponentially with N. The interest charges follow the law of compounded interest.

For the economic analysis of energy efficiency projects, the interest rate is typically assumed to be constant throughout the lifetime of the projects. Therefore, it is common to use average interest rates when an economic analysis is performed for energy efficiency projects. Table 3.1 provides historical data for long term interest rates for selected countries.

Table 3.1: Average long term interest rates for selected countries (*Source:* OCDE, 1999)

Period/Year	FRANCE	GERMANY	JAPAN	UNITED STATES
Period				
1961-1973	6.9	7.2	7.0	5.3
1974-1980	11.2	8.1	8.0	8.6
1981-1990	12.0	7.8	6.5	10.3
1990-1995	8.5	7.5	5.1	7.2
Year				
1995	7.7	6.9	3.4	6.6
1996	6.5	6.2	3.1	6.4
1997	5.7	5.6	2.3	6.4
1998	5.1	5.1	1.8	5.9
1999*	5.5	5.5	2.0	6.1

*: based on predictions

Example 3.1: *A building owner has $10,000 available and has the option to invest this money in either (i) a bank that has an annual interest rate of 7% or (ii) buying a new boiler for the building. If he decides to invest all the money in the bank, how much will the building owner have after 10 years? Compare this amount if simple interest had been paid.*

Solution: if the interest is compounded, using Eq. (3.6) with P=$10,000; N=10; and i=0.07; the investment will accumulate to the total amount F:

$$F = \$10,000 * (1 + 0.07)^{10} = \$19,672$$

Thus, the building owner's original investment will have almost doubled over the 10-year period.

If simple interest had been paid, the total amount that would have accumulated is slightly less and is determined from Eq. (3. 3):

$$F = \$10,000 * (1 + 0.07 * 10) = \$17,000$$

3.2.2   Inflation Rate

Inflation occurs when the cost of goods and services increases from one period to the next. While the interest rate, i, defines the cost of money, the inflation rate, λ, measures the increase in the cost of goods and services. Therefore, the future cost of a commodity, FC, is higher than the present cost, PC, of the same commodity:

$$FC = PC(1+\lambda) \tag{3.7}$$

Over a lifetime, N, the future cost of a commodity increases exponentially:

$$FC = PC(1+\lambda)^N \tag{3.8}$$

Similar to the interest rates, the inflation rates are typically assumed to be constant over the lifetime of the energy retrofit project. Table 3.2 presents historical data for inflation rates for selected countries.

Table 3.2: Average inflation rates for selected countries (*Source:* OCDE, 1999)

PERIOD/YEAR	FRANCE	GERMANY	JAPAN	UNITED STATES
Period				
1971-1980	9.8	5.0	8.8	7.1
1981-1990	6.2	2.5	2.1	4.7
1990-1995	1.9	2.8	1.0	2.6
Year				
1991	3.2	3.7	2.5	9.2
1992	2.4	4.7	1.9	3.3
1993	2.2	4.1	1.2	2.7
1994	2.1	3.0	0.7	2.4
1995	1.6	1.7	-0.5	2.6
1996	1.8	2.0	0.1	2.4
1997	1.2	1.9	1.7	2.0
1998*	1.0	1.7	0.5	1.0

*: based on predictions

The escalation of energy cost is an important factor to consider in evaluating energy retrofit projects. The energy escalation rate can be considered as one form of inflation rate.

If the interest charges are compounded at the same period during which inflation occurs, the future worth can be determined from the present value, P, as follows

$$F = P \frac{(1+i)^N}{(1+\lambda)^N}$$ (3.9)

The expression above for F can be rearranged as follows:

future
value
$$F = P \left(\frac{1+i}{1+\lambda} \right)^N = P \left(1 + \frac{i-\lambda}{1+\lambda} \right)^N$$ (3.10)
real interest

A composite interest rate, θ, can be defined to account for the fact that inflation decreases the buying power of money due increases in the cost of commodities:

$$\theta = \frac{i-\lambda}{1+\lambda}$$ (3.11)

It should be mentioned that theoretically the composite interest rate can be negative. In this case, the money loses its value with time.

Example 3.2: *Determine the actual value of the $10,000 investment for the building owner of example 3.1 if the economy experiences an annual inflation rate of 4%.*

Solution: The composite interest rate can be determined using Eq. (3.11):

$$\theta = \frac{i-\lambda}{1+\lambda} = \frac{0.07 - 0.04}{1 + 0.04} = 0.02885$$

Using Eq. (3.6) with P=$10,000; N=10; and i=0.02885; the investment will accumulate to the total amount F:

$$F = \$10,000*(1+0.02885)^{10} = \$13,290$$

3.2.3   Tax Rate

In most economies, the interest that is received from an investment is subject to taxation. If this taxation has a rate, t, over a period that coincides with the interest period, then the amount of taxes, T, to be collected from an investment, P, with an interest rate, i, is determined as follows:

$$i = \frac{I}{P}$$

$$T = tiP \tag{3.12}$$

Therefore, the net return from the investment, P, to the investor after tax deductions is:

$$I' = I - T = (1-t)iP \qquad = I - tiP \\ = iP - tiP \tag{3.13} \\ = (1-t)ip$$

An effective interest rate, i′, can then be defined to account for the loss of income due to taxation:

$$i' = (1-t)i \tag{3.14}$$

Therefore, the composite interest rate defined by Eq. (3.11) can be generalized to account for both inflation and tax rates related to present and future values:

$$\theta = \frac{(1-t)i - \lambda}{1 + \lambda} \tag{3.15}$$

Example 3.3: *If the building owner is in the 28% tax bracket, determine the actual value of his $10,000 investment considered in example 3.1 if the economy experiences an annual inflation rate of 4%.*

Solution: The composite interest rate can be determined using Eq. (3.15):

$$\theta = \frac{(1-t)i - \lambda}{1 + \lambda} = \frac{(1 - 0.28) * 0.07 - 0.04}{1 + 0.04} = 0.01$$

Using Eq. (3.6) with P=$10,000; N=10; and i=0.01; the investment will accumulate to the total amount F:

$$F = \$10,000 * (1 + 0.01)^{10} = \$11,046$$

$$F = P * \left\{ 1 + \left(\frac{(1-t)i - \lambda}{1 + \lambda} \right) \right\}^N$$

3.2.4   Cash Flows

Comp. int Rate

In evaluating energy efficiency projects, it is important to account for the total cash receipts and disbursements due to an implementation of an energy conservation measure (such as the installation of a new boiler) for each period during the entire lifetime of the project. The difference between the total cash receipts (*inflows*) and total cash disbursements (*outflows*) for a given period of time is called a *cash flow*.

Over the lifetime of project, an accurate accounting of all the cash flows should be performed. For energy efficiency improvement projects, the cash flow accounting can be in a tabular format as illustrated in Table 3.3 that lists the cash flows attributed to the installation of a new steam boiler in a hospital. Table 3.3 accounts for the cost related to the initial cost of a new boiler installation (counted as a disbursement for Year 0) and the cost savings due to the higher energy efficiency of the new boiler (counted as receipts in Year 1 through Year 10). The reduction in the yearly receipts is attributed to the aging of the equipment.

Table 3.3: Cash flows for an installation of a new boiler over a lifetime of 10 years.

END OF YEAR	TOTAL CASH RECEIPTS	TOTAL CASH DISBURSEMENTS	TOTAL CASHFLOWS	COMMENTS
0	$ 0	$ 400,000	– $ 400,000	Installation cost of a new boiler
1	$ 40,000	$ 0	+ $ 40,000	Net cost savings
2	$ 38,000	$ 0	+ $ 38,000	" "
3	$ 36,000	$ 0	+ $ 36,000	
4	$ 34,000	$ 0	+ $ 34,000	
5	$ 33,000	$ 0	+ $ 33,000	
6	$ 32,000	$ 0	+ $ 32,000	
7	$ 31,000	$ 0	+ $ 31,000	
8	$ 30,500	$ 0	+ $ 30,500	
9	$ 30,000	$ 0	+ $ 30,000	
10	$ 29,500	$ 0	+ $ 29,500	

Note that the cash flows are positive when they represent inflows (i.e., receipts) and are negative when they are outflows (i.e., disbursements). To better visualize the evolution of the cash flows over time, *a cash flow diagram* as depicted in Figure 3.1 is used. Note that in Fig. 3.1, the initial cash flow, $C_0 = -$ \$400,000 (disbursement), is represented by a downward-pointing arrow. Meanwhile, the cash flows occurring later, C_1 through C_N with $N=10$, are receipts and are represented by upward-pointing arrows.

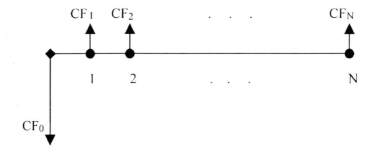

Figure 3.1: Typical cash flow diagram

It should be noted again that the cash flows cannot simply be added since the value of money changes from one period to the next. In the next section, various factors are defined to correlate cash flows occurring at different periods.

3.3 Compounding Factors

Two types of payment factors are considered in this section. These payment factors are useful in the economical evaluation of various energy audit projects. Without loss of generality, the interest period is assumed in the remainder of this chapter to be one year. Moreover, a nominal discount rate, d, is used throughout this chapter. This discount rate is an effective interest rate that includes various effects such as inflation and taxation discussed above.

3.3.1 Single Payment

In this case, an initial payment is made to implement a project by borrowing an amount of money, P. If this sum of money earns interest at a discount rate , d, then the value of the payment, P, after N years, is provided by Eq. (3.6). The ratio F/P is often called single payment compound-amount factor (SPCA). The SPCA factor is a function of i and N and is defined as:

$$SPCA(d, N) = F / P = (1 + d)^N \qquad (3.16)$$

Using the cash flow diagram of Figure 3.1, the single payment represents the case where $C_0 = P$, $C_1 = \ldots = C_{N-1} = 0$, and $C_N = F$ as illustrated in Figure 3.2.

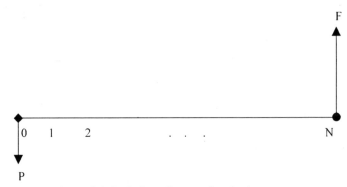

Figure 3.2 Cash flow diagram for single payment

The inverse ratio P/F allows one to determine the value of the cash flow P needed to attain a given amount of cash flow F after N years. The ratio P/F is called single payment present worth (SPPW) factor and is equal to:

$$SPPW(d, N) = P/F = (1+d)^{-N} \qquad (3.17)$$

3.3.2   Uniform-Series Payment

In the vast majority of energy retrofit projects, the economic benefits are estimated annually and are obtained after a significant initial investment. It is hoped that during the lifetime of the project, the sum of all the annual benefits can surpass the initial investment.

Consider then an amount of money, P,   that represents the initial investment, and a receipt of an amount A that is made each year and represents the cost savings due to the retrofit project. To simplify the analysis, the amount A is assumed to be the same for all the years during the lifetime of the project. Therefore, the cash flows – using the diagram of Fig. 3.1 – are $C_0=P$, $C_1=...=C_N=A$ as depicted in Figure 3.3.

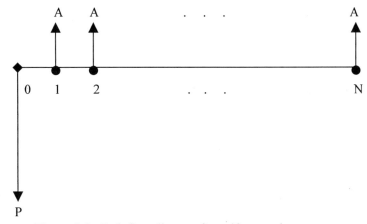

Figure 3.3: Cash flow diagram for uniform-series payment

To correlate between P and A, we note that for any year k, the present worth, P_k, of the receipt A can be determined by using Eq. (3.17):

$$P_k = A(1+d)^{-k} \tag{3.18}$$

By summing the present worth values for all the annual receipts A, the result should equal to the cash flow, P:

$$P = \sum_{k=1}^{N} P_k = \sum_{k=1}^{N} A(1+d)^{-k} \tag{3.19}$$

The sum can be rearranged to obtain a geometric series that can be evaluated as shown below:

$$P = \frac{A}{(1+d)^N} \sum_{k=1}^{N} (1+d)^{N-k} = \frac{A}{(1+d)^N} \sum_{k=0}^{N-1} (1+d)^{k} = \frac{A[(1+d)^N - 1]}{d(1+d)^N} \tag{3.20}$$

The ratio A/P is called uniform-series capital recovery factor (USCR). This USCR factor can be determined as a function of both d and N:

$$USCR(d,N) = A/P = \frac{d(1+d)^N}{(1+d)^N - 1} = \frac{d}{1-(1+d)^{-N}} \tag{3.21}$$

The uniform-series present worth factor (USPW), that allows us to calculate the value of P knowing the amount A, is the ratio P/A and can be expressed as follows:

$$USPW(d,N) = P/A = \frac{(1+d)^N - 1}{d(1+d)^N} = \frac{1-(1+d)^{-N}}{d} \qquad (3.22)$$

Example 3.4: Find the various compounding factors for N=10 years and d=5%.

Solution: The values of the compounding factors for d=0.05 and N= 10 years are summarized below:

Compounding factor	Equation Used	Value
SPCA	Eq. (3.16)	1.629
SPPW	Eq. (3.17)	0.613
USCR	Eq. (3.21)	0.129
USPW	Eq. (3.22)	7.740

3.4   Economic Evaluation Methods Among Alternatives

To evaluate the cost-effectiveness of energy retrofit projects, several evaluation tools can be considered. The basic concept of all these tools is to compare among the alternatives the net cash flow that results during the entire lifetime of the project. As discussed earlier, a simple addition of all the cash flows such as those represented in Figure 3.1 is not possible. However, by using compound factors discussed in section 3.3, "conversion" of the cash flows from one period to another is feasible. This section provides a brief description of the common evaluation methods used in engineering projects.

3.4.1   Net Present Worth

The basic principle of this method is to evaluate the present worth of the cash flows that occur during the lifetime of the project. Referring to the cash flow diagram of Fig. 3.1, the sum of all the present worth of the cash flows can be obtained by using the single payment present worth factor defined in Eq. (3.17):

$$NPW = -CF_0 + \sum_{k=1}^{N} CF_k * SPPW(d,k) \qquad (3.23)$$

Note that the initial cash flow is negative (a capital cost for the project) while the cash flows for the other years are generally positive (revenues).

In the particular but common case of a project with a constant annual revenue (due to energy operating cost savings), $CF_k = A$, the net present worth is reduced to:

$$NPW = -CF_0 + A * USPW(d, N) \qquad (3.24)$$

For the project to be economically viable, the net present worth has to be positive or at worst zero ($NPW \geq 0$). Obviously, the higher is the NPW, the more economically sound is the project.

The net present worth value method is often called the net savings method since the revenues are often due to the cost savings from implementing the project.

3.4.2   Rate of Return

In this method, the first step is to determine the specific value of the discount rate, d', that reduces the net present worth to zero. This specific discount rate is called the rate of return (ROR). Depending on the case, the expression of NPW provided in Eq. (3.23) or Eq. (3.24) can be used. For instance, in the general case of Eq. (3.23), the rate of return, d', is solution of the following equation:

$$-CF_0 + \sum_{k=1}^{N} CF_k * SPPW(d', k) = 0 \qquad (3.25)$$

To solve accurately this equation, any numerical method (such Newton-Raphson iteration method) can be used. However, an approximate value of d' can be obtained by trial and error. This approximate value can be determined by finding the two d-values for which the NPW is slightly negative and slightly positive, and then interpolate linearly between the two values. It is important to remember that a solution for ROR may not exist.

Once the rate of return is obtained for a given alternative of the project, the actual market discount rate or the minimum acceptable rate of return is compared to the ROR value. If the value of ROR is larger (d'>d), the project is cost-effective.

3.4.3   Benefit-Cost Ratio

The benefit-cost ratio (BCR) method is also called the savings-to-investment ratio (SIR) and provides a measure of the net benefits (or savings) of the project relative to its net cost. The net values of both benefits (B_k) and costs (C_k) are computed relative to a base case. The present worth of all the cash flows are typically used in this method. Therefore, the benefit-cost ratio is computed as follows:

$$BCR = \frac{\sum_{k=0}^{N} B_k * SPPW(d,k)}{\sum_{k=0}^{N} C_k * SPPW(d,k)} \qquad (3.26)$$

The alternative option for the project is considered economically viable relative to the base case when the benefit-cost ratio is greater than one (BCR>1.0).

3.4.4   Payback Period

In this evaluation method, the period Y (expressed typically in years) required to recover an initial investment is determined. Using the cash flow diagram of Fig. 3.1, the value of Y is solution of the following equation:

$$CF_0 = \sum_{k=1}^{Y} CF_k * SPPW(d,k) \qquad (3.27)$$

If the payback period Y is less than the lifetime of the project N (Y<N), then the project is economically viable. The value of Y obtained using Eq. (3.27) is typically called discounted payback period (DPB) since it includes the value of money.

In the vast majority of applications, the time value of money is neglected in the payback period method. In this case, Y is called simple payback period (SPB) and is solution of the following equation:

$$CF_0 = \sum_{k=1}^{Y} CF_k \qquad (3.28)$$

In the case where the annual net savings are constant ($CF_k = A$), the simple payback period can be easily calculated as the ratio of the initial investment over the annual net savings:

$$Y = \frac{CF_0}{A} \qquad (3.29)$$

The values for the simple payback period are shorter than for the discounted payback periods since the undiscounted net savings are greater than their discounted counterparts. Therefore, acceptable values for simple payback periods are typically significantly shorter than the lifetime of the project.

3.4.5   Summary of Economic Analysis Methods

Table 3.4 summarizes the basic characteristics of the economic analysis methods used to evaluate single alternatives of an energy retrofit project.

Table 3.4: Summary of the basic criteria for the various economic analysis methods for energy conservation projects.

EVALUATION METHOD	EQUATION	CRITERIO
Net Present Worth (NPW)	$NPW = -CF_0 + \sum_{k=1}^{N} CF_k * SPPW(d,k)$	NPW > 0
Rate of Return (ROR)	$-CF_0 + \sum_{k=1}^{N} CF_k * SPPW(d',k) = 0$	d' > d
Benefit Cost Ratio (BCR)	$BCR = \dfrac{\sum_{k=0}^{N} B_k * SPPW(d,k)}{\sum_{k=0}^{N} C_k * SPPW(d,k)}$	BCR > 1
Discounted Payback Period (DPB)	$CF_0 = \sum_{k=1}^{Y} CF_k * SPPW(d,k)$	Y < N
Simple Payback Period (SPB)	$Y = \dfrac{CF_0}{A}$	Y << N

It is important to note that the economic evaluation methods described above provide an indication of whether or not a single alternative of a retrofit project is cost-effective. However, these methods cannot be used or relied on to compare and rank various alternatives for a given retrofit project. Only the life-cycle cost (LCC) analysis method is appropriate for such endeavor.

> **Example 3.5**: *After finding that the old boiler has an efficiency of only 60% while a new boiler would have an efficiency of 85%, a building owner of Example 3.1 has decided to invest the $10,000 in getting a new boiler. Determine whether or not this investment is cost-effective if the lifetime of the boiler is 10 years and the discount rate is 5%. The boiler consumes 5,000 gallons per year at a cost of $1.20 per gallon. An annual maintenance fee of $150 is required for the boiler (independently of its age). Use all the five methods summarized in Table 3.4 to perform the economic analysis.*

Solution: The base case for the economic analysis presented in this example is the case where the boiler is not replaced. Moreover, the salvage value of the boiler is assumed insignificant after 10 years. Therefore, the only annual cash flows (A) after the initial investment on a new boiler are the net savings due to higher boiler efficiency as calculated below:

$$A = Fuel - Use_{before} * (1 - \eta_{before} \Big/ \eta_{after}) * Fuel - \cos t \,/\, gallon$$

Thus,

$$A = 5000 * (1 - 0.60 \Big/ 0.85) * \$1.20 = \$1,765$$

The cost-effectiveness of replacing the boiler is evaluated as indicated below:

(1) *Net Present Worth*. For this method $CF_0=\$10,000$ and $CF_1=...=CF_{10}=A$, $d=0.05$, and $N=10$ years. Using Eq. (3.24) with USPW=7.740 (see example 3.4):

$$NPW=\$3,682$$

Therefore, the investment in purchasing a new boiler is cost-effective.

(2) *Rate of return*. For this method also $CF_0=\$10,000$ and $CF_1=...=CF_{10}=A$, while SPPW(d',k) is provided by Eq. (3.17). By trial and error, it can be shown that the solution for d' is:

$$d'=12.5\%$$

Since $d'>d=5\%$, the investment in replacing the boiler is cost-effective.

(3) *Benefit Cost Ratio*. In this case, $B_0=0$ and $B_1=...=B_{10}=A$ while $C_0=\$10,000$ and $C_1=...=C_{10}=0$.

Note that since the maintenance fee is applicable to both old and new boiler, this cost is not accounted for in this evaluation method (only the benefits and costs relative to the base case are considered).

Using Eq. (3.26):

$$BCR = 1.368$$

Thus, the benefit cost ratio is greater than unity (BCR >1) and the project of getting a new boiler is economically feasible.

(4) *Compounded Payback Period.* For this method, $CF_0=\$10,000$ and $CF_1=\ldots=CF_{10}=A$. Using Eq. (3.27), Y can be solved Y = 6.9 year.

Thus, the compounded payback period is shorter than the lifetime of the project (Y>N=10years) and therefore replacing the boiler is cost-effective.

(5) *Simple Payback Period.* For this method, $CF_0=\$10,000$ and A=\$1,765. Using Eq. (3.29), Y can be easily determined: Y = 5.7 years.

Thus, the simple payback period method indicates that the boiler retrofit project can be cost-effective.

3.5 Life-Cycle Cost Analysis Method

The Life-Cycle Cost (LCC) analysis method is the most commonly accepted method used to assess the economic benefits of energy conservation projects over their lifetime. Typically, the method is used to evaluate at least two alternatives of a given project (for instance, evaluate two alternatives for the installation of a new HVAC system: a VAV system or a heat pump system to condition the building).   Only one alternative will be selected for implementation based on the economic analysis.

The basic procedure of the LCC method is relatively simple since it seeks to determine the relative cost effectiveness of the various alternatives. For each alternative including the base case, the total cost is computed over the project lifetime. The cost is commonly determined using one of two approaches: the present worth or the annualized cost estimate. Then, the alternative with the lowest total cost (or LCC) is typically selected.

Using the cash flow diagram of Figure 3.1, the LCC amount for each alternative can be computed by projecting all the costs (including costs of acquisition, installation, maintenance, and operating of the energy systems related to the energy-conservation project) on either :

(i)   One single present value amount. This single cost amount can be computed as follows:

$$LCC = \sum_{k=0}^{N} CF_k * SPPW(d,k) \tag{3.30}$$

This is the most commonly used approach in calculating LCC in energy retrofit projects.

Or

(ii)   Multiple annualized costs over the lifetime of the project:

$$LCC_a = USCR(d,N)*\left[\sum_{k=1}^{N}CF_k*SPPW(d,k)\right]$$

(3.31)

Note that the two approaches for calculating the LCC values are equivalent.

Example 3.6: *The building owner of Example 3.5 has three options to invest his money as briefly described below*

(A) Replace the entire older boiler (including burner) with a more efficient heating system. The old boiler/burner system has an efficiency of only 60% while a new boiler/burner system has an efficiency of 85%. The cost of this replacement is $10,000.
(B) Replace only the burner of the old boiler. This action can increase the efficiency of the boiler/burner system to 66%. The cost of the burner replacement is $2,000.
(C) Do nothing and replace neither the boiler nor the burner

Determine the best economical option for the building owner. Assume that the lifetime of the retrofit project is 10 years and the discount rate is 5%. The boiler consumes 5,000 gallons per year at a cost of $1.20 per gallon. An annual maintenance fee of $150 is required for the boiler (independently of its age). Use all the life cycle cost analysis method to determine the best option.

Solution: The total cost of operating the boiler/burner system is considered for the three options. In this analysis, the salvage value of the boiler or burner is neglected. Therefore, the only annual cash flows (A) after the initial investment on a new boiler are the maintenance fee and the net savings due to higher boiler efficiency. To present the calculations for LCC analysis, it is recommended to present the results in a tabular format and proceed as shown below:

Cost Item	Option A	Option B	Option C
Initial Investment			
(a) Replacement Cost ($)	10,000	2,000	0
Annual Operating Costs:			
(b) Fuel Use (gallons)	3,530	4,545	5,000
(c) Fuel Cost ($) [$1.2*(b)]	4,236	5,454	1,000
(d) Maintenance fee ($)	150	150	150
(f) Total Operating Cost ($)			
[(c)+(d)]	4,386	5,594	6,150

USPW factor
[d=5%, N=10, Eq. (3.26)]   7.740   7.740   7.740

Present Worth ($)
[(a)+USPW*(f)]   43,948   45,298   47,601

Therefore, the life cycle cost for option A is the lowest. Thus, it is recommended for the building owner to replace the entire boiler/burner system.

This conclusion is different from that obtained by using the simple payback analysis [Indeed, the payback period for option A – relative to the base case C – is SPB(A)=($10,000)/($1,765)=5.66 years; while for option B, SPB(B)=($2,000)/ ($ 546) = 3.66 years].

Note: If the discount rate was d=10% (which is unusually high for most markets), the USPW would be equal to USPW=6.145 and the life-cycle cost for each option will be

LCC(A)= $36,952   LCC(B)=$36,375   LCC(C)=$37,791

Therefore, Option B will become the most effective economically and will be the recommended option to the building owner.

3.6   General Procedure for an Economic Evaluation

It is important to remember that the recommendations for energy conservation projects that stem from an energy audit should be based on an economically sound analysis. In particular, the auditor should ask several questions before making the final recommendations such as:

(a) Will project savings exceed costs?
(b) Which design solution will be most cost-effective?
(c) What project size will minimize overall building costs?
(d) Which combination of interrelated projects will maximize net savings?
(e) What priority should projects be given if the owner has limited investment capacity?

As alluded to earlier, the best suitable economic assessment method is the LCC method described in section 3.6. Before the application of the LCC, several data are needed to perform an appropriate and meaningful economic analysis. To help the auditor in gathering the required information and in the application of the LCC method, the following systematic approach in any economic evaluation is proposed:

1) Define the problem that the proposed retrofit project is attempting to address and state the main objective of the project. [For instance, a building has an old boiler that does not provide enough steam to heat the entire building. The project is to replace the boiler with main objective to heat all the conditioned spaces within the building].

2) Identify the constraints related to the implementation of the project. These constraints can vary in nature and include financial limitations or space requirements. [For instance, the new boiler cannot be gas fired since there is no supply of natural gas near the building].

3) Identify technically sound strategies and alternatives to meet the objective of the project. [For instance, three alternatives can be considered for the old boiler replacement: (i) a new boiler with the burner of the old boiler, (ii) a new boiler/burner system, and (iii) a new boiler/burner system with an automatic air-fuel adjustment control].

4) Select a method of economic evaluation. When several alternatives exist including the base case (which may consist of the alternative of "doing nothing"), the LCC method is preferred for energy projects. When a preliminary economic analysis is considered, the simple payback period method can be used. As mentioned earlier, the payback period method is not accurate and should be used with care.

5) Compile data and establish assumptions. The data include the discount rates, the energy costs, the installation costs, operating costs, and maintenance costs. Some of these data are difficult to acquire and some assumptions or estimations are required. For instance an average discount rate over the life cycle of the project may be assumed, based on a historical data.

6) Calculate indicators of economic performance. These indicators depend on the economic evaluation method selected. The indicators are the life-cycle costs (LCCs) for the LCC method.

7) Evaluate the alternatives. This evaluation can be performed by simply comparing the values of LCC obtained for various alternatives.

8) Perform sensitivity analyses. Since the economic evaluation performed in step 6 is typically on some assumed values (for instance the annual discount rate), it is important to determine whether or not the results of the evaluation performed in step 7 depend on some of these assumptions. For this purpose, the economic evaluation is repeated for all alternatives using different plausible assumptions.

9) Take into account unqualified effects. Some of the alternatives may have effects that cannot be included in the economic analysis but may be determining factors in decision making. For instance, the environmental impact (emission of pollutants) can disqualify an otherwise economically sound alternative.

10) Make recommendations. The final selection will be based on the findings of the three previous steps (i.e., steps 7, 8, and 9). Typically, the alternative with the lowest LCC value will be recommended.

Once the project for energy retrofit is selected based on an economic analysis, it is important to decide on the financing options to actually carry out the project and implement the measures that allow a reduction in energy cost of operating the facility. The next section discusses the common payment and financing options typically available for energy retrofit projects.

3.7   Financing Options

There are several alternatives that the owner or the facility manager can use to finance an energy retrofit project. These alternatives can be found under three main categories:

(a) Direct Purchasing
(b) Leasing
(c) Performance Contracting

Each of the above listed financing options is briefly described in the following sections with some indications of their advantages and disadvantages.

3.7.1   Direct Purchasing

This category includes all the financing alternatives where the facility (through its representatives) purchases the equipment and the services required to implement the energy retrofit projects either using its own money or through a loan.

Typically, the facility uses its own money to purchase the equipment and services when it has strong cash reserves and/or when the measures to be implemented are simple and are inexpensive with a payback period less than one year. The advantages of using its own funds are that (i) the organization benefits directly from any cost savings realized from the retrofit project, and (ii) the depreciation value of the purchased equipment can be deducted from the taxes. The main disadvantage is a loss of capital available for other investment opportunities.

The facility can take a loan to finance the retrofit project. Generally, the lending companies provide only part of the funds required for the implementation of the project. Therefore, the facility has to finance the other part of the project initial cost through its own capital and has to provide assets as security for the loan. The loan payments are generally structured to be lower than estimated energy cost savings. However, the facility bears all the risks associated with the project such as lower energy savings than predicted.

Therefore, only projects with low payback periods (less than one year) are considered for this financing option.

If the management of the operation and maintenance is performed by an external organization (such as the "P4 exploitants" that serves large facilities in France), the financing of the energy profit can be carried out by this external organization. In this case, the facility purchases and gets the ownership of the installed equipment.

3.7.2   Leasing

Instead of purchasing, the facility can lease the equipment required for the energy retrofit project. The laws and regulations for equipment leasing are generally complex and change frequently. It is therefore recommended that a financial expert and/or an attorney be consulted before finalizing any lease agreement. The advantage of leasing is that the payments are typically lower than loan payments. Two types of leasing are commonly available: (i) capital leasing, and (ii) operating leasing.

The capital leasing is one that meets one or more of the following criteria defined by the Financial Accounting Standard (FASB):

- The lease transfers ownership of the property to customer at end of lease term.
- The lease includes a bargain purchase option (i.e. the purchase amount is less than fair market value).
- The lease term covers 75% or more of the estimated economic life of the leased equipment.
- The present value of the minimum lease equals or exceeds 90% of the fair market value of the leased equipment.

Therefore, the capital lease does not require an initial capital payment from the facility. Moreover, the capital lease is recorded on the facility's balance sheet as an asset and a liability.

If a lease does not meet any of the criteria listed above for a capital lease, it is considered an operating lease. Thus, the equipment is leased to the facility for a fixed monthly or annual fee during the contract period. At the end of the contract, the facility has only three option: remove the equipment, renegotiate the lease, or purchase the equipment but only at fair market value. The operating lease is recorded only as a periodic expense for the facility and is not booked on the facility's balance sheet.

3.7.3   Performance Contracting

Performance contracting has become an increasingly common financing option for energy retrofit projects during the last five years. Typically, a third party such as an energy services company (ESCO) obtains financing and assumes the performance risks associated with the energy retrofit project. Generally, the facility does not assume or provide any up-front investment but is rather guaranteed a certain amount from the energy cost savings. The financing organization owns the equipment during the term of the contract. Thus, the equipment asset and the associated debt do not appear on the facility's balance sheet.

The performance contracts are sometimes referred to as "shared savings" or "paid from savings" contracts since the incurred savings due to the energy retrofit project is distributed between the ESCO and the facility based on an agreement that is documented in the performance contract. At the end of the contract, ownership of the equipment is transferred to the facility based on terms specified in the contract. In general, performance contracts are inherently complex and take a long time to negotiate. Indeed, these contracts typically involve:

- Detailed specifications of the work to be performed for each facility.
- Large sums of capital to implement the agreed-on energy retrofit measures.
- Long periods, since the contracts may span over 5 years.
- A wide range of contingencies.
- High level of expertise in various disciplines including engineering, finances, and law.

Therefore, the energy retrofit project has to have high potential for energy savings for the ESCO and the financial institution to commit to performance contracting. Therefore, the ESCO typically spends a significant amount of time, effort, and resources to perform the energy audit and establish a sound baseline from which energy savings can be estimated.

3.8   Summary

An energy audit of commercial and industrial buildings encompasses a wide variety of tasks and requires expertise in a number of areas to determine the best energy conservation measures suitable for an existing facility. This chapter described a general but a systematic approach to perform the economic evaluation of various alternatives of energy retrofit projects.

The analysis methods (such as the net present worth, the benefit cost ratio, the rate of return, and the payback period) suitable for evaluating single alternatives cannot be used to rank alternatives. Only the life-cycle cost (LCC) analysis method should be used to select the most economical option among alternatives of the same project. Under certain market conditions, the simple payback method can lead to erroneous conclusions and thus should be used only to provide an indication of the cost-effectiveness of an energy retrofit project.

For sound economic analysis, the simple payback analysis method should be used.

Problems

3.1 A Company is considering the purchase of a new machine that will last 8 years and cost $90,000 the first year, decreasing by $1,000 each year to $2,000 the eighth year. Determine how much money the company should set aside for this machine:

      (a) If the interest rate is 4% per year, compounded annually.

      (b) If the interest rate is 7% per year, compounded annually.

3.2 An energy audit of a residential heating system reveals that the boiler-burner efficiency is only 60%. In addition, the energy audit showed that an electric water heater is used. The house uses 1,500 gallons of oil annually at a cost of $1.40 per gallon. The total electric bill averaged $84.50 per month for an average monthly consumption of 818 kWh. Of this total, about 35% was for domestic water heating. It suggested that the owner of the house equip the boiler with a tank-less domestic water heater, replacing the existing electric water heater. The existing water heater is 12 years old and has no resale value and little expected life. The owner of the house is expected to spend $300 within the coming year to replace the electric water heater. The cost of the tank-less heater is $400.

To improve heating efficiency, it is suggested to

      (i) Replace the existing burner with a new one (this burner replacement will improve efficiency to 65%) costing $520.

      (ii) Replace the entire heating plant with higher efficiency (85%) with a cost of $2,000.

For two discount rates (5% and 10%), provide the LCC analysis of the following options and make the appropriate recommendations:

- Keep the boiler-burner and replace the electric water heater with a like system.
- Replace burner and electric water heater with like systems.
- Replace boiler-burner with efficient boiler and replace electric heater with a like system.
- Replace existing boiler-burner and electric water heater with a new boiler/take-less domestic water heater.

Note: State all the assumptions made in your calculations.

3.3 An electrical energy audit indicates that the motor control center consumption is 8×10^6 kWh per year. By using high efficiency motors, a savings of 15% can be achieved. The additional cost for these motors is about $80,000. Assuming that the average energy charge is $0.08 per kWh, is the expenditure justified based on a minimum rate of return of 18% before taxes? Assume a 20-year life cycle and use the present worth, annual cost, and rate of return methods.

3.4 A chiller consumes 9.5×10^5 kWh annually with an overall efficiency of 0.90 kW/ton. If this chiller is replaced by a more energy efficient chiller (0.75 kW/ton) at a cost of $95,000. In particular, identify

(a) Determine the simple payback period of the chiller replacement.

(b) If the expected life of the old chiller is 15 years. Is cost-effective to replace the chiller in an economy with a discount rate of 7%?

Assume that the average cost of electricity is $0.06/kWh.

4

ENERGY ANALYSIS
TOOLS

Abstract

This chapter provides a brief overview of various energy simulation tools available to energy auditors to estimate savings incurred from energy conservation measures. The chapter focuses on simplified calculation methods, inverse models, and computer simulation programs. In particular, the basic features of each simulation tool are discussed. The application of the simulation tools to estimate energy savings for energy conservation measures is also outlined.

4.1   Introduction

To analyze energy consumption and estimate the cost-effectiveness of energy conservation measures, an auditor can use a myriad of calculation methods and simulation tools. The existing energy analysis methods vary widely in complexity and accuracy. To select the appropriate energy analysis method, the auditor should consider several factors including speed, cost, versatility, reproducibility, sensitivity, accuracy, and ease of use (Sonderegger, 1985). There are hundreds of energy analysis tools and methods that are used worldwide to predict the potential savings of energy conservation measures. In the US, DOE provides an up-to-date listing of selected building energy software (DOE, 2000).

Generally, the existing energy analysis tools can be classified into either forward or inverse methods. In the forward approach as depicted in Figure 4.1, the energy predictions are based on a physical description of the building systems such as geometry, location, construction details, and HVAC system type and operation. Most of the existing detailed energy simulation tools such as DOE-2, TRNSYS, ENERGYPlus, and BLAST follow the forward modeling approach. In the inverse approach illustrated in Figure 4.2, the energy analysis model attempts to deduce representative building parameters [such as the building load coefficient (BLC), the building base-load, or the building time constant] using existing energy use, weather, any relevant performance data. In general, the inverse models are less complex to formulate than the forward models. However, the flexibility of inverse models is typically limited by the formulation of the representative building parameters and the accuracy of the building performance data. Most of the existing inverse models rely on regression analysis [such as the variable-base degree-day models (Fels 1986) or the change-point models (Kissock et al.1998)] tools or connectionist approach (Kreider et al., 1997) to identify the building parameters.

It should be noted that tools based on the forward or inverse approaches are suitable for other applications. Among the common applications are verification of energy savings actually incurred from energy conservation measures (for more details about this application, the reader is referred to Chapter 16), diagnosis of equipment malfunctions, and efficiency testing of building energy systems.

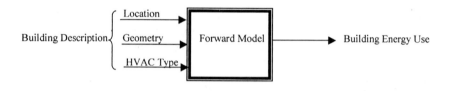

Figure 4.1: Basic approach of a typical forward energy analysis model

Energy analysis tools can also be classified based on their ability to capture the dynamic behavior of building energy systems. Thus, energy analysis tools can use either steady-state or dynamic modeling approaches. In general, the steady-state models are sufficient to analysis seasonal or annual building energy performance. However, dynamic models may be required to assess the transient effects of building energy systems such as those encountered for thermal energy storage systems and optimal start controls.

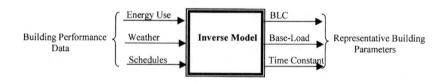

Figure 4.2: Basic approach of a typical inverse energy analysis model

In this chapter, selected energy analysis tools commonly used in the US and Europe are described. These tools are grouped into three categories:

- Ratio-based methods: which are pre-audit analysis approaches that rely on building energy/cost densities to quickly evaluate the building performance.
- Inverse methods: using both steady-state and dynamic modeling approaches including variable-base degree-day methods.
- Forward methods: include either steady-state or dynamic modeling approaches and are often the basis of detailed energy simulation computer programs.

4.2   Ratio-Based Methods

4.2.1   Introduction

The ratio-based methods are not really energy analysis tools but rather pre-audit analysis approaches to determine a specific energy or cost indicators for the building. These energy/cost building indicators are then compared to reference performance indices obtained from several other buildings with the same attributes. The end-use energy or consumption ratios can provide useful insights on some potential problems within the building such as leaky steam pipes, inefficient cooling systems, or high water usage. In particular, the building energy densities or ratios can be useful:

- to determine if the building has high energy consumption and to assess if an energy audit of the building would be beneficial.
- to assess if a pre-set energy performance target has been achieved for the building. If not, the energy ratio can be used to determine the magnitude of the required energy use reduction to reach the target.
- to estimate typical consumption levels for fuel, electricity, and water to be expected for new buildings.
- to monitor the evolution of energy consumption of buildings and estimate the effectiveness and profitability of any energy management program carried out following an audit.

To estimate meaningful reference ratios, large databases have to be collected. Typically, data for thousands of similar buildings and facilities are required to estimate reference indicators. These data have to be screened in order to eliminate any erroneous or implausible data points. A number of statistical analysis procedures can be carried out to screen the databases. Common statistical methods include:

i- Elimination of any data point that has a value which is not part of an interval bounded by 2.5 times the standard deviation centered on the average of the entire data set.

ii- Elimination of any data point that has a value below the 10th percentile or above the 90th percentile.

When new data points are added to the database, the screening process of the data should be repeated. In particular, any data points previously eliminated should be reinstated in the database. To obtain up-to-date reference ratios, the databases have to be regularly screened and updated. It is recommended that the databases be updated at least once every 5 years for energy ratios, and once every two years for the cost ratios.

The ratio-based methods are approaches used only for pre-audit energy diagnostic or energy screening. These methods do not provide sufficient information to conduct a complete and a detailed energy analysis of the building and its systems. However, energy screening tools using ratio-based methods can be useful to quickly assess the energy efficiency of buildings, if properly used. Examples of screening tools developed in the US include BEST (Building Energy Screening Tool) or Scheduler (developed for the Energy Star Buildings program).

4.2.2   Types of Ratios

Energy or cost ratios are typically computed as a fraction made up of a numerator and a denominator. A set of variables is used in the numerator. Another set of specific parameters is suitable for the denominator. For energy ratios, the variables that are typically used for the numerator include:

- Total building energy use (i.e., including all end-uses). This total energy use can be expressed in kWh or MMBtu.
- Building energy consumption by end-use (i.e., heating, cooling, and lighting
- Energy demand (in kW)

For the cost ratios, a monetary value (specifically for the energy expenditure or for the overall building operation) is typically used for the numerator.

For the denominator, several variables can be used depending on the building type and the main goal intended for the computed ratios. Some of the common variables, used in the denominator of energy and cost ratios, are:

- Surface area or space volume (such as heating area or conditioned volume in offices)
- Building users (in collective buildings such as hotels, schools)
- Degree-day [with generally 65°F (18°C) as a base temperature]
- Units of production (especially for manufacturing facilities, restaurants)

In general annual or seasonal values are used to obtain the energy or cost ratios. However, daily or monthly ratios can be considered. The monthly variations of energy ratios are often referred to as building signatures.

4.2.3   Examples of Energy Ratios

Generally, meaningful energy ratios require careful analysis and screening of the data. For instance, it is important to consider effects such the climate and the building function when estimating energy ratios. Sources of building energy data are typically difficult to obtain. In the US, the Energy Information Administration (EIA) provides annually statistical data about the energy use of various building types. In other countries, it is generally very difficult to obtain such data even through governmental agencies. Table 4.2 provides some energy ratios for selected commercial and institutional building types in the US and France. Table 4.3 lists the energy ratios for some US industrial facilities.

It should be stressed that these energy ratios such as those provided in Tables 4.2 and 4.3 should be used only as generic indicators of typical energy use for the listed buildings or facilities.   Energy ratios specified by climate zone, type of HVAC system, and/or building size may be required for an adequate energy screening or pre-audit energy analysis.

Table 4.1: Energy Ratio (or Energy Intensity) by Principal Building Activity in kWh/m^2

Major Building Activity	FRANCE[1]	US[2]
Office	395	300
Education	185	250
Health Care	360	750
Lodging	305	395
Food Service	590	770
Mercantile and Service	365	240
Sports	405	NA[3]
Public Assembly	NA	375
Warehouse and Storage	NA	125

(1) Source: CEREN based on 1993 energy consumption (CEREN, 1997)
(2) Source: EIA using 1995 data (EIA, 1996)
(3) Not Available

Table 4.2: Energy Operating Ratios by Industry Group in the US (*Source:* EIA, 1994)

Industry Group	Energy Use Per Employee (10^3 kWh)	Energy Use per Dollar of value added (kWh)	Energy Use per Dollar of Shipments (kWh)
Food and Kindred Products	236.2	1.9	0.8
Textile Mill Products	151.5	2.8	1.1
Apparel and Other textile Products	13.9	0.3	0.1
Lumber and Wood Products	198.0	2.9	1.3
Furniture and Fixtures	34.3	0.6	0.3
Paper and Allied Products	1,289.0	12.7	5.6
Printing and Publishing	24.9	0.3	0.2
Chemicals and Allied Products	1,197.8	5.6	3.0
Petroleum and Coal Products	8,436.0	33.1	6.5
Rubber and Misc. Plastics Products	93.1	1.2	0.6
Stone, Clay and Glass Product	595.9	6.7	3.9
Primary Metal Industries	1158.3	11.7	4.7
Fabricated Metal Products	82.9	1.2	0.6
Industrial Machinery and equipment	43.1	0.5	0.2
Electronic and Other Electric Equipment	48.5	0.5	0.3
Transportation Equipment	67.8	0.7	0.3
Instruments and Related Products	39.5	0.3	0.2
Misc. Manufacturing Industries	41.0	0.6	0.3

4.3   Inverse Modeling Methods

As discussed in the introduction, methods using the inverse modeling approach rely on existing building performance data to identify a set of building parameters. The inverse modeling methods can be valuable tools in improving the building energy efficiency. In particular, the inverse models can be used to:

- help detect malfunctions by identifying time periods or speficic systems with abnormally high energy consumption,
- provide estimates of expected savings from a defined set of energy conservation measures, and
- verify the savings achieved by energy retrofits.

Typically, regression analyses are used to estimate the representative parameters for the building and/or its systems (such as building load coefficient or heating system efficiency) using measured data. In general, steady-state inverse models are based on monthly and/or daily data and include one or more independent variables. Dynamic inverse models are usually developed using hourly or sub-hourly data to capture any significant transient effect such as the case where the building has a high thermal mass to delay cooling or heating loads.

4.3.1   Steady-State Inverse Models

These models generally attempt to identify the relationship between the building energy consumption and selected weather-dependent parameters such as monthly or daily average outdoor temperatures, degree-hours, or degree days. As mentioned earlier, the relationship is identified using statistical methods (based on linear regression analysis). The main advantages of the steady-state inverse models are:

■ Simplicity: steady-state inverse models can be developed based on a small data set such as energy data obtained from utility bills.
■ Flexibility: steady-state inverse models have a wide range of applications. They are particularly valuable in predicting the heating, cooling energy end-uses for both residential and small commercial buildings.

However, steady-state inverse models have some limitations since they cannot be used to analyze transient effects such as thermal mass effects and seasonal changes in the efficiency of the HVAC system.

Steady-state inverse models are especially suitable for measurement and verification (M&V) of energy savings accrued from energy retrofits. Chapter 16 provides a more detailed discussion about the inverse models that are commonly used for M&V applications. In this section, only simplified methods based on steady-inverse modeling are briefly presented. These simplified models have been used to determine the energy impact of selected energy efficiency measures and are based on the degree-day methods. Two simplified energy analysis approaches are briefly outlined:

■ Cumulative (or aggregated) degree-day method that consists of correlating – using a linear regression analysis – the cumulative building energy use to the cumulative degree-days (using a reference temperature of 18°C [65°F]). Figure 4.3 illustrates the basic concept of the cumulative degree-day method.

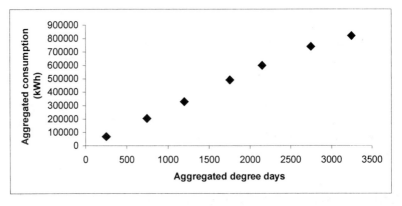

Figure 4.3: Typical application of cumulative (or aggregated) degree-day method.

This method is used in some European countries to monitor the variation of building energy use throughout the heating season. In particular, the cumulative degree-day approach helps easily visualize any changes in building energy use pattern attributed to energy retrofit measure through the slope of the regression line. Any improvement in the building thermal performance (such as addition of thermal insulation of increase in the efficiency of the heating system) will reduce the slope.

■ Variable-base degree-day method that uses a linear regression analysis to estimate the building balance temperature. Chapter 6 discusses the details of the variable-base degree-method. Several energy analysis tools and software have been developed using one form of the variable-base degree-day method. Among these tools are the PRISM (Princeton Scorekeeping Method) used in the US (Fels, 1986) to analyze energy use for residential and small commercial buildings, and the ANAGRAM ("ANalyse GRAphique Mensuelle des consommations") software developed in France by GDF, (1985) specifically to estimate monthly heating energy use for buildings. Both of these programs are briefly described below.

ANAGRAM method

Using the ANAGRAM approach, the annual building heating energy use is first calculated as follows:

$$E_H = 24 \times \frac{\text{BLC}_V}{\eta_H} \times V_B \times \text{DD}_H \times I \qquad (4.1)$$

Where,
- E_H is the annual building heating energy consumption (in kWh)
- BLC_V is the building loss coefficient based on the building volume (in $kW/m^3 \cdot °C$)
- η_H is the average seasonal energy efficiency of the heating system
- V_B is the heated building volume (m^3)
- DD_H is the heating degree-days (based on 18°C)
- I is a correction factor to account for the effects of night setback coefficient in reducing the building heating load (if there is no night setback, I=1).

Then, ANAGRAM proceeds to calculate the monthly heating energy use using the following expression:

$$E_{H,m} = 24 \times \frac{\text{BLC}_V}{\eta_H} \times V_B \times I \times [\text{DD}_{H,m} - (18 - T_b) \times 30] \qquad (4.2)$$

Where:
- $E_{H,m}$ is the month building heating energy use (in kWh)
- $\text{DD}_{H,m}$ is the heating degree-days (based on 18°C)
- T_b is the building balance temperature. T_b is defined as the outdoor temperature for which the building does not need any heating.
- BLC_V, I, V_B, η_H have the same definition as in Eq. (4.1)
- 30 is the number of days in a month (which corresponds the number of days included in the GDF utility bills).

Thus, a linear regression is carried out to correlate the monthly building energy use to the monthly degree-days (based on 18°C) using only heating season data. This regression analysis provides an estimation of the building balance temperature, T_b, and the ratio of BLC_V/η_H and is the average seasonal energy efficiency of the heating system. Example 4.1 illustrates how the ANAGRAM approach is used to analyze monthly building energy use.

Example 4.1: *Illustration of the ANAGRAM method.*

Using the ANAGRAM approach, analyze the heating energy use data for a building having a heated volume of 15,000 m^3. Monthly energy consumption and degree-days data are provided below for two years, 0 and 1.

	Monthly Degree Days	Monthly Consumption (MWh) Year 0	Monthly Consumption (MWh) Year 1
September (S)	115	0	0
October (O)	235	80	20
November (N)	375	200	80
May (My)	185	60	10
March (M)	465	260	110
June (Ju)	105	0	0
January (J)	580	340	150
Febuary (F)	535	310	135
December (D)	485	280	120
April (A)	310	140	50

The results of the linear regression analysis of the monthly heating energy consumption with degree days are provided in the graph below. The regression analysis is based on the model described by Eq. (4.2) with I=1 (no temperature setback).

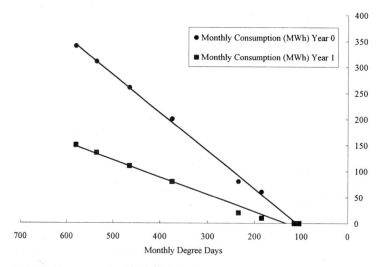

For year 0, the regression line intersects the X-axis at degree-days DD$_{H.,m}$= 120 °C-days. Thus, the balance temperature, T$_b$, can be estimated as estimated as follows: [DD$_{H,m}$-(18-T$_b$)x30]=0. Therefore, for year 0, the balance temperature is:

$$T_b=18-120/30=14°C$$

Using the slope of the regression line (which is 570 kWh/°C-days), the ratio BLC_V/η_H can be estimated using the fact that:

$$570 \text{ kWh/°C-days} = 24*BLC_V/\eta_H *V_B,$$

Since $V_B = 15,00 \text{ m}^3$, it is estimated that:

$$BLC_V/\eta_H = 1.58 \text{ W/m}^3.°C$$

Similar analysis can be carried out for year 1. It is found that between year 0 and 1, the balance temperature has decreased from 14°C to 13.5°C, while the ratio BLC_V/η_H is reduced from 1.58 W/m^3.°C to 0.70 W/m^3.°C. It is most likely that this reduction in both T_b and BLC_V/η_H is attributed to improvement in the energy efficiency of the building envelope.

4.3.2   PRISM method

PRISM correlates the building energy use per billing period to heating or cooling degree-days (obtained for the billing period). Thus the energy consumption is estimated for each billing period using the following expression:

$$E_{H/C} = 24 \times \frac{BLC}{\eta_{H/C}} \times DD_{H/C}(T_{b,H/C}) + E_{base,H/C} \qquad (4.3)$$

Where,

- $E_{H/C}$ represents the annual building energy use during heating or cooling season
- BLC is the building loss coefficient
- $\eta_{H/C}$ is the average seasonal energy efficiency of the heating or the cooling system
- $T_{b,H/C}$ is the building balance temperature for heating or cooling energy use.
- DD_H is the heating or cooling degree-days (based on the balance temperature). Chapter 6 discusses in more details how variable base degree-days are defined. For heating, for instance, the heating degree days is estimated as follows: $DD_H(T_b) = \sum_j [T_b - T_{o,j}]^+$ with $T_{o,j}$ is the outdoor air temperature at day j.
- $E_{base,H/C}$ is the base-load for building energy use. It represents the non-heating or non-cooling energy use.

Through regression analysis, the balance temperature and the building load coefficient (assuming the heating or cooling system efficiency is known) can be determined. With these two parameters determined, PRISM tool can be used to establish an energy use model for the building and to determine any energy savings attributed to measures that affect one of the three parameters, balance temperature, building load coefficient, or heating/cooling system efficiency. One variation of the PRISM approach represented by Eq. (4.3) is to use the average outdoor temperature instead of the variable-base degree-day. This method is described in Chapter 6.

4.3.3   Dynamic models

Steady-state inverse models are only suitable for predicting long-term building energy use. Therefore, energy use data is collected for a relatively long time period (at least one season or one year) to carry out the regression analysis. On the other hand, dynamic inverse models can be used to predict short-term building energy use variations using data collected for a short period of time such as one week. Generally, a dynamic inverse model is based on a building thermal model that uses a specific set of parameters. These building model parameters are identified using typically some form of a regression analysis. Chapter 16 discusses other approaches to develop dynamic inverse models suitable for predict building energy use.

An example of a dynamic model relating building cooling energy use to the outdoor air temperatures at various time steps (typically hours) is presented by Eq. (4.4):

$$E_C^n + b_1 E_C^{n-1} + ... + b_N E_C^{n-N} = a_0 T_o^n + a_1 T_o^{n-1} + ... + a_M T_0^{n-M} \quad (4.4)$$

Other examples of dynamic inverse models include equivalent thermal network analysis, Fourier series models, and artificial neural networks. These models are capable of capturing dynamic effects such as building thermal mass dynamics. The main advantages of the dynamic inverse models include the ability to model complex systems that depend on several independent parameters. Their disadvantages include their complexity and the need for more detailed measurements to fine-tune the model. Unlike steady-state inverse models, dynamic inverse models usually require a high degree of user interaction and knowledge of the modeled building or system.

4.4   Forward modeling methods

Forward modeling methods are generally based on a physical description of the building energy systems. Typically, forward models can be used to determine the energy end-uses as well as predict any energy savings incurred from energy conservation measures. Selected existing US energy analysis tools that use the forward modeling approach are described in the following sections. For a more detailed discussion, the reader is referred to ASHRAE Handbook of Fundamentals (ASHRAE, 1997).

4.4.1 Steady-state methods

Steady-state energy analysis methods that use the forward modeling approach are generally easy to use since most of the calculations can be performed by hand or using spreadsheet programs. Two types of steady-state forward tools can be distinguished: degree-day methods and bin methods.

4.4.2 Degree Day Methods

The degree-day methods use seasonal degree-days computed at a specific set-point temperature (or balance temperature) to predict the energy use for building heating. Typically, these degree-day methods are not suitable for predicting building cooling loads. In the US, the traditional degree method day using a base temperature of 65°F has been replaced by the variable-base degree-days method and is applied mostly to residential buildings. In Europe, heating degree-days using 18°C as the base temperature is still used for both residential and commercial buildings.

The variable-base degree-day methods predict seasonal building energy used for heating with one variation of the following formulation:

$$FU = \frac{24.BLC.f.DD_H(T_b)}{\eta_H} \tag{4.5}$$

Where:

- FU represents the fuel use (gas, fuel oil, or electricity depending on the heating system).
- BLC is the building loss coefficient including transmission and infiltration losses through the building envelope. Chapter 6 indicates how BLC can be computed.
- f is a correction factor to include various effects such the part-load performance of the heating system, night set-back effects, and free heat gains.
- T_b is the building heating balance temperature. Again, the reader is referred to Chapter 6 which describes the procedure to calculate the balance temperature based on the building description and indoor temperature settings.
- $DD_H(T_b)$ is the heating degree days calculated at the balance temperature T_b.

Variable-base degree-day methods provide generally good predictions of the fuel use for buildings dominated by transmission loads (i.e., low-rise buildings). However, they are not recommended for buildings dominated by internal loads and/or with involved HVAC system operation strategies.

4.4.3   Bin Methods

Another energy analysis method that uses the forward approach but is based on steady-state modeling of building energy systems is the bin method (Knebel et al. 1983). The bin method is similar to the variable-base degree-day method but relies on bin weather data to estimate total building heating and/or cooling energy consumption. In the US, a number of HVAC engineers use the bin method to perform a variety of energy analyses. Moreover, computer energy simulation tools based on the bin methods have been developed such as ASEAM which is a public domain tool and can be acquired from DOE (DOE, 2000). The simulation tools based on the bin method are typically appropriate for residential or small commercial buildings.

In the classical bin method, only the outdoor temperatures are grouped into bins of equal size, typically 5°F (2.8°C) bins. The number of hours of occurrence is determined for each bin. For other weather variables, only average values coincident to each temperature bin are determined. The resulting weather data from the classical bin method is often referred to as one-dimensional bin weather data. Table 4.3 illustrates one-dimensional weather data obtained for Atlanta, GA. In addition to the outdoor dry-bulb temperature bins, Table 4.3 provides the average values for coincident humidity ratio. The current version of ASEAM software uses one-dimensional bin weather data.

Table 4.3: Classical (or one-dimensional) bin weather data for Atlanta, GA.

Average of Outdoor Dry-Bulb Temperature Bin (^{o}F)	Number of Hours of Occurrence	Average Coincident Humidity ratio (lb/lb)
15	1	0.0020
20	42	0.0020
25	154	0.0020
30	291	0.0025
35	354	0.0031
40	641	0.0038
45	623	0.0044
50	665	0.0053
55	741	0.0065
60	882	0.0083
65	905	0.0100
70	1225	0.0128
75	1000	0.0133
80	672	0.0137
85	421	0.0143
90	133	0.0156
95	19	0.0170

The accuracy of the classical bin method is adequate for only buildings dominated by sensible heat loads and with no significant thermal mass effects. However, the classical bin method may not provide accurate energy predictions for buildings with high latent heat loads as reported by Harriman et al. (1999) and Cohen and Kosar (2000). To improve the accuracy of the bin method especially for buildings with significant latent loads, two-dimensional weather data bins are introduced by ASHRAE (1997). The two-dimensional (also referred to as joint-frequency) weather data bins are generated based on bins obtained for two variables (such as the dry-bulb temperature and humidity ratio) as presented in Table 4.4 using partial data set for Atlanta, GA.

Table 4.4: Partial listing of two-dimensional bin weather data for Atlanta, GA.

Average of Humidity Ratio Bin (lb/lb)	Average of Dry-Bulb Temperature Bin (oF)								
	50	55	60	65	70	75	80	85	90
0	0	0	0	0	0	0	0	0	0
0.0020	69	48	29	14	0	0	0	0	0
0.0040	229	183	123	54	35	11	1	0	0
0.0060	238	178	91	75	39	37	10	0	0
0.0080	129	210	196	122	71	85	28	14	0
0.0100	0	122	335	228	140	111	72	40	3
0.0120	0	0	108	312	202	131	156	53	5
0.0140	0	0	0	100	486	245	156	131	42
0.0160	0	0	0	0	237	289	184	142	57
0.0180	0	0	0	0	15	89	62	37	19
0.0200	0	0	0	0	0	2	3	4	7
0.0220	0	0	0	0	0	0	0	0	0

The two-dimensional weather bin data can be created using hourly data such as TMY-2 files. A number of software is available to create these bins including the ASHRAE Weather Data Viewer developed by Colliver et al. (1998).

4.4.4   Dynamic methods

Dynamic analytical models use numerical or analytical methods to determine energy transfer between various building systems. These models consist generally of simulation computer programs with hourly or sub-hourly time steps) to estimate adequately the effects of thermal inertia due for instance to energy storage in the building envelope and/or its heating system. The important characteristic of the simulation programs is their capability to account for several parameters that are crucial to accurately energy use especially for buildings with significant thermal mass, thermostat setbacks or setups, explicit energy storage, or predictive control strategies. A typical calculation flow chart of detailed simulation programs is presented in Figure 4.4.

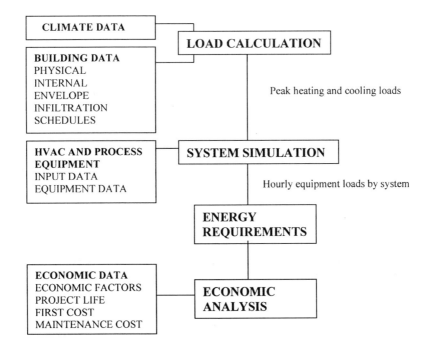

Figure 4.4: Flow chart of complete building model

Detailed computer programs require a high-level of expertise and are generally suitable to simulate large buildings with complex HVAC systems and involved control strategies that are difficult to model by simplified energy analysis tools.

In general, an energy simulation program requires a detailed physical description of the building (including building geometry, building envelope construction details, HVAC equipment type and operation, and occupancy schedules). Thermal load calculations are based on a wide range of algorithms depending on the complexity and the flexibility of the simulation program. To adequately estimate energy savings from energy efficiency measures, energy simulation tools have to be calibrated using existing measured energy data (utility bills, for instance). A basic calibration procedure is discussed in details in Chapter 16.

While energy simulation programs are generally capable of modeling most of the building energy systems, they are often not sufficiently flexible and have inherent limitations. To select the appropriate energy simulation program, it is important that the user be aware of the capabilities of each simulation available to him/her. Some of the well-known simulations programs are briefly presented below:

■ DOE-2 (version DOE-2.1). DOE-2 was developed at the Lawrence Berkeley National Laboratory (LBNL) by the US Department of Energy and is widely used because of its comprehensiveness. It can predict hourly, daily, monthly, and/or annual building energy use. DOE-2 is often used to simulate complex buildings. Figure 4.4 illustrates a typical zoning scheme used to model an office building with DOE-2. Typically, significant efforts are required to create DOE-2 input files using a programming language called Building Description Language (BDL). Several tools are currently available to facilitate the process of developing DOE-2 input files. Among energy engineers and professionals, DOE-2 has become a standard building energy simulation tool in the US and several other countries.

■ BLAST (Building Loads Analysis and Systems Thermodynamics) program enables the user to predict the energy use of a whole building under design conditions or for long-term periods. The heating/cooling load calculations implemented in BLAST are based on heat balance approach (instead of the transfer function technique adopted by DOE-2). Therefore, BLAST can be used to analyze systems such radiant heating or cooling panels that cannot be adequately modeled by DOE-2.

■ ENERGYPlus builds on the features and capabilities of both DOE-2 and BLAST. Its first version is expected to be issued in year 2000. ENERGPlus uses new integrated solution techniques to correct one of the deficiencies of both BLAST and DOE-2 – the inaccurate prediction of space temperature variations. Accurate prediction of space temperatures is crucial to properly analyze energy efficient systems. For instance, HVAC system performance and occupant comfort are directly affected by space temperature fluctuations. Moreover, ENERGYPlus has several features that should aid engineers and architects to evaluate a number of innovative energy efficiency measures that cannot be simulated adequately with either DOE-2 or BLAST. These features include:

  - Free cooling operation strategies using outdoor air,
  - Realistic HVAC systems controls,
  - Effects of moisture adsorption in building elements,
  - Indoor air quality with a better modeling of contaminant and air flows within the building.

■ HAP (1.17 or 3.05) is software developed by Carrier and is currently used to calculate cooling and heating loads in order to design air conditioning systems in buildings with multiple zones (up to 20 zones). It can also perform seasonal and annual energy analysis and evaluate potential energy savings from several energy conservation measures.

- TRNSYS provides a flexible energy analysis tool to simulate a number of energy systems using user-defined modules. A good knowledge of computer programming (Fortran) is required to properly use TRNSYS simulation tool.

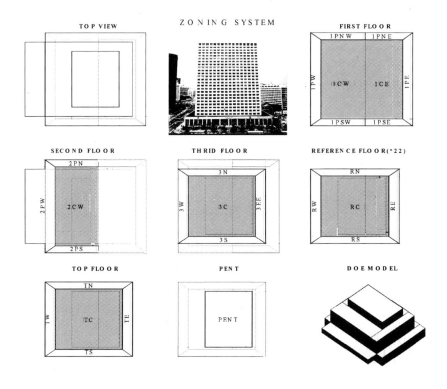

Figure 4.5: Space zoning used to model and office building using DOE-2

4.5   Summary

In this chapter, selected energy analysis tools are described with a brief discussion of the general analysis procedures performed with these tools. The advantages as well as the limitations of the presented modeling approaches are outlined. The auditor should select the proper tool to carry out the energy analysis of the building and to estimate the potential energy and cost savings for retrofit measures. Throughout the following chapters, simplified analysis methods are presented to estimate energy savings.

5

ELECTRICAL SYSTEMS

Abstract

This chapter describes some energy efficiency measures regarding building electrical systems. First, a review of basic characteristics of an electric system operating under alternating current is provided. Then, electric equipment commonly used in HVAC systems such as motors are described. Finally, design procedure of electrical distribution systems specific to motors is illustrated with step by step approach using specific examples. Throughout this chapter, several measures of improving the energy efficiency of electrical systems are provided. Moreover, simplified calculation methods are presented to evaluate of the cost-effectiveness of the proposed energy efficiency measures.

5.1   Introduction

In most buildings and industrial facilities, electric systems consume a significant part of the total energy use. Table 5.1 compares the part of electricity consumption in three sectors (residential, commercial, and industrial) for both the US and France which is representative of most western European countries. It is clear that in the US, electric energy is used more significantly in commercial and residential buildings than the industrial facilities where fossil fuels (such as coal, oil, and natural gas) are predominantly used.

Table 5.1: Percentage share of electricity in the total energy use in three sectors for US[1] and France[2]

SECTOR	US	FRANCE
Residential buildings	61 %	52 %
Commercial buildings	52 %	68 %
Industrial facilities	12 %	52 %

[1] *Source:* Office of Technology Assessment (1995)
[2] *Source*: Electricite de France (1997)

For residential buildings, lighting and heating, ventilating, and air conditioning (HVAC) account each for approximately 20% of total US electricity use. Refrigerators represent another important energy end-use in the residential sector with about 16% of electricity. For the commercial sector as a whole, lighting accounts for over 40% while HVAC accounts for 11% of the total electricity use. However, for commercial buildings with space conditioning, HVAC is one of the major electricity end-uses and can be more energy intensive than lighting. Moreover, computers and other office equipment (such as printers, copiers, and facsimile machines) are becoming an important electric energy end-use in office buildings.

In this chapter, a brief description of the electric energy end-use such as motors with focus on some measures to reduce the electrical energy use. First, a brief review of basic characteristics of an electric system is provided to highlight the major issues that should be considered when designing, analyzing, or retrofitting an electric system.

5.2   Review of Basics

5.2.1   Alternating Current Systems

For a linear electrical system subject to an alternating current (AC), the time variation of the voltage and current can be represented as a sine function:

$$v(t) = V_m \cos \omega t \tag{5.1}$$

$$i(t) = I_m \cos(\omega t - \phi) \tag{5.2}$$

Where:

- V_m and I_m are the maximum instantaneous values of respectively, voltage and current. These maximum values are related to the effective or root mean square (rms) values as follow:

$$V_m = \sqrt{2} * V_{rms} = 1.41 * V_{rms}$$

$$I_m = \sqrt{2} * I_{rms} = 1.41 * I_{rms}$$

In the US, the values of V_{rms} are typically 120 V for residential buildings or plug-load in the commercial buildings, 277 V for lighting systems in commercial buildings, 480 V for motor loads in commercial and industrial buildings. Higher voltages can be used for some power-intensive industrial applications.

- ω is the angular frequency of the alternating current and is related to the frequency f as follows:

$$\omega = 2\pi f$$

In the US, the frequency f is 60 Hz, that is 60 pulsations or oscillations in one second. In other countries, the frequency of the alternating current is f=50 Hz.

- ϕ is the phase lag between the current and the voltage. In the case, the electrical system is a resistance (such as an incandescent lamp), the phase lag is zero and the current is on phase with the voltage. If the electrical system consists of a capacitance load (such as a capacitor or a synchronous motor), the phase lag is negative and the current is in advance relative to the voltage. Finally, when the electrical system is dominated by an inductive load (such a fluorescent fixture or an induction motor), the phase lag is positive and the current lags the voltage.

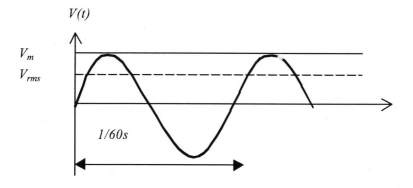

Figure 5.1: Illustration of the voltage waveform and the concept of V_{rms}.

Figure 5.1 illustrates the time variation of the voltage for a typical electric system. The concept of root mean square (also called effective value) for the voltage, V_{rms}, is also indicated in Fig.5.1. It should be noted that in the US the cycle for the voltage waveform repeats itself every 1/60 s (since the frequency is 60 Hz).

The instantaneous power, $p(t)$, consumed by the electrical system operated on one-phase AC power supply can be calculated using the ohm law:

$$p(t) = v(t).i(t) = V_m I_m \cos\omega t.\cos(\omega t - \phi) \tag{5.3}$$

The above equation can be rearranged using some basic trigonometry and the definition of the rms values for voltage and current:

$$p(t) = V_{rms}.I_{rms}(\cos\phi.(1 + \cos 2\omega t) + \sin\phi.\sin 2\omega t) \tag{5.4}$$

Two types of power can be introduced as function of the phase lag angle ϕ, the real power P_R, and the reactive power P_X, as defined below:

$$P_R = V_{rms}.I_{rms}\cos\phi \tag{5.5}$$

$$P_X = V_{rms}.I_{rms}\sin\phi \tag{5.6}$$

Note that both types of power are constant and are not a function of time. To help understand the meaning of each power, it is useful to note that the average of the instantaneous power actually consumed by the electrical system over one period is equal to P_R:

$$\overline{p} = \frac{1}{T}\int_0^T p(t)dt = P_R \tag{5.7}$$

Therefore, P_R is the actual or real power consumed by the electrical system over its operation period (which consists typically of a large number of periods $T = \frac{1}{2\pi f}$). P_R is typically called real power and is measured in kW. Meanwhile, P_X is the power required to produce a magnetic field to operate the electrical system (such as an induction motors). P_X is stored and then released by the electrical system; this power is typically called reactive power and is measured in kVAR. A schematic diagram is provided in Figure 5.2 to help illustrate the meaning of each type of power.

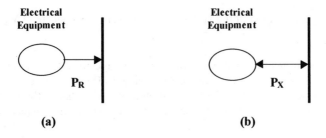

Figure 5.2: Illustration of the direction of electricity flow for (a) real power, and (b) reactive power.

While the user of the electrical system actually consumes only the real power, the utility or the electricity provider has to make available to the user both the real power, P_R, and the reactive power, P_X. The algebraic sum of P_R and P_X constitutes the total power, P_T. Therefore, the utility has to know in addition to the real power needed by the customer, the magnitude of the reactive power, and thus the total power.

As mentioned earlier, for a resistive electrical system, the phase lag is zero and thus the reactive power is also zero (see Eq. 5.6). Unfortunately, for commercial buildings and industrial facilities, the electrical systems are not often resistive and the reactive power can be significant. In fact, the higher the phase lag angle ϕ, the larger is the reactive power P_X. To illustrate the importance of the reactive power relative to the real power P_R and the total power P_T consumed by the electrical system, a power triangle is typically used to represent the power flow as shown in Figure 5.3.

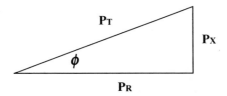

Figure 5.3: Power Triangle for an electrical system

In Figure 5.3, it is clear that the ratio of the real power to the total power represents the cosine of the phase lag. This ratio is widely known as the power factor, *pf*, of the electrical system:

$$pf = \frac{P_R}{P_T} = \cos\phi \qquad (5.8)$$

Ideally, the power factor has to be as close to unity as possible (i.e., *pf* =1.0). Typically however, power factors above 90% are considered to be acceptable. If the power factor is low, that is if the electrical system has a high inductive load, capacitors can be added in parallel to reduce the reactive power as illustrated in Figure 5.4.

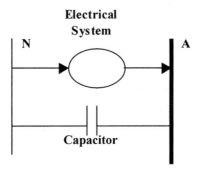

Figure 5.4: The addition of capacitor can improve the power factor of an electrical system

5.2.2   Power Factor Improvement

As mentioned in the previous section, the reactive power has to be supplied by the utility even though it is not actually recorded by the power meter (as real power used). The magnitude of this reactive power increases as the power factor decreases. To account for the loss of energy due to the reactive power, most utilities have established rate structures that penalize any user that has low power factor. The penalty is imposed through a power factor clause set in the utility rate structures as discussed in Chapter 2. Significant savings in the utility costs can be achieved by improving the power factor. As illustrated in Figure 5.4, this power factor improvement can be obtained by adding a set of capacitors connected in parallel to the electrical system. The size of these capacitors, P_C, is typically measured in KVAR (the same unit as the reactive power) and can be determined as indicated in Figure 5.5 using the power triangle analysis:

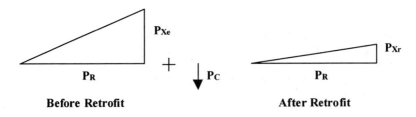

Figure 5.5: Effect of adding capacitors on the power triangle of the electrical system

$$P_C = P_{Xe} - P_{Xr} = P_R.(\tan\phi_e - \tan\phi_r) \qquad (5.9)$$

where,
- P_{Xe} and P_{Xr} are reactive power respectively before retrofit (existing conditions) and after retrofit (retrofitted conditions).
- P_C is the reactive power of the capacitor to be added.
- ϕ_e and ϕ_r are phase lag angles respectively before retrofit (existing conditions) and after retrofit (retrofitted conditions).

Using the values of power factor before and after the retrofit, the size of the capacitors can be determined:

$$P_C = P_R \left[\tan(\cos^{-1} pf_e) - \tan(\cos^{-1} pf_r) \right] \qquad (5.10)$$

The calculations of the cost savings incurred from power factor improvement depend on the utility rate structure. In most all the rate structures, one of three options is used to assess the penalty for low power factors. These three options are summarized below. Basic calculation procedures are typically performed to estimate the annual cost savings in the utility bills:

(1) Modified Billing Demand: In this case, the demand charges are increased in proportion with a fraction by which the power factor is less than a threshold value. The size for the capacitors should be selected so the system power factor reaches at least the defined threshold value.
(2) Reactive Power Charges: In this case, charges for reactive power demand are included as part of the utility bills. In this option, the size of the capacitors should be ideally determined to eliminate this reactive power (so that the power factor is unity).
(3) Total Power Charges: This rate, the penalty charges are imposed based on the total power required by building/facility. Again, capacitors should be sized so the power factor is equal to unity.

The calculations of the cost savings due to power factor improvement are illustrated in example 5.1.

Example 5.1:

Problem: *Consider a building with a total real power demand of 500 kW with a power factor of $pf_e=0.70$. Determine the required size of a set of capacitors to be installed in parallel with the building service entrance so that the power factor becomes at least $pf_r=0.90$*

Solution:

The size in kVAR of the capacitor is determined using Eq. (5.10):

$$P_C = 500.\left[\tan(\cos^{-1} 0.70) - \tan(\cos^{-1} 0.90)\right] = 268 \;\; kVAR$$

Thus a capacitor rated at 275 KVAR can be selected to ensure a power factor for the building electrical system to be higher than 0.90.

5.3   Electrical Motors

5.3.1   Introduction

In the United States, there were 125 million operating motors in the range of 1 to 120 horsepower in 1991. These motors consumed approximately 55% of the electricity generated in the US (Andreas, 1992). In large industrial facilities, motors can account for as much as 90% of the total electrical energy use. In commercial buildings, motors can account for more than 50% of the building electrical load.

Motors convert electrical energy to mechanical energy and are typically used to drive machines. The driven machines can serve a myriad of purposes in the building including moving air (supply and exhaust fans), moving liquids (pumps), moving objects or people (conveyors, elevators), compressing gases (air compressors, refrigerators), and producing materials (production equipment). To select the type of motor to be used for a particular application, several factors have to be considered including:

(a) The form of the electrical energy that can be delivered to the motor: direct current (DC) or alternating current (AC), single or three phase.
(b) The requirements of the driven machine such as motor speed and load cycles.
(c) The environment in which the motor is to operate: normal (where a motor with an open-type ventilated enclosure can be used), hostile (where a totally enclosed motor must be used to prevent outdoor air from infiltrating inside the motor), or hazardous (where a motor with an explosion-proof enclosure must be used to prevent fires and explosions)

The basic operation and the general characteristics of AC motors are discussed in the following sections. In addition, simple measures are described to improve the energy efficiency of existing motors.

5.3.2   Overview of Electrical Motors

There are basically two types of electric motors used in buildings and industrial facilities: (i) induction motors, and (ii) synchronous motors. Induction motors are the more common type, accounting for about 90% of the existing motor horsepower. Both types use a motionless stator and a spinning rotor to convert electrical energy into mechanical power. The operation of both types of motor is relatively simple and is briefly described below.

Alternating current is applied to the stator, which produces a rotating magnetic field in the stator. A magnetic field is also created in the rotor. This magnetic field causes the rotor to spin in trying to align with the rotating stator magnetic field. The rotation of the magnetic field of the stator has an angular speed that is a function of both the number of poles, N_P, and the frequency, f, of the AC current as expressed in Equation (5.11):

$$\omega_{mag} = \frac{4\pi \cdot f}{N_P} \tag{5.11}$$

The above expression is especially useful to explain the operation of variable frequency drives for motors with variable loads, as will be discussed in this Chapter.

One main difference between the two motor types (synchronous vs. induction) is the mechanism by which the rotor magnetic field is created. In an induction motor, the rotating stator magnetic field induces a current, and thus a magnetic field, in the rotor windings which are typically of the squirrel-cage type. In a magnetic motor, the rotor cannot rotate at the same speed as the magnetic field (if the rotor spins with the same speed as the magnetic field, no current can be induced in the rotor since effectively the stator magnetic field remains at the same position relative to the rotor). The difference between the rotor speed and the stator magnetic field rotation is called the slip factor.

In a synchronous motor, the magnetic field is produced by application of direct current through the rotor windings. Therefore, the rotor spins at the same speed as the rotating magnetic field of the stator and thus the rotor and the stator magnetic fields are synchronous in their speed.

Because of their construction characteristics, the induction motor is basically an inductive load and thus has a lagging power factor while the synchronous motor can be set so it has a leading power factor (i.e., acts like a capacitor). Therefore, it is important to remember that a synchronous motor can be installed to both provide mechanical power and improve the power factor for a set of induction motors. This option may be more cost-effective than just adding a bank of capacitors.

Three parameters are typically used to characterize an electric motor during full-load operation. These parameters include:

(a) The mechanical power output of the motor, P_M. This power can be expressed in kW or horsepower (1 Hp = 0.746 kW). The mechanical power is generally the most important parameter in selecting a motor.
(b) The energy conversion efficiency of the motor, η_M. This efficiency expresses the mechanical power as a fraction of the real electric power consumed by the motor. Due to various losses (such as friction, core losses due to the alternating of the magnetic field, and resistive losses through the windings), the motor efficiency is always less than 100%. Typical motor efficiencies range from 75% to 95% depending on the size of the motor.
(c) The power factor of the motor, pf_M. As indicated earlier in this chapter, the power factor is a measure of the magnitude of the reactive power needed by the motor.

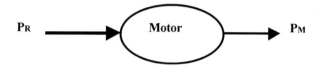

P_R **Motor** P_M

Figure 5.6: Definition of the efficiency of a motor

Using the schematic diagram of Figure 5.6, the real power used by the motor can be calculated as follows:

$$P_R = \frac{1}{\eta_M}.P_M \qquad (5.12)$$

Therefore, the total power and the reactive power needed to operate the motor are respectively:

$$P_T = \frac{P_R}{pf_M} = \frac{1}{pf_M.\eta_M}.P_M \qquad (5.13)$$

$$P_X = P_R \tan\phi = \frac{1}{\eta_M}.P_M.\tan(\cos^{-1} pf) \qquad (5.14)$$

5.3.3 Energy-efficient Motors

General description

Based on their efficiency, motors can be classified into two categories: (i) standard-efficiency motors, and (ii) high or premium-efficiency (i.e., energy-efficient) motors. The energy-efficient motors are 2 to 10 percentage points more efficient than standard-efficiency motors depending on the size. Table 5.2 summarizes the average efficiencies for both standard and energy-efficient motors that are currently available commercially. The improved efficiency for the high or premium-motors are mainly due to better design and use of better materials to reduce losses. However, this efficiency improvement comes with a higher price of about 10 to 30% more than standard-efficiency motors. These higher prices may be the main reason that only one-fifth of the motors sold in the US is energy-efficient.

However, the installation of premium-efficiency motors is becoming a common method of improving the overall energy efficiency of buildings. The potential for energy savings from premium-efficiency motor retrofits is significant. In the United States alone, there were in 1991 about 125 million operating motors which consumed approximately 55% of the electrical energy generated in the US (Andreas, 1992). It was estimated that replacing all these motors with premium-efficiency models would save approximately 60 TWh of energy per year (Nadel et al., 1991).

To determine the cost-effectiveness of motor retrofits, there are several tools available including the MotorMaster developed by the Washington State Energy Office (WSEO, 1992). These tools have the advantage of providing large databases for cost and performance information for various motor types and sizes.

Adjustable Speed Drives (ASDs)

With more emphasis on energy efficiency, an increasing number of designers and engineers are recommending the use of variable speed motors for various HVAC systems. Indeed, the use of adjustable speed drives (ASDs) is now becoming common especially for supply and return fans in Variable Air Volume (VAV) systems and for hot and chilled water pumps in central heating and cooling plants.

Electronic ASDs convert the fixed-frequency AC power supply (50 or 60 Hz) first to a DC power and then to a variable frequency AC power as illustrated in Figure 5.7. Therefore, the ASDs can change the speed of AC motors with no moving parts, presenting high reliability and low maintenance requirements.

Table 5.2: Typical Motor Efficiencies (Adapted from Hoshide, 1994)

MOTOR MECHANICAL POWER OUTPUT KW (HP)	AVERAGE NOMINAL EFFICIENCY FOR STANDARD-EFFICIENCY MOTOR	AVERAGE NOMINAL EFFICIENCY FOR PREMIUM-EFFICIENCY MOTOR
0.75 (1.0)	0.730	0.830
1.12 (1.5)	0.750	0.830
1.50 (2.0)	0.770	0.830
2.25 (3.0)	0.800	0.865
3.73 (5.0)	0.820	0.876
5.60 (7.5)	0.840	0.885
7.46 (10)	0.850	0.896
11.20 (15)	0.860	0.910
14.92 (20)	0.875	0.916
18.65 (25)	0.880	0.926
22.38 (30)	0.885	0.928
29.84 (40)	0.895	0.930
37.30 (50)	0.900	0.932
44.76 (60)	0.905	0.933
55.95 (75)	0.910	0.935
74.60 (100)	0.915	0.940
93.25 (125)	0.920	0.942
111.9 (150)	0.925	0.946
149.2 (200)	0.930	0.953

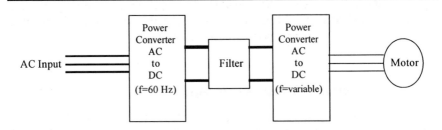

Figure 5.7: Basic concept of ASD power inverter.

In order to achieve the energy savings potential for any HVAC application, the engineer needs to know the actual efficiency of the motors. For ASD applications, it is important to separate the losses between the drive and the motor to achieve an optimized HVAC system with the lowest operating cost. Specifically, the engineer would need to know the loss distribution including the iron losses, copper losses, friction and winding losses, and the distribution of losses between stator and rotor. To determine these losses and thus the motor efficiency, accurate measurements are needed. Unfortunately, existent power measurement instruments are more suitable for sinusoidal rather than distorted waveforms (which are typical for ASD applications).

The use of low-cost solid-state power devices and integrated circuits to control the speed have made most of the commercially available ASDs draw power with extremely high harmonic content. Some investigators (Dominijan et al. 1995; Czarkowski and Domijan, 1997) have studied the behavior of commercially available power measurement instruments subjected to voltage and current waveforms typical of ASD-motor connections. In particular, three drive technologies used in HVAC industry were investigated by Czarkowski and Domijan, 1997; and by Domijan et al., 1996) used by namely PWM Induction, Switched Reluctance, and Brushless DC drives. The main findings of their investigation is that existing power instruments fail to accurately measure power losses (and thus motor efficiency) due to the high harmonic content in the voltage spectra especially for Brushless DC and Switched Reluctance motors which represent a substantial portion of the HVAC market.

Typically, the motor losses are measured indirectly by monitoring the power input and the power output and taking the difference. This traditional approach requires an extremely high accuracy for the power measurements to achieve "reasonable" estimation of motor losses especially for premium-efficiency motors (with low losses). However, a new approach to measure the ASD-motor losses directly and with better accuracy has been proposed (Fuchs et al., 1996).

Energy Savings Calculations

There are three methods to calculate the energy savings due to energy-efficient motor replacement. These three methods are outlined below:

Method 1: *Simplified Method*
This method has been and is still being used by most energy engineers to determine the energy and cost savings incurred by motor replacement. Inherent to this method, two assumptions are made: (i) the motor is fully loaded, and (ii) the change in motor speed is neglected.

The electric power savings due to the motor replacement is first computed as follows:

$$\Delta P_R = P_M \left(\frac{1}{\eta_e} - \frac{1}{\eta_r} \right)$$ (5.15)

Where,
- P_M is the mechanical power output of the motor
- η_e is the design (i.e., full-load) efficiency of the existing motor (e.g. before retrofit)

- η_r is the design (i.e., full-load) efficiency of the energy-efficient motor (e.g. after retrofit)

The electric energy savings incurred from the motor replacement is thus:

$$\Delta kWh = \Delta P_R . N_h . LF_M \qquad (5.16)$$

Where,
- N_h is the number of hours per year during which the motor is operating.
- LF_M is the load factor of the motor's operation during one year.

Method 2: *Mechanical Power Rating Method*
In this method, the electrical peak demand of the existing motor is assumed to be proportional to its average mechanical power output:

$$P_{R,e} = \frac{P_M}{\eta_{op,e}} . LF_{M,e} . PDF_{M,e} \qquad (5.17)$$

With,
- $\eta_{op,e}$ is the motor efficiency at the operating average part-load conditions.

  To obtain this value, the efficiency curve for the motor can be used. If the efficiency curve for the specific existing motor is not available, a generic curve can be used.
- $LF_{M,e}$ is the load factor of the existing motor and is the ratio between the average operating load of the existing motor and its rated mechanical power. In most applications, the motor is oversized and operates at less its capacity.
- $PDF_{M,e}$ is the peak demand factor and represents the fraction of typical motor load that occurs at the time of the building peak demand. In most applications, $PDF_{M,e}$ can be assumed to be unity since the motors often contribute to the total peak demand of the building .

Since the mechanical load does not change after installing an energy-efficient motor, it is possible to consider a smaller motor with a capacity $P_{M,r}$ if the existing motor is oversized with a rating of $P_{M,e}$. In this case, the smaller energy-efficient motor can operate at a higher load factor than the existing motor. The new load factor, LF_r, of the energy-efficient motor can be calculated as follows:

$$LF_r = LF_e . \frac{P_{M,r}}{P_{M,e}} \qquad (5.18)$$

Moreover, the energy-efficient motors often operate at a higher speed than the standard motors they replace since they have lower internal losses. This higher speed actually has a negative impact since it reduces the effective efficiency of the energy-efficient motor by a factor called the slip penalty. The slip penalty factor, $SLIP_P$, is defined as shown in Eq. (5.19):

$$SLIP_p = \left(\frac{\omega_{M,r}}{\omega_{M,e}} \right)^3 \tag{5.19}$$

Where,

- $\omega_{M,e}$ is the rotation speed of the existing motor

- $\omega_{M,r}$ is the rotation speed of the energy-efficient motor

Using similar equation than Eq. (5.17), the peak electrical demand for the retrofitted motor (e.g. energy-efficient motor) can be determined:

$$P_{R,r} = \frac{P_{M,r}}{\eta_{op,r}} . LF_{M,r} . PDF_{M,r} . SLIP_P \tag{5.20}$$

The electrical power savings due to the motor replacement can thus be estimated:

$$\Delta P_R = P_{R,e} - P_{R,r} \tag{5.21}$$

The electric energy savings can be therefore calculated using Eq. (5.16).

Method 3: *Field Measurement Method*

In this method, the motor electrical power demand is measured directly on-site. Typically, current, I_M, voltage, V_M, power factor, pf_M, readings are recorded for the existing motor to be retrofitted. For three-phase motors (which is common in industrial facilities and in most HVAC systems for commercial buildings), the electrical power used by the existing motor can be either directly measured or calculated from current, voltage, and power factor readings as follows:

$$P_{R,E} = \sqrt{3}.V_M.I_M.pf_M \tag{5.22}$$

The load factor of the existing motor can be estimated by taking the ratio of the measured current over the nameplate full-load current, I_{FL}, as expressed by Eq. (5.23):

$$LF_{M,E} = \frac{I_M}{I_{FL}} \tag{5.23}$$

A study by Biesemeyer and Jowett (1996) has indicated that Eq. (5.23) has much higher accuracy to estimate the motor load ratio than an approach based on the ratio of the motor speeds (i.e. measured speed over nominally rated speed) used by BPA (1990) and Lobodovsky (1994). It should be noted that Eq. (5.23) is recommended for load ratios that are above 50% since for these load ratios, a typical motor draws electrical current that is proportional to the imposed load.

The methodology for the calculation of the electrical power and energy savings is the same as described for the Mechanical Power Rating Method using Eq. (5.18) through Eq. (5.23).

5.4   Lighting Systems

5.4.1   Introduction

Lighting accounts for a significant portion of the energy use in commercial buildings. For instance, in office buildings, 30% to 50% of the electricity consumption is used to provide lighting. In addition, heat generated by lighting contributes to additional thermal loads that need to be removed by the cooling equipment. Typically, energy retrofits of lighting equipment are very cost-effective with payback periods of less than 2 years in most applications.

In the US, lighting energy efficiency features are the most often considered strategies to reduce the energy costs in commercial buildings as shown in Table 5.3. The data for Table 5.3 is based on the results of a survey (EIA, 1995) to determine the participation level of commercial buildings in a variety of specific types of conservation programs and energy technologies.

To better understand the retrofit measures that need to be considered in order to improve the energy-efficiency of lighting systems, a simple estimation of the total electrical energy use due to lighting is first considered:

$$Kwh_{Lit} = \sum_{j=1}^{J} N_{Lum,j} . WR_{Lum,j} . N_{h,j} \qquad (5.24)$$

Where,
- $N_{Lum,j}$ is the number of lighting luminaires of type j in the building to be retrofitted. Recall that a luminaire consists of the complete set of a ballast, electric wiring, housing, and lamps.
- $WR_{Lum,j}$ is the wattage rating for each luminaire of type j. In this rating the energy use due to both the lamps and ballast should be accounted for.
- $N_{h,j}$ is the number of hours per year when the luminaires of type j are operating.
- J is the number of luminaire types in the building.

Table 5.3: Level of Participation in Lighting Conservation Programs by US Commercial Buildings (*Source*: EIA, 1995)

LIGHTING RETROFIT	PERCENT PARTICIPATION IN NUMBER OF BUILDINGS	PERCENT PARTICIPATION IN FLOOR AREA OF SPACES
Energy Efficient Lamps and Ballasts	31	49
Specular Reflectors	18	32
Time Clock	10	23
Manual Dimmer Switches	10	23
Natural Lighting Control Sensors	7	13
Occupancy Sensors	5	11

It is clear from Eq. (5.24) that there are three options to reduce the energy use attributed to lighting systems as briefly discussed below:

(a) Reduce the wattage rating for the luminaires including both the lighting sources (e.g. lamps) and the power transforming devices (e.g. ballasts) [therefore, decrease the term $WR_{Lum,j}$ in Eq. (5.24)]. In the last decade, technological advances such as compact fluorescent lamps and electronic ballasts have increased the energy efficiency of lighting systems.

(b) Reduce the time of use of the lighting systems through lighting controls [therefore, decrease the term $N_{h,j}$ in Eq. (5.24)]. Automatic controls have been developed to decrease the use of a lighting system so illumination is provided only during times when it is actually needed. Energy-efficient lighting controls include the occupancy sensing systems and light dimming controls through the use of daylighting.

(c) Reduce the number of luminaires [therefore, decrease the term $N_{Lum,j}$ in Eq. (5.24)]. This goal can be achieved only in cases where delamping is possible due to over-illumination.

In this section, only measures related to the general actions described in items (a) and (b) are discussed. To estimate the energy savings due to any retrofit measure for the lighting system, Eq (5.24) can be used. The energy use due to lighting has to be calculated before and after the retrofit and the difference between the two estimated energy uses represents the energy savings.

5.4.2   Energy-Efficient Lighting Systems

Improvements in the energy-efficiency of lighting systems have provided several opportunities to reduce electrical energy use in buildings. In this section, the energy savings calculations for the following technologies are discussed:

- High efficiency fluorescent lamps
- Compact fluorescent lamps
- Compact halogen lamps
- Electronic ballasts

First, a brief description is provided for the factors that an auditor should consider to achieve and maintain an acceptable quality and level of comfort for the lighting system. Second, the design and the operation concepts are summarized for each available lighting technology. Then, the energy savings that can be expected from retrofitting existing lighting systems using any of the new technologies are estimated and discussed.

Typically, three factors determine the proper level of light for a particular space. These factors include: age of the occupants, speed and accuracy requirements, and background contrast (depending on the task being performed). It is a common misconception to consider that overlighting a space provides higher visual quality. Indeed, it has been shown that overlighting can actually reduce the illuminance quality and the visual comfort level within a space in addition to wasting energy. Therefore, it is important when upgrading a lighting system to determine and maintain the adequate illuminance level as recommended by the appropriate authorities. Table 5.4 summarizes the lighting levels recommended for various activities and applications in selected countries including the US based on the most recent illuminance standards.

High Efficiency Fluorescent Lamps:

Fluorescent lamps are the most commonly used lighting systems in commercial buildings. In the US, fluorescent lamps illuminate 71% of the commercial space. Their relatively high efficacy, diffuse light distribution, and long operating life are the main reasons for their popularity.

A fluorescent lamp consists generally of a glass tube with a pair of electrodes at each end. The tube is filled at very low pressure with a mixture of inert gases (primarily argon) and with liquid mercury. When the lamp is turned on, an electric arc is established between the electrodes. The mercury vaporizes and radiates in the ultraviolet spectrum. This ultraviolet radiation excites a phosphorous coating on the inner surface of the tube which emits visible light. High-efficiency fluorescent lamps use a krypton-argon mixture which increases the efficacy output by 10 to 20% from a typical efficacy of 70 lumens/watt to about 80 lumens/watt. Improvements in phosphorous coating can further increase the efficacy to 100 lumens/watt.

Table 5.4: Recommended lighting levels for various applications in selected countries (in Lux maintained on horizontal surfaces).

Application	France AEF (92&93)	Germany DIN5035 (90)	Japan JIS (89)	US/Canada IESNA (93)
Offices				
General	425	500	300-750	200-500
Reading Tasks	425	500	300-750	200-500
Drafting (detailed)	850	750	750-1500	1000-2000
Classrooms				
General	325	300-500	200-750	200-500
Chalkboards	425	300-500	300-1500	500-1000
Retail Stores				
General	100-1000	300	150-750	200-500
Tasks/till Areas	425	500	750-1000	200-500
Hospitals				
Common Areas	100	100-300	--	--
Patient Rooms	50-100	1000	150-300	100-200
Manufacturing				
Fine Knitting	850	750	750-1500	1000-2000
Electronics	625-1750	100-1500	1500-300	1000-2000

It should be mentioned that the handling and the disposal of fluorescent lamps is highly controversial due to the fact that mercury inside the lamps can be toxic and hazardous to the environment. A new technology is being tested to replace the mercury with sulphur to generate the radiation that excites the phosphorous coating of the fluorescent lamps. The sulphur lamps are not hazardous and would present an environmental advantage to the mercury-containing fluorescent lamps.

The fluorescent lamps come in various shapes, diameters, lengths, and ratings. A common labeling used for the fluorescent lamps is

$$F.S.W.C - T.D$$

Where:

- F stands for the fluorescent lamp.
- S refers to the style of the lamp. If the glass tube is circular, then the letter C is used. If the tube is straight, no letter is provided.
- W is the nominal wattage rating of the lamp (it can be 4, 5, 8, 12, 15, 30, 32, 34, 40, etc.)

- *C* indicates the color of the light emitted by the lamp: W for White, CW for Cool White, BL for Black Light
- *T* refers to tubular bulb.
- D indicates the diameter of the tube in eighths of one inch (1/8 in = 3.15 mm) and can be for instance 12 (D=1.5 in = 38 mm) for the older and less energy efficient lamps and 8 (D= 1.0 in = 31.5 mm) for more recent and energy efficient lamps.

Thus, F40CW-12 designates a fluorescent lamp that has a straight tube, uses 40 W electric power, provides cool white color, and is tubular with 38 mm (1.5 inches) in diameter.

Among the most common retrofit in lighting systems is the upgrade of the conventional 40 W T12 fluorescent lamps to more energy efficient lamps such 32 W T8 lamps. For a lighting retrofit, it is recommended that a series of tests be conducted to determine the characteristics of the existing lighting system. For instance, it is important to determine the illuminance level at various locations within the space especially in working areas such as benches and/or desks.

Compact Fluorescent Lamps:

These lamps are miniaturized fluorescent lamps with small diameter and shorter length. The compact lamps are less efficient that full size fluorescent lamps with only 35 to 55 lumens/Watt. However, they are more energy efficient and have longer life than incandescent lamps. Currently, compact fluorescent lamps are being heavily promoted as energy savings alternatives to incandescent lamps even though they may have some drawbacks. In addition to their high cost, compact fluorescent lamps are cooler and thus provide less pleasing contrast than incandescent lamps.

Compact Halogen Lamps:

The compact halogen lamps are adapted for use as direct replacements for standard incandescent lamps. Halogen lamps are more energy-efficient, produce whiter light, and last longer than incandescent lamps. Indeed, incandescent lamps typically convert only 15% of their electrical energy input into visible light since 75% is emitted as infrared radiation and 10% is used by the filament as it burns off. In halogen lamps, the filament is encased inside a quartz tube which is contained in a glass bulb. A selective coating on the exterior surface of the quartz tube allows visible radiation to pass through but reflects the infrared radiation back to the filament. This recycled infrared radiation permits the filament to maintain its operating temperatures with 30% less electrical power input.

The halogen lamps can be dimmed and present no power quality or compatibility concerns as can be the case for the compact fluorescent lamps.

Electronic Ballasts:

Ballasts are integral parts of fluorescent luminaires since they provide the voltage level required to start the electric arc and regulate the intensity of the arc. Before the development of electronic ballasts in early 1980's, only magnetic or "core and coil" ballasts are used to operate fluorescent lamps. While the frequency of the electrical current is kept at 60 Hz (in countries other than the US, the frequency is set at 50 Hz) by the magnetic ballasts, electronic ballasts use solid-state technology to produce high-frequency (20 – 60 MHz) current. The use of high-frequency current increases the energy efficiency of the fluorescent luminaires since the light is cycling more quickly and appears brighter. When used with high-efficiency lamps (T8 for instance), electronic ballasts can achieve 95 lumens/watts as opposed to 70 watts/lumens for conventional magnetic ballasts. It should be mentioned however that efficient magnetic ballasts can achieve similar lumen/watt ratios as electronic ballasts.

Other advantages that electronic ballasts have relative to their magnetic counterparts include:

- Higher power factor. The power factor of electronic ballasts are typically in the 0.90 to 0.98 range. Meanwhile, the conventional magnetic ballasts have low power factor (less than 0.80) unless a capacitor is added as discussed in section 5.2.
- Less flicker problems. Since the magnetic ballasts operate at 60 Hz current, they cycle the electric arc about 120 times per second. As a result, flicker may be perceptible especially if the lamp is old during normal operation or when the lamp is dimmed to less than 50% capacity. However, electronic ballasts cycle the electric arc several thousands of times per second and flicker problems are avoided even when the lamps are dimmed to as low as 5% of capacity.
- Less noise problems. The magnetic ballasts use electric coils and generate audible hum which can increase with age. Such noise is eliminated by the solid-state components of the electronic ballasts.

5.4.3   Lighting Controls

As illustrated by Eq. (5.24), energy savings can be achieved by not operating the lighting system in cases when illumination becomes unnecessary. The control of the operation of lighting system can be achieved by several means including manual on/off and dimming switches, occupancy sensing systems, and automatic dimming systems using daylighting controls.

While energy savings can be achieved by manual switching and manual dimming, the results are typically unpredictable since they depend on the occupant behavior. Scheduled lighting controls provide a more efficient approach to energy savings but can also be affected by the frequent adjustments from occupants. Only automatic light switching and dimming systems can respond in real-time to changes in occupancy and climatic changes. Some of the automatic controls available for lighting systems are briefly discussed below.

Occupancy Sensors

Occupancy sensors save energy by automatically turning off the lights in spaces that are not occupied. Generally, occupancy sensors are suitable for most lighting control applications and should be considered for lighting retrofits. It is important to properly specify and install the occupancy sensors to provide reliable lighting during periods of occupancy. Indeed, most failed occupancy sensor installations result from inadequate product selection and improper placement. In particular, the auditor should select the proper motion sensing technology used in occupancy sensors. Two types of motion sensing technologies are currently available in the market:

i.   *Infrared sensors*: which register the infrared radiation emitted by various surfaces in the space including the human body. When the controller connected to the infrared sensors receives a sustained change in the thermal signature of the environment (as is the case when an occupant moves), it turns the lights on. The lights are kept on until the recorded changes in temperature are not significant. The infrared sensors operate adequately only if they are in direct line-of-sight with the occupants and thus must be used in smaller enclosed spaces with regular shapes and without partitions.

ii. *Ultrasound sensors:* operate on a sonar principle like the submarines and airport radars. A device emits a high frequency sound (25-40 KHz) so it is beyond the hearing range of humans. This sound is reflected by the surfaces inside a space (including furniture and occupants) and is sensed by a receiver. When people move inside the space, the pattern of sound wave changes. The lights remain on until no movement is detected for a preset period of time (after 5 minutes). Unlike infrared radiation, sound waves are not easily blocked by obstacles such wall partitions. However, the ultrasound sensors may not operate properly in large spaces which tend to produce weak echoes.

Based on a study by EPRI, Table 5.5 provides typical energy savings to be expected from occupancy sensor retrofits. As shown in Table 5.5, significant energy savings can be achieved in spaces where occupancy is intermittent such as conference rooms, rest rooms, storage areas, and warehouses.

Table 5.5: Energy Savings Potential with Occupancy Sensor Retrofits

SPACE APPLICATION	RANGE OF ENERGY SAVINGS
Offices (Private)	25-50 %
Offices (Open Space)	20-25 %
Rest Rooms	30-75 %
Conference Rooms	45-65 %
Corridors	30-40 %
Storage Areas	45-65 %
Warehouses	50-75 %

Light Dimming Systems

Dimming controls allow the variation of the intensity of lighting system output based on natural light level, manual adjustments, and occupancy. A smooth and uninterrupted decrease in the light output is defined as a continuous dimming as opposed to stepped dimming in which the lamp output is decreased in stages by preset amounts.

To accurately estimate the energy savings from dimming systems that use natural light controls (e.g. daylighting), computer software exists such as RADIANCE (LBL, 1991). With this computer tool, an engineer can predict the percentage of time when natural light is sufficient to meet all lighting needs.

Example 5.2 provides a simple calculation procedure to estimate the energy savings from a lighting retrofit project.

Example 5.2:

Problem: *Consider a building with total 500 luminaires of four 40 Watt lamps/luminaire. Determine, the energy saving after replacing those with four 32 Watt high efficacy lamps/ luminaire. This building is operated 8 hours/day, 5 days/week, 50 weeks/year.*

Solution:

The energy saving in KWh is

$$\Delta KWh = 500 \cdot (4*40 - 4*32) \cdot 8.5.50 \cdot \frac{1}{1000} = 32,000 \ kWh \ / \ yr$$

Thus, the energy saving is 116,800 KWh/year.

5.5   Electrical Appliances

5.5.1 Office Equipment

Over the last decade, the energy used by office equipment has increased significantly and accounts currently for more than 7% of the total commercial sector electricity use. Recognizing this problem, the US Environmental Protection Agency (EPA) in cooperation with the US Department of Energy (DOE) has launched the Energy Star Office Program to increase the energy efficiency of commonly used office equipment such computers, fax machines, printers and scanners. To help the consumers in identifying energy-efficient office equipment, Energy Star labels are provided to indicate the energy savings features of the products.

It is estimated that Energy Star labeled products can save as much as 75% of the total electricity use depending on the type and the usage pattern of the office equipment. Almost all office equipment manufacturers currently integrate power management features in their products. For instance, computers can enter a low-power "sleep" mode when idled for a specific period of time. Similarly, copiers can go into a low-power mode of only 15-45 watts after 30-90 minutes of activity. Table 5.6 summarizes the Energy Star features of common office equipment.

Table 5.6: Summary of the Energy Star specifications for office equipment

Product	Watts in Low-Power State	Power Management Preset Default Times
Computers	< 30 W (for power supply < 200 W)	15-30 minutes
	< 15% of power supply (for power supply > 200 W)	15-60 minutes
Monitors	< 15 W (for sleep mode)	15-30 minutes
	< 8 W (for deep-sleep mode)	< 70 minutes
Copiers	3.85*cpm+5 (ppm: copies per minute)	15 minutes
Scanners	12 W	15 minutes
Fax Machines (Specifications depend on the ppm: pages per minute)	15 W (ppm < 7)	5 minutes
	30 W (7 < ppm < 14)	5 minutes
	45 W (ppm > 14)	15 minutes
Printers (Specifications depend on the ppm: pages per minute)	15 W (ppm < 7)	15 minutes
	30 W (7 < ppm < 14)	30 minutes
	45 W (ppm > 14)	60 minutes

It is estimated that the energy efficiency features proposed by the Energy Star Program can reduce by year 2010 the annual pollution generated within the US by an amount equivalent to that generated – annually – by about 6.5 millions cars. In addition, Energy Star labeled office equipment in one home (equipped with one computer, one monitor, one printer, and one fax machine) can save enough electricity to light an entire home for the duration of one year (EPA, 1999).

5.5.1   Residential Appliances

Appliances account for a significant part of the energy consumption in buildings which used about 41% of electricity generated world-wide in 1990

(IPCC, 1996). Table 5.7 illustrates a forecast of world demand by region for white goods which include refrigerators, clothes washers, dishwashers, clothes dryers, and cooking products (Euromonitor, 1994). The greatest growth rate from 1992-2000 is predicted to be in South America and the Middle East. The increase in the use of appliances is anticipated to require a significant increase in electricity demand.

Moreover, the operating cost of appliances during their lifetime (typically 10 to 15 years) far exceeds their initial purchase price. However, consumers – especially in the developing countries where no labeling programs for appliances are enacted – do not generally consider energy efficiency and operating cost when making purchases since they are not well informed.

Recognizing the significance and the impact of appliances on the national energy requirements, a number of countries have established energy efficiency programs. In particular, some of these programs target improvements of energy efficiency for residential appliances. Methods to achieve these improvements include energy efficiency standards and labeling programs.

Table 5.7: World Market for White Goods in Percent Market value (*source:* Euromonitor, 1994)

Region	1992	2000	% Growth
Western Europe	39.9	38.4	10
South East Asia	20.2	23.6	35
North America	25	21.1	−3
Eastern Europe	6.8	7.2	21
South America	1.8	2.6	71
Middle East	1.9	2.5	49
Australasia/Pacific Islands	1.6	1.6	15
Other	2.8	3.0	7

Minimum efficiency standards for residential appliances have been implemented in some countries for a number of residential end-uses. The energy savings associated with the implementation of these standards are found to be substantial. For instance, studies have indicated that in the UK, the energy consumption of average new refrigerators and freezers in 1993 was about 60% of the consumption in 1970. Similar improvements have been obtained in Germany (Waide et al., 1997). In the US, the savings due to the standards are estimated to be about 0.7 exajoules per year during the period extending from 1990 to 2010 (1 exajoule = 10^{18} joules = 1 quadrillion of Btu = 10^{15} Btu).

Energy standards for appliances in the residential sector have been highly cost-effective. In the US, it is estimated that the average benefit/cost ratios for promoting energy efficient appliances are about 3.5. In other terms, each US dollar of federal expenditure on implementing the standards is expected to contribute $165 of net present-valued savings to the economy over the period of 1990 to 2010. In addition to energy and cost savings, minimum efficiency standards reduce pollution with significant reduction in carbon emissions. In the period of 2000 to 2010, it is estimated that energy efficiency standards will result in an annual carbon reduction of 4% (corresponding to 9 million metric tons of carbon/year) relative to the 1990 level.

Currently, energy efficiency standards are utilized with various degrees of comprehensiveness, enforcement, and adoption in a limited number of countries as summarized in Table 5.8. However, several other countries are in the process of developing national standards or labeling programs to promote energy efficient residential appliances.

Minimum efficiency standards have been in use in the United States and Canada for more than two decades and cover a wide range of products. In the last few years, standards programs have spread to other countries like Brazil, China, Korea, Mexico, and the Philippines. Most of these countries have currently one or two products subject to standards such as the case of European countries, Korea, and Japan. Other countries are in the process of implementing or considering energy efficiency standards for household appliances. Among these countries are Colombia, Denmark, Egypt, Indonesia, Malaysia, Pakistan, Singapore, and Thailand.

Many of the currently existing standards are mandatory and prohibit the manufacture and/or sale of non-complying products. However, there are other standards that are voluntary and thus are not mandatory, such as the case of product quality standards established in India.

Table 5.8: Status of International Residential Appliance Energy Efficiency Standards (*Source*; Turiel, 1997).

Country/region	Compliance Status	Products
Australia	Mandatory	R, FR, WH
Brazil	Voluntary	R, FR
Canada	Mandatory	All
China	Mandatory	R, CW, RAC
European Union	Mandatory	R, FR
India	Voluntary	R, RAC, A/C
Japan	Voluntary	A/C
Korea	Mandatory	R, A/C
Mexico	Mandatory	R, FR, RAC
Philippines	Mandatory	A/C
US	Mandatory	All

Notes: Products are: Refrigerators (R), Freezers (FR), Clothes Washers (CW), Dishwashers (DW), Clothes Dryers (CD), Cooking, Water Heaters (WH), Room Air Conditioners (RAC), and Central Air Conditioning (A/C).

As indicated in Table 5.8, most countries have established minimum efficiency standards for refrigerators and freezers since this product type has one of the highest growth rates both in terms of sales value and volume. The existing international energy efficiency standards for refrigerators and freezers set a limit on the energy use over a specific period of time (generally, one month or one year). This energy use limit may vary depending on the size and the configuration of the product. Table 5.9 shows the maximum limits for the allowable annual energy use for US refrigerators and freezers. Two standards are shown in Table 5.9: the standards that are currently effective since 1993 and the updated standards to be effective in July 2001. It should be noted that models with higher energy efficiencies than those listed in Table 5.9 do exist and are sold in the US market. To keep up with the technology advances, US standards are typically amended or changed periodically. However, any new or amended US standard for energy efficiency has to be based on improvements that are technologically feasible and economically justified. Typically, energy-efficient designs with payback periods of less than three years can be incorporated into new US standards.

Table 5.9: Maximum allowable annual energy use (kWh/yr) for refrigerators and freezers sold and manufactured in the US (*Source*: Turiel, 1997).

Product Category	1993 Standard	2001 Standard
Manual defrost R/FR	299+0.48AV	248+0.31AV
Partial auto-defrost R/FR	398+0.37AV	248+0.31AV
Top-mount auto-defrost R/FR	355+0.57AV	276+0.35AV
Top-mount auto-defrost with through the door features R/FR	391+0.616AV	356+0.36AV
Side-mount auto-defrost R/FR	501+0.413AV	508+0.17AV
Side-mount auto-defrost with through the door features R/FR	527+0.571AV	406+0.36AV
Bottom-mount auto-defrost R/FR	367+0.578AV	459+0.16AV
Upright manual FR	264+0.361AV	258+0.27AV
Upright auto-defrost FR	391+0.522AV	326+0.44AV
Chest RF	160+0.385AV	144+0.35AV

Notes: R = Refrigerators
FR = Freezers
AV = adjusted volume = Volume of R + 1.63 x Volume of FR (for R/FR)
= 1.73 x Volume of FR   (for R)

In addition to standards, labeling programs have been developed to inform consumers about the benefits of energy efficiency. There is a wide range of labels used in various countries to promote energy efficiency for appliances. These labels can be grouped into three categories:

- Efficiency type labels used to allow consumers to compare the performance of different models for a particular product type. For instance, a common label used for refrigerators indicates the energy use and operating cost over a specific period such as one month or one year. Other labels such as labels for clothes washers show the efficiency expressed in kWh of energy used per pound of clothes washed. Another feature of efficiency labels is the ability to provide consumers with a comparative evaluation of the product models by showing the energy consumption or efficiency of a particular model on a scale of the lowest (or highest) energy use (or efficiency) models.

- Eco-labels provide information on more than one aspect (i.e., energy efficiency) of the product. Other aspects include noise level, waste disposal, and emissions. Green Seal in the US is an example of an eco-label program that certifies that the products are designed and manufactured in an environmentally responsible manner (Green Seal, 1993). Certification standards have been established for refrigerators, freezers, clothes washers, clothes dryers, dishwashers, and cooktops/ovens.

- Efficiency seals of approval, such as the Energy Star program in the US, are labels that indicate that a product has met a set of energy efficiency criteria but do not quantify the degree by which the criteria were met. The Energy Star label, established by the US Environmental Protection Agency, indicates for instance that a computer monitor is capable of reducing its standby power level when not in use for some time period (Johnson and Zoi, 1992).

In recent years, labeling of appliances is becoming a popular approach around the world in order to inform consumers about the energy use and energy cost of purchasing different models of the same product. Presently, Australia, United States, and Canada have the most comprehensive and extensive labeling programs. Other countries such as the European Union, Japan, Korea, Brazil, Philippines, and Thailand, have developed labels for a few products.

In addition to energy efficiency, standards have been developed to improve the performance of some appliances in conserving water. For instance, water-efficient plumbing fixtures and equipment have been developed in the US to promote water conservation.

The reduction of water use by some household appliances can also increase their energy efficiency. Indeed, a large fraction of the electrical energy used by both clothes washers and dishwashers is attributed to heating the water (85% for clothes washers and 80% for dishwashers). Chapter 15 provides more detailed discussion on water and energy performance of conventional and energy/water efficient models currently available for residential clothes washers and dishwashers.

5.6   Electrical Distribution Systems

5.6.1   Introduction

All electrical systems have to be designed in order to provide electrical energy to the utilization equipment as safely and reliably as economically possible. Figure 5.8 shows a typical one-line diagram of an electrical system for a commercial building. The main distribution panel includes the switchgear-breakers to distribute the electric power and the unit substation to step down the voltage. The unit substation consists typically of a high voltage disconnect switch, a transformer, and a set of low-voltage breakers. The circuit breakers for lighting and plug-connected loads are housed in lighting panelboards while the protective devices for motors are assembled typically into motor control centers (MCCs). Specifically, a MCC consists generally of the following equipment:

- Overload relays to prevent the current from the motor from exceeding any dangerous level.
- Fuse disconnect switches or breakers to protect the motor from short-circuit currents.

An important part of any electrical system is the electrical wiring that connects all the system components. Three types of connecting wires can be identified:

- Service entrance conductors are those electrical wires that deliver electricity from the supply system to the facility. For large facilities, electricity is typically supplied by an electric utility at a relatively high voltage (13.8 kV) requiring a transformer (part of a unit substation) to step down the voltage to the utilization level.
- Feeders are the conductors that deliver electricity from the service entrance equipment location to the branch-circuits. Two types of feeders are generally distinguished: the main feeders that originate at the service entrance (or main distribution panel) and the sub-feeders that originate at distribution centers (lighting panelboards or motor control centers).
- Branch circuits are the conductors that deliver electricity to the utilization equipment from the point of the final over-current device.

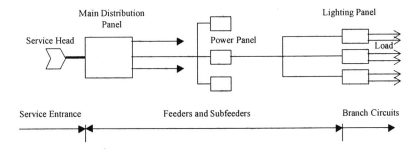

Figure 5.8: A schematic one-line diagram for a basic electrical distribution system within a building.

5.6.2   Transformers

The transformer is the device that changes the volt level of an alternating current. In particular, it is common to use transformers at generating stations to increase the transmission voltages to high levels (13800 volt) and near or inside buildings to reduce the distribution voltages to low levels for utilization (480 or 208 volt).

A typical transformer consists of two windings: primary and secondary windings. The primary winding is connected to the power source while the secondary winding is connected to the load. Between the primary and the secondary windings, there is no electrical connection. Instead, the electric energy is transferred by induction within the core which is generally made up of laminated steel. Therefore, transformers operate only on alternating current.

There are basically two types of transformers (i) liquid-filled transformers and (ii) dry-type transformers. In liquid-filled transformers, the liquid acts as a coolant and as insulation dielectric. Dry-type transformers are constructed so that the core and coils are open to allow for cooling by free movement of air. In some cases, fans may be installed to increase the cooling effect. The dry-type transformers are widely used because of their lighter weight and simpler installation compared to liquid-filled transformers.

A schematic diagram for a single-phase transformer is illustrated in Figure 5.9. A three-phase transformer can be constructed from a set of three single-phase transformers electrically connected so that the primary and the secondary windings can be either wye or delta configurations. For buildings, delta-connected primary and wye-connected secondary is the most common arrangement for transformers. It can be shown that the primary and the secondary voltages, V_p and V_s, are directly proportional to the respective number of turns, N_p and N_s, in the windings:

$$\frac{V_p}{V_s} = \frac{N_p}{N_s} = a \qquad (5.25)$$

where, a is the turns ratio of the transformer. As indicated by Eq. (5.25), the turns ratio can be determined directly from the voltages without the need to know the actual number of turns on the transformer windings.

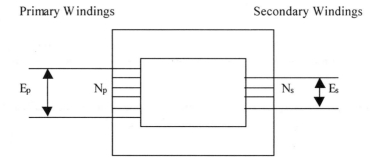

Figure 5.9: Simplified Model for a single-phase transformer.

Transformers are rated by their volt-ampere capacity from the secondary windings. For large transformers, the power output in kilo volt-ampere, or kVA, rating is generally used as expressed by Eq. (5.26):

$$kVA = \frac{\sqrt{3}.V_s.I_s}{1000} \qquad (5.26)$$

where, V_s and I_s are respectively the rated line-to-line voltage and the rated line current of the secondary.

Transformers are typically very efficient with energy losses (in the core and windings) representing only 1-2% of the transformer capacity. It may be cost-effective to invest on more energy efficient transformers especially if they are used continuously at their rated capacity as illustrated in Example 5.3.

Example 5.3:

Problem:
Determine the cost-effectiveness of selecting a unit with an efficiency of 99.95% rather than 99.90% for a 500-kVA rated transformer. Assume that:

- *The cost of electricity is $0.10/kWh.*

- The installed costs of 99.0% and 99.5% efficient transformers are respectively $7,000 and $9,000.
- The average power factor of the load 0.90.
- The no-load losses are the same for both transformers.

For the analysis consider two case for the length of time during which the transformer is used at its rated capacity:
(a) 10 hours/day and 250 days/year,
(b) 16 hours/day and 300 days/year.

Solution:

To determine the cost-effectiveness of installing an energy efficient transformer, a simplified economic analysis is used to estimate the simple payback period. The savings in energy losses in kWh for the high-efficiency transformer can be calculated as follows:

$$kWh_{saved} = N_h.kVA.pf\left(\frac{1}{\eta_{std}} - \frac{1}{\eta_{eff}}\right)$$

with:
✓ N_h is the total number of hours (per year) during which the transformer is operating at full-load [for case (a) N_h = 10*250=2500 hrs/yr; for case (b) N_h = 16*300=4800 hrs/yr].
✓ kVA is the rated transformer power output [500 kVA]
✓ pf is the annual average power factor of the load [pf=0.90]
✓ η_{std} and η_{eff} are the efficiency of respectively the standard transformer and the efficient transformer [0.990 and 0.995].

The energy savings and the simple payback period for each case are presented above.

For case (a) N_h=2500 hrs/yr

The energy savings in kWh is calculated as follows:

$$kWh_{saved} = 2500*500*0.90*\left(\frac{1}{0.990} - \frac{1}{0.995}\right) = 5710kWh/yr$$

Therefore, the simple payback period, SPB, for investing on the efficient transformer is:

$$SPB = \frac{\$9,000 - \$7,000}{5710kWh * \$0.07 / kWh} = 5.0 years$$

For case (a) N_h=4800 hrs/yr

The energy savings in kWh is calculated as follows:

$$kWh_{saved} = 4800 * 500 * 0.90 * \left(\frac{1}{0.990} - \frac{1}{0.995} \right) = 10964kWh / yr$$

Thus, the simple payback period, SPB, for investing on the efficient transformer is:

$$SPB = \frac{\$9,000 - \$7,000}{10964kWh * \$0.07 / kWh} = 2.6 years$$

It is clear that it can be cost-effective to consider investing in a more energy-efficient transformer especially when the load is supplied during longer periods of time. It should be noted that additional energy savings can be expected during no-load conditions for the energy-efficient transformer.

5.6.3   Electrical Wires

The term electrical wire is actually generic and refers typically to both a conductor and a cable. The conductor is the copper or aluminum wires that actually carries electrical current. The cable refers generally to the complete wire assembly including the conductor, the insulation, and any shielding and/or protective covering. A cable can have more than one conductor, each with its own insulation.

The size of an electrical conductor represents its cross-sectional area. In the US, two methods are used to indicate the size of a conductor: the American Wire Gage (AWG) for small sizes, and thousand of circular mils (MCM) for larger sizes. For the AWG method, the available sizes are from number 18 to number 4/0 with the higher number representing the smaller conductor size. For buildings, the smallest size of copper conductor that can be used is number 14 which is rated for a maximum loading of 15 amperes. The AWG size designation became inadequate soon after its implementation in early 1900's due to the ever increasing electrical loads in buildings. For larger conductors, the cross-sectional area is measured in circular mils. A circular mil corresponds to the area of a circle that has diameter of 1 mil or 1/1000th of an inch. For instance, a conductor with a diameter of ½ inch (500 mils) has a circular mil area of 250,000 which is designated by 250 MCM.

To determine the correct size of conductors to be used for feeders and branch circuits in buildings, three criteria need generally to be considered:

- The rating of the continuous current under normal operating conditions. The National Electric Code (NEC, 1996) refers to the continuous current rating as the ampacity of the conductor. The main parameters that affect the ampacity of a conductor include the physical characteristics of the wire such as its cross-sectional area (or size) and its material and the conditions under which the wire operates such as the ambient temperature and the number of conductors installed in the same cable. Table 5.11 indicates the ampacity rating of copper and aluminum conductors with various sizes. Various derating and correction factors may need to be applied to the ampacity of the conductor to select its size.

- The rating of short-circuit current under fault conditions. Indeed, high short-circuit currents can impose significant thermal or magnetic stresses not only on the conductor but also on all the components of the electrical system. The conductor has to withstand these relatively high short-circuit current since the protective device requires some finite time before detecting and interrupting the fault current.

- The maximum allowable voltage drop across the length of the conductor. Most electrical utilization equipment are sensitive to the voltage applied to them. It is therefore important to reduce the voltage drop that occurs across the feeders and the branch circuits. The NEC recommends a maximum voltage drop of 3% for any one feeder or branch circuit with a maximum voltage drop from the service entrance to the utilization outlet of 5%.

For more details, the reader is referred to section 220 of the NEC that covers the design calculations of both feeders and branch circuits.

Two conductor materials are commonly used for building electrical systems: copper and aluminum. Because of its highly desirable electrical and mechanical properties, copper is the preferred material used for conductors of insulated cables. Aluminum has some undesirable properties and its use is restricted. Indeed, an oxide film which is not a good conductor can develop on the surface of aluminum and can cause poor electrical contact especially at the wire connections. It should be noted that aluminum can be considered in cases when cost and weight are important criteria for the selection of conductors. However, it is highly recommended even in these cases to use copper conductors for the connections and the equipment terminals to eliminate poor electrical contact.

To protect the conductor, several types of insulation materials are used. The cable (which is the assembly that includes the conductor, insulation, and any other covering) is identified by letter designations depending on the type of insulation material and the conditions of use. In buildings, the following letter designations are used:

- For the insulation material type: A (asbestos), MI (mineral insulation), R (rubber), SA (silicone asbestos), T (thermoplastic), V (varnished cambric), and X (cross-linked synthetic polymer).
- For the conditions of use: H (heat up to 75 °C), HH (heat up to 90 °C), UF (suitable for underground), W (moisture resistant).

Thus, the letter designation THW refers to a cable that has a thermoplastic insulation rated for maximum operating temperature of 75 °C and suitable for use in dry as well as wet locations.

Moreover, some types of electrical cables have outer coverings that provide mechanical/corrosion protection such as lead sheath (L), nylon jacket (N), armored cable (AC), metal-clad cable (MC), and NM (nonmetallic sheath cable).

For a full description and all types of insulated conductors, their letter designations, and their uses, the reader is referred to the NEC, article 310 and Table 310-13.

In general, the electrical cables are housed inside conduits for additional protection and safety. The types of conduit commonly used in buildings are listed below:

- Rigid Metal Conduit (RMC) can be of either steel or aluminum and has the thickest wall of all types of conduit. Rigid metal conduit is used in hazardous locations such as high exposure to chemicals.
- Intermediate Metal Conduit (IMC) has a thinner wall than the rigid metal conduit but can be used in the same applications.
- Electrical Metallic Tubing (EMT) is a metal conduit, but with a very thin wall. The NEC restricts the use of EMT to locations where it is not subjected to severe physical damage during installation or after installation.
- Electrical Nonmetallic Conduit (ENC) is made of nonmetallic material such as fiber or rigid PVC (polyvinyl chloride). Generally, rigid nonmetallic conduit cannot be used where subject to physical damage.
- Electrical Nonmetallic Tubing (ENT) is a pliable corrugated conduit that can be bent by hand. Electrical nonmetallic tubing can be concealed within walls, floors, and ceilings.
- Flexible Conduit can be readily flexed and thus is not affected by vibration. Therefore, a common application of the flexible conduit is for the final connection to motors or recessed lighting fixtures.

It should be noted that the number of electrical conductors that can be installed in any one conduit is restricted to avoid any damage of cables (especially when the cables are pulled through the conduit). The NEC restricts the percentage fill to 40% for three or more conductors. The percentage fill is defined as the fraction of the total cross-sectional area of the conductors – including the insulation- over the cross-sectional area of the inside of the conduit.

When selecting the size of the conductor, the operating costs and not only the initial costs should be considered. As illustrated in Example 5.4, the cost of energy encourages the installation of larger conductors than are required typically by the NEC especially when smaller size conductors are involved (i.e., numbers 14, 12, 10 and 8). Unfortunately, most designers do not consider the operating costs in their design, for several reasons including interest in lower first costs and uncertainty in electricity prices.

Example 5.4:

Problem:
Determine if it is economically feasible to install a number 10 (AWG) copper conductor instead of a number 12 (AWG) on 400 feet branch circuit that feeds a load of 16 amperes. Assume that
- The load is used 10 hours/day and 250 days/year.
- The cost of electricity is $0.10/kWh.
- The installed costs of No. 12 and No. 10 conductors are respectively $60.00 and $90.00 per 1000 ft long cable.

Solution:

In addition to the electric energy used to meet the load, there is an energy loss in the form of heat generated by the flow of current, I, through the resistance of the conductor, R. The heat loss in *Watts* can be calculated as follows:

$$Watts = R.I^2$$

Using the information by the NEC (Table 8), the resistance of both conductors No. 12 and 10 can be determined to be respectively: 0.193 ohm and 0.121 ohm per 100 feet. Thus, the heat loss for the 400-ft branch circuit if No. 12 conductor is used can be estimated as follows:

$$Watts_{12} = 0.193*400/100.(16)^2 = 197.6 \quad W$$

Similarly the heat loss for the 400-ft branch circuit when No. 10 conductor is used is found to be:

$$Watts_{10} = 0.121*400/100.(16)^2 = 123.9 \quad W$$

The annual cost of copper losses for both cases can be easily calculated:

$Cost_{12} = 197.6W * 250days/yr * 10hrs/Day * 1kW/1000W * \$0.10/kWh = \$49.4/yr$

$Cost_{10} = 123.9W * 250days/yr * 10hrs/Day * 1kW/1000W * \$0.10/kWh = \$31.0/yr$

Therefore, if No. 10 is used instead of No. 12, the simple payback periods, SPB, for the higher initial cost for the branch circuit conductor is:

$$SPB = \frac{(\$90/1000ft - \$60/1000ft) * 400ft}{(\$49.4 - \$31.0)} = 0.68yr = 8 \quad months$$

The savings in energy consumption through the use of larger conductors can thus be cost-effective. Moreover, it should be noted that the larger size conductors reduce the voltage drop across the branch circuit which permits the connected electrical utilization equipment to operate more efficiently. However, the applicable code has to be carefully consulted to determine if larger size conduit is required when larger size conductors are used.

Table 5.10: Ampacity of selected insulated conductors used in buildings (adapted from NEC Table 310-16)

CONDUCTOR SIZE (AWG OR MCM)	THW (COPPER)	THHN (COPPER)	THW (ALUMINUM)	THHN (ALUMINUM)
18	-	14	-	-
16	-	18	-	-
14	20	25	-	-
12	25	30	20	25
10	35	40	30	35
8	50	55	40	45
6	65	75	50	60
4	85	95	65	75
3	100	110	75	85
2	115	130	90	100
1	130	150	100	115
1/0	150	170	120	135
2/0	175	195	135	150
3/0	200	225	155	175
4/0	230	260	180	205
250	255	290	205	230
300	285	320	230	255
350	310	350	250	280
400	335	380	270	305
500	380	430	310	350
600	420	475	340	385
700	460	520	375	420
750	475	535	385	435
800	490	555	395	450
900	520	585	425	480
1000	545	615	445	500
1250	590	665	485	545
1500	625	705	520	585
1750	650	735	545	615
2000	665	750	560	630

5.7 Power Quality

5.7.1 Introduction

Under ideal operation conditions, the electrical current and voltage vary as a sine function of time. However, problems due a utility generator or distribution system such as voltage drops, spikes, or transients can cause fluctuations in the electricity which can reduce the life of electrical equipment including motors and lighting systems. Moreover, an increasing number of electrical devices operating on the system can cause distortion of the sine waveform of the current and/or voltage. This distortion leads to poor power quality which can waste energy and harm both electrical distribution and devices operating on the systems.

5.7.2   Total Harmonic Distortion

Power quality can be defined as the extent to which an electrical system distorts the voltage or current sine waveform. The voltage and current for an electrical system with ideal power quality vary as a simple sine function of time, often referred to as the fundamental harmonic, and are expressed by Eq. (5.1) and Eq. (5.2), respectively. When the power is distorted, due for instance to electronic ballasts (which change the frequency of the electricity supplied to the lighting systems), several harmonics need to be considered in addition to the fundamental harmonic to represent the voltage or current time-variation as shown in Eq. (5.27) and Eq. (5.28):

$$v(t) = \sum_{k=1}^{N_V} V_k \cos(k\omega - \theta_k) \tag{5.27}$$

$$i(t) = \sum_{k=1}^{N_I} I_k \cos(k\omega - \phi_k) \tag{5.28}$$

Highly distorted waveforms contain numerous harmonics. While the even harmonics (i.e., second, fourth, etc.) tend to cancel each other's effects, the odd harmonics (i.e., third, fifth, etc.) have their peaks coincide and significantly increase the distortion effects. To quantify the level of distortion for both voltage and current, a dimensionless number referred to as the total harmonic distortion (THD) is determined through a Fourier series analysis of the voltage and current waveforms. The THD for voltage and current are respectively defined as follow:

$$THD_V = \sqrt{\frac{\sum_{k=2}^{N_V} V_k^2}{V_1^2}} \tag{5.29}$$

$$THD_I = \sqrt{\frac{\sum_{k=2}^{N_V} I_k^2}{I_1^2}} \tag{5.30}$$

Table 5.11 provides current THD for selected but specific lighting and office equipment loads (NLPID, 1995). Generally, it is found that devices with high current THD contribute to voltage THD in proportion to their share of the total building electrical load. Therefore, the engineer should consider higher-wattage devices before lower-devices to reduce the voltage THD for the entire building or facility. Example 5.5 shows a simple calculation procedure that can be used to assess the impact of an electrical device on the current THD. Thus, the engineer can determine which devices need to be corrected first to improve the power quality of the overall electric system. Typically, harmonic filters are added to electrical devices to reduce the current THD values.

Example 5.5:

Problem:
Assess the impact on the current THD of a building of two devices: 13-W compact fluorescent lamp (CFL) with an electronic ballast and a laser printer while printing. Use the data provided in Table 5.16.

Solution:
Both devices have an rms voltage of 120 V (i.e., V_{rms} = 120 V); their rms current can be determined using the real power used and the the power factor given in Table 5.16 and Eq. (5.5):

$$I_{rms} = \frac{P_R}{V_{rms} \cdot pf}$$

The above equation gives an rms current of 0.22 A for the CFL and 6.79 A for the printer. These values correspond actually to the rms of each device's fundamental current waveform and can be used in the THD equation, Eq (5.29), to estimate the total harmonic current of each device:

$$I_{tot} = I_{rms} \cdot THD_I$$

The resultant values of 0.33 A for the CFL and 1.02 A for the printer show that although the printer has relatively low current THD (15%), the actual distortion current produced by the printer is more than three times that of the CFL because the printer uses more power.

IEEE (1992) recommends a maximum allowable voltage THD of 5% at the building service entrance (i.e., point where the utility distribution system is connected to the building electrical system). Based on a study by Verderber et al. (1993), the voltage THD reaches the 5% limit when about 50% of the building electrical load has a current THD of 55% or when 25% of the building electrical load has a current THD of 115%.

It should be noted that when the electrical device has a power factor of unity (i.e., pf=1), there is little or no current THD (i.e., THD_I= 0%) since the device has only a resistive load and effectively converts input current and voltage into useful electric power. As shown in Table 5.11, the power factor and the current THD are interrelated and both define the characteristics of the power quality. In particular Table 5.11 indicates that lighting systems with electronic ballasts have typically high power factor and low current THD. This good power quality is achieved using capacitors to reduce the phase lag between the current and voltage (thus improving the power factor as discussed in section 5.2) and filters to reduce harmonics (and therefore increase the current THD value).

Table 5.11: Typical power quality characteristics (power factor and current THD) for selected electrical loads (*Source*: Adapted from NLPIP, 1995)

ELECTRICAL LOAD	REAL POWER USED (W)	POWER FACTOR	CURRENT THD (%)
Incandescent Lighting Systems			
100-W incandescent lamp	101	1.0	1
Compact fluorescent lighting systems			
13-W lamp w/ magnetic ballast	16	0.54	13
13-W lamp w/electronic ballast	13	0.50	153
Full-size fluorescent lighting systems (2 lamps per ballast)			
T12 40-W lamp w/ magnetic ballast	87	0.98	17
T12 40-W lamp w/ electronic ballast	72	0.99	5
T10 40-W lamp w/ magnetic ballast	93	0.98	22
T10 40-W lamp w/ electronic ballast	75	0.99	5
T8 32-W lamp w/ electronic ballast	63	0.98	6
High-intensity discharge lighting systems			
400-W high-pressure sodium lamp w/ magnetic ballast	425	0.99	14
400-W metal halide lamp w/magnetic ballast	450	0.94	19
Office equipment			
Desktop computer w/o monitor	33	0.56	139
Color monitor for desktop computer	49	0.56	138
Laser printer (in standby mode)	29	0.40	224
Laser printer (printing)	799	0.98	15
External fax/modem	5	0.73	47

The possible problems that have been reported due to poor power quality include:

(1) Overload of neutral conductors in three-phase with four wires. In a system with no THD, the neutral wire carries no current if the system is well balanced. However, when the current THD becomes significant, the currents due to the odd harmonics do not cancel each other and rather add up on the neutral wire which can overheat and cause a fire hazard.
(2) Reduction in the life of transformers and capacitors. This effect is mostly caused by distortion in the voltage.
(3) Interference with communication systems. Electrical devices that operate with high frequencies such as electronic ballasts (that operate at frequencies ranging from 20 to 40 kHz) can interfere and disturb the normal operation of communication systems such radios, phones, and energy management systems (EMS).

5.8 Summary

In this chapter, an overview is provided for the basic characteristics of electrical systems in HVAC applications for buildings. In particular, the operation principles of motors are emphasized. Throughout the chapter, several measures are described to improve the energy performance of existing or new electrical installations. Moreover, illustrative examples are presented to evaluate the cost-effectiveness of selected energy efficiency measures. For instance, it was shown that the use of larger conductors for branch circuits can be justified based on the reduction of energy losses and thus operating costs. Moreover, the chapter provided suggestions to improve the power quality, increase the power factor, and reduce lighting energy use in buildings. These suggestions are presented to illustrate the wide range of issues that an engineer should address when retrofitting electrical systems for buildings.

Problems

5.1 Provide a simple payback period analysis of lighting controls in a 10,000 square-foot office area comprised of 7,200 square-feet of open landscaped office space and twenty (20) enclosed perimeter offices (2,800 square-feet). The total lighting load used for the space is 17 kW and the office was operated 260 days per year. The open space is monitored by 24 ceiling-mounted motion sensors. The perimeter offices are monitored with wall-mounted sensors (one for each office). The installed cost of a ceiling sensor is $110 with a $64 utility rebate. The wall sensor costs $75 with a $32 utility rebate.

Determine the simple payback periods with and without rebate. Assume the additional off-time (hours/day) is 1, 2, 4, 6, and 8. Two scenarios for electricity cost: $0.07/kWh and $0.14/kWh. Show your results in a tabular format.

5.2 Two motors operate 5,000 h/hr at full load. One is 40 HP with an efficiency of 0.75 and a power factor of 0.65 while the other is 100 HP with an efficiency of 0.935 and a power factor of 0.85. Determine:

(a) The overall power factor.
(b) The simple payback period of replacing each motor by and energy-efficient motor. Assume that the cost of electricity is $0.07/kWh.

5.3 Determine the capacitor ratings (in kVAR) to add to a 80 HP motor with an efficiency of 0.85 to increase its power factor from 0.80 to 0.85, 0.90, and 0.95, respectively.

5.4 Consider an aluminum fabrication plant with 400 employees that runs on three shifts for a total of 8,760 hours per year. The company spent $1.1 million in electric energy bills last year. The plant uses numerous motors to drive process-related equipment such as shell presses and compound presses. Determine if it is cost-effective to replace the existing motors with high efficient motors under the following three rate structures:

For all the rates, the utility company chares the facility and average cost of electricity –without demand- of $0.05/kWh. The demand charges are:
- Rate-1: 7.02/kVA
- Rate-2: $5.00/kW (billed demand)
- Rate-3: $5.00/kW plus $0.75 for excess KVAR above 6-percent of real demand.

The motors that will be replaced are listed below.

HP	N	LF	eff-s	eff-e	pf-s	pf-e
100	4	0.75	0.919	0.950	0.872	0.905
60	2	0.75	0.916	0.940	0.854	0.861
40	13	0.50	0.908	0.934	0.797	0.805
20	2	0.50	0.886	0.923	0.759	0.833
15	12	0.50	0.875	0.916	0.681	0.774
10	5	0.50	0.864	0.910	0.714	0.770
7.5	57	0.50	0.846	0.902	0.683	0.731
5	43	0.50	0.839	0.890	0.687	0.714
2	14	0.50	0.791	0.864	0.516	0.540
1.5	2	0.50	0.780	0.852	0.546	0.580

Note: LF indicates the load factor of the motor.

5.5 The nameplate of a motor provides the following information:

Full Load HP	100
Volts	440/220
Amperes	123/246
Full Load RPM	1,775

During an audit, measurements on the motor indicated that:

Average Volts	440
Amperes	116
Full Load RPM	1,779

(a) Calculate the following parameters: the slip RPM, the efficiency, and the power factor of the motor

(b) Determine if it is cost-effective to replace this motor with a more efficient motor (You need to document the reference for the motor cost). Perform the calculations with and without adjusting the speed of the energy efficient motor.

5.6 Consider a three-phase 1,500 kVA transformer to step-down the volt from 13.8 kV to 480Y/277 volt. The transformer is old and needs to be replaced. Two options are available: (i) replace it by the same transformer with an efficiency of 98.6% and an installed cost of $45,000 and (ii) replace it by more energy efficient transformer with an efficiency of 99.0% and an installed cost of $60,000.

(a) Determine the best option for the transformer replacement considering a lifetime of 25 years, and electricity price of $0.06, a discount rate of 4.1%, and an average load factor of 50%.

(b) Determine the electricity price for which the energy efficient motor is not cost-effective using the same assumptions of question (a).

(c) Determine the size of the main secondary feeders for both the standard and the energy efficient transformers. Calculate the energy savings if the next higher feeder size is used.

6

BUILDING ENVELOPE

Abstract

This chapter provides some energy conservation measures related to the building thermal envelope. The measures discussed in this chapter tend to improve the comfort level within the buildings as well as the energy efficiency of the building envelope. Specifically, this chapter provides the auditor with selected measures to improve the thermal performance of building envelope components such as adding thermal insulation to roofs and walls, weather-stripping to improve the leakage characteristics of the building shell, and installing energy efficiency features for windows.

6.1   Introduction

Generally, the envelope of a structure is designed by architects to respond to many considerations including structural and esthetics. Before the oil crisis of 1973, the energy efficiency of the envelope components was rarely considered as an important factor in the design of a building. However, since 1973 several standards and regulations have been developed and implemented to improve the energy efficiency of various components of building envelopes. For energy retrofit analysis, it is helpful to determine if the building was constructed or modified to meet certain energy-efficiency standards. If it is the case, retrofitting of the building envelope may not be cost-effective especially for high-rise commercial buildings. However, improvements to building envelope can be cost-effective if the building or industrial facility was built without any concern for energy efficiency such as the case with structures constructed with no insulation provided in the walls or roofs.

Moreover, the building envelope retrofit should be performed after careful assessment of the building thermal loads. For instance in low-rise buildings such as residential and small commercial buildings or warehouses, the envelope transmission losses and infiltration loads are dominant and the internal loads within these facilities are typically low. Meanwhile in high-rise commercial, industrial, and institutional facilities, the internal heat gains due to equipment, lighting, and people are typically dominant and the transmission loads affect only the perimeter spaces.

157

The accurate assessment of the energy savings incurred by building envelope retrofits generally requires detailed hourly simulation programs since the heat transfer in buildings is complex and involves several mechanisms. In this chapter, only simplified calculation methods are presented to estimate the energy savings for selected improvements of the building envelope commonly proposed to improve not only the energy efficiency of the building but also the thermal comfort of its occupants and the structural integrity of its shell.

6.2   Basic Heat Transfer Concepts

The heat transfer from the building envelope can occur by various mechanisms including conduction, convection, and radiation. In this section, various fundamental concepts and parameters are briefly reviewed. These concepts and parameters are typically used to characterize the thermal performance of various components of the building envelope and are useful to estimate the energy use savings accrued by retrofits of building envelope.

6.2.1   Heat Transfer from Walls and Roofs

In buildings, the heat transfer through walls and roofs is dominated by conduction and convection. Typically, one-dimensional heat conduction is considered to be adequate for above-grade building components unless significant thermal bridges exist such as at the wall corners or at the slab edges. Specifically, the heat transfer from a homogeneous wall or roof layer illustrated in Figure 6.1 can be calculated as follows using the Fourier law:

$$\dot{q} = \frac{k}{d} . A(T_i - T_o) \qquad (6.1)$$

Where,

- A is the area of the wall
- T_i is inside wall surface temperature
- T_o is outside wall surface temperature
- k is thermal conductivity of the wall
- d is the thickness of the wall

To characterize the heat transfer presented by Eq. (6.1), a thermal resistance R-value or a U-value is defined for the layer as shown below:

$$R = \frac{d}{k} = \frac{1}{U} \qquad (6.2)$$

Where,

- h is the convective heat transfer coefficient of the surface

The concept of thermal resistance can be extended to convection heat transfer that occurs at the outer or inner surfaces of the building envelope:

$$R_{conv} = \frac{1}{h} \qquad (6.3)$$

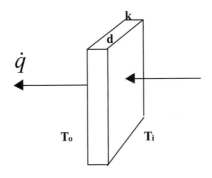

Figure 6.1: Conduction heat transfer through one-layer wall

In buildings, a wall or a roof consists of several layers of homogeneous materials as illustrated in Figure 6.2:

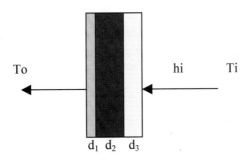

Figure 6.2: Heat Transfer from a multi-layered wall

The heat transfer from a multi-layered wall or roof can be found by determining first its overall R-value:

$$R_T = \sum_{j=1}^{N_L} R_j \qquad (6.4)$$

Where:

- R_j is the R-value of each homogeneous layer part of the construction of the wall or roof assembly. It includes the R-value due to convection at both inner and outer surfaces of the wall or roof obtained by Eq. (6.3).

- N_L is the number of layers (including the convection boundary layers) that are part of the wall or roof assembly. For instance, in the wall assembly presented in Figure 6.2, N_L=5 (3 conductive layers and 2 convective layers).

The overall U-value of the wall or roof can be defined simply as the inverse of the overall R-value:

$$U_T = \frac{1}{R_T} \tag{6.5}$$

It should be noted that practitioners usually prefer to use R-values rather than U-values since the U-values are small, especially when insulation is added to the wall or roof assembly. For doors and windows, the use of U-values is more common since these components have low R-values.

From Eq. (6.1), it is clear that in order to reduce the heat transfer from the above-grade building envelope components, its R-value should be increased or its U-value decreased. To achieve this objective, a thermal insulation can be added to the building envelope. In the next section, calculation methods of the energy savings due to addition of insulation are presented to determine the cost-effectiveness of such measure.

To characterize the total heat transmission of the entire building, a building load coefficient (BLC) is defined to account for all the above grade building envelope components (roofs, walls, doors, and windows):

$$BLC = \sum_{i=1}^{N_E} A_i . U_{T,i} = \sum_{i=1}^{N_E} \frac{A_i}{R_{T,i}} \tag{6.6}$$

With, A_i is the area of each element of the above-grade building envelope including walls, roofs, windows, and doors.

6.2.2   Infiltration Heat Loss/Gain

Air can flow in or out of the building envelope through leaks. This process is often referred to as air infiltration or ex-filtration. Thus, infiltration (and ex-filtration) is rather an uncontrolled flow of air unlike ventilation (and exhaust) for which air is moved by mechanical systems. Generally, air infiltration occurs in all buildings but is more important for smaller buildings such as detached residential buildings. In larger buildings, air infiltration is typically less significant for two reasons:

a) The volume over envelope surface area (from which air leakage occurs) is small for larger buildings.
b) The indoor pressure is generally maintained higher than outdoor pressure by mechanical systems in larger buildings.

Typically, infiltration is considered significant for low-rise buildings and can affect energy use, thermal comfort, and especially structural damages through rusting and rotting of the building envelope materials due to the humidity transported by infiltrating/exfiltrating air. Without direct measurement, it is difficult to estimate the leakage air flow through the building envelope. There are two basic measurement techniques that allow estimation of the infiltration characteristics for a building. These measurement techniques include fan pressurization/depressurization techniques and tracer gas techniques.

Fan pressurization/depressurization techniques are commonly known as blower door tests and allow one to determine the volumetric air flow rate variation with the pressure difference between outdoors and indoors of a building. Several pressure differential values are typically considered and a correlation is found in the form of:

$$\dot{V} = C.\Delta P^n \qquad (6.7)$$

Where, C and n are correlation coefficients determined by fitting the measured data of pressure differentials and air volumetric rates. Using the correlation of Eq. (6.7), an effective leakage area (ELA) can be determined as follows:

$$ELA = \dot{V}_{ref} \cdot \sqrt{\rho / 2.\Delta P} \qquad (6.8)$$

with $\dot{V}_{ref}$ is the reference volume air rate through the building at a reference pressure difference (between indoors and outdoors), of typically 4 Pa and is obtained by extrapolation from Eq. (6.7). The ELA provides an estimate of the equivalent area of holes in the building envelope through which air leaks can occur.

To determine the building air infiltration rate under normal climatic conditions (due to wind and temperatures effects), the LBL infiltration model developed by Shermann and Grimsrud (1980) is commonly used:

$$\dot{V} = ELA \cdot \left(f_s.\Delta T + f_w.v_w^2 \right)^{1/2} \qquad (6.9)$$

with ΔT is the indoor-outdoor temperature difference and v_w is the period-average wind speed, and f_s and f_w are the stack and wind coefficients, respectively. Table 6.1 provides the crack coefficients for three levels of building heights. Table 6.2 lists the wind coefficients for various shielding classes and building heights.

Blower door tests are still being used to find and repair leaks in low-rise buildings. Typically, the leaks are found by holding a smoke source and watching where the smoke exits the house. Several weather-stripping methods are available to reduce air infiltration through building envelope including caulking, weather-stripping, landscaping around the building to reduce the wind effects and installing air-barriers to tighten the building envelope.

It should be mentioned however, that the blower door technique cannot be used to determine accurately the amount of fresh air supplied to the building through either infiltration or ventilation. For this purpose, it is recommended to use the tracer gas techniques described below.

In a typical blower test, the house should first be prepared. In particular, windows are closed, interior doors that are normally open are kept open, and the fire place ash is cleaned. The main entrance door is generally used to place the blower fan to either introduce air (for pressurization test) or extract air (for depressurization test). The air flow rate is generally measured using a pressure gage attached to the blower set-up. The pressure gage should be first checked to make sure that it reads zero with the fan off. An additional pressure gage is used to measure the differential in pressure between the inside and outside of the house. Figure 6.3 shows the setup for both the depressurization and pressurization tests. Example 6.1 illustrates how the results of blower door tests can be used to determine the infiltration rate in a house. The results and the analysis presented in Example 6.1 are based on actual tests performed by Azerbegi et al. (2000).

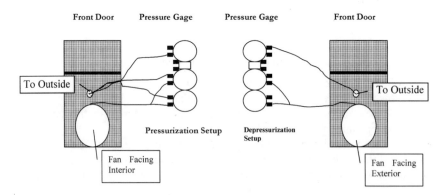

Figure 6.3: Typical Blower door set-up for both pressurization and depressurization tests.

Table 6.1: Stack Coefficient, f_s (*Source*: ASHRAE, <u>Handbook of Fundamentals</u>, 1997. With permission.)

	IP Units			SI Units		
	House Height (Stories)			**House Height (Stories)**		
	One	**Two**	**Three**	**One**	**Two**	**Three**
Stack Coefficient	0.0156	0.0313	0.0471	0.000145	0.000290	0.000435

IP Units for f_s: $(\text{ft}^3/\text{min})^2 \cdot \text{in}^4 \cdot {}^\circ\text{F}$
SI Units for f_s: $(\text{L/sft})^2 \cdot \text{cm}^4 \cdot {}^\circ\text{C}$

Table 6.2: Wind Coefficients and Shielding-Class Descriptions (*Source*: ASHRAE, <u>Handbook of Fundamentals</u>, 1997. With permission.)

	Wind Coefficient, f_w					
	IP Units			**SI Units**		
Shielding	**House Height (Stories)**			**House Height (Stories)**		
Class[a]	**One**	**Two**	**Three**	**One**	**Two**	**Three**
1	0.0119	0.0157	0.0184	0.000319	0.000420	0.000494
2	0.0092	0.0121	0.0143	0.000246	0.000325	0.000382
3	0.0065	0.0086	0.0101	0.000174	0.000231	0.000271
4	0.0039	0.0051	0.0060	0.000104	0.000137	0.000161
5	0.0012	0.0016	0.0018	0.000032	0.000042	0.000049

IP Units for f_s: $(\text{ft}^3/\text{min})^2 \cdot \text{in}^4 \cdot \text{mph}$
SI Units for f_s: $(\text{L/sft})^2 \cdot \text{cm}^4 \cdot (\text{m/s})^2$
[a]Descriptions of shielding classes:
1. No obstructions or local shielding
2. Light local shielding: few obstructions, few trees, or small shed.
3. Moderate local shielding: some obstructions within two house height, thick hedge, solid fence, or one neighboring house.
4. Heavy shielding: obstructions around most of perimeter, buildings or trees within 30 ft (10 m) in most directions; typical suburban shielding.
5. Very heavy shielding: Large obstructions surrounding perimeter within two house heights; typical downtown shielding.

Example 6.1: *A blower door test has been performed in a house located in Evergreen, CO. The results of the pressurization and depressurization tests are summarized in Figure 6.4 based on the results of the blower door. Determine the leakage areas and ACHs for both pressurization and depressurization tests.*

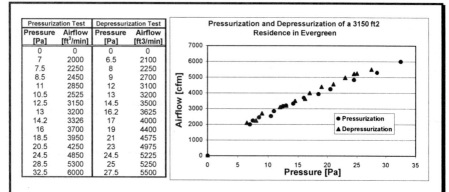

| Pressurization Test | | Depressurization Test | |
Pressure [Pa]	Airflow [ft³/min]	Pressure [Pa]	Airflow [ft3/min]
0	0	0	0
7	2000	6.5	2100
7.5	2250	8	2250
8.5	2450	9	2700
11	2850	12	3100
10.5	2525	13	3200
12.5	3150	14.5	3500
13	3200	16.2	3625
14.2	3326	17	4000
16	3700	19	4400
18.5	3950	21	4575
20.5	4250	23	4975
24.5	4850	24.5	5225
28.5	5300	25	5250
32.5	6000	27.5	5500

Figure 6.4: Summary of the blower door results for both pressurization and depressurization tests (Azerbegi et al. 2000)

Solution: To determine the air leakage characteristics of a house using blower door tests, the following procedure is used:

(i) First, the data consisting of pressure differential (ΔP [Pa]) and airflow rate ($\dot{V}$ [CFM]) presented in Figure 6.4 are plotted in a log-log scale as illustrated in Figure 6.5.

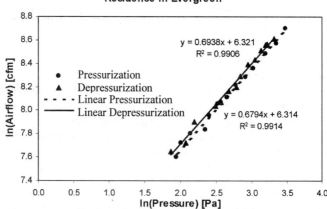

Figure 6.5: Flow rate as a function of the pressure difference in a Log-Log scale

(ii) Then, a linear regression analysis is used to determine the coefficients C and n of Eq. (6.7):

$$\dot{V} = C.\Delta P^n$$

- For the pressurization test, the coefficients C and n are found to be respectively: C = 552.25 and n = 0.679
- For the depressurization test, the coefficients C and n are found to be: C = 556.13 and n = 0.694

It should be noted that the values of C and n provided above are valid only if the airflow rate $\dot{V}$ and pressure differential ΔP are expressed respectively in cfm (cubic feet per minute) and Pa.

(iii) Based on the correlation and the coefficients provided above, the infiltration and exfiltration rates under a normal pressure differential ($\Delta P_{ref} = 4$ Pa) can be calculated:
- For pressurization: $\dot{V}_{ref}$ = 552.25*(4)$^{0.679}$ = 1455 cfm
- For depressurization: $\dot{V}_{ref}$ = 556.13*(4)$^{0.694}$ = 1416 cfm

The leakage areas for both pressurization and depressurization tests can be obtained using Eq. (6.8):

$$ELA = \dot{V}_{ref} \cdot \sqrt{\rho / 2.\Delta P}$$

For Evergreen CO, the air density should be adjusted for altitude and is found to be: $\rho = 0.063$ lbm/ft^3

- For pressurization: ELA = 369.6 in^2
- For depressurization ELA = 379.7 in^2

Therefore, the average leakage area for the house based on both pressurization and depressurization tests is:

ELA(average) = 2.60 ft^2 = 374.4 in^2

(iv) The annual average leakage air flow rate expressed in air change per hour (ACH) can be determined using the LBL model expressed by Eq. (6.9):

$$\dot{V} = ELA \cdot \left(f_s.\Delta T + f_w.v_w^2 \right)^{1/2}$$

To determine the average ACH for the house, the average ELA of 374.4 in^2 is used. Moreover, the coefficients f_s = 0.0266 and f_w = 0.0051 need to used for two story building for a shielding class of 4 (refer to Tables 6.1 and 6.2). The annual average wind speed is V_w (for Denver) = 8.5 mi/hr, and the annual average outdoor temperature tout = 46.17 °F (for Evergreen). To account for the effect of indoor temperature setback (during the winter season), an annual average indoor temperature of 60 °F is considered to calculated the annual average leakage air flow rate $\dot{V}$.

$$\dot{V} = 374.9 * \left(0.026 * [60 - 46.2) + 0.0051 * (8.5)^2\right)^{1/2} = 331\ cfm$$

Then, the volume flow rate $\dot{V}$, is divided by the conditioned volume of the house (in this case, 28,350 ft^3) to obtain the annual average ACH:

$$ACH = \frac{331\,cfm * 60\ min/\ hr}{28,350\ ft^3} = 0.70$$

It should be noted that ASHRAE recommends a leakage of 0.35 ACH for proper ventilation of a house with minimum heat loss.

Tracer gas techniques are commonly used to measure the ventilation rates in buildings. By monitoring the injection and the concentration of a tracer gas (a gas that is inert, safe, and mixes well with air), the exchange of air through the building can be estimated. For instance, in the decay method, the injection of tracer gas is performed for a short time and then stopped. The concentration of the decaying tracer gas is then monitored over time. The ventilation is measured by the air change rate (ACH) within the building and is determined from time variation of the tracer gas concentration:

$$c(t) = c_0 . e^{-ACH . t} \qquad (6.10)$$

or

$$ACH = \frac{\dot{V}}{V_{bldg}} = \frac{1}{t} . Ln\left[\frac{c_0}{c(t)}\right] \qquad (6.11)$$

where C_0 is the initial concentration of the tracer gas, and V_{bldg} is the volume of the building.

The thermal load of air infiltration is rather difficult to assess. It has traditionally thought that the sensible thermal load due to infiltration air is simply calculated as follows:

$$E_{inf} = \rho.c_{p,a}.\dot{V}_{inf}.(T_i - T_o) = \dot{m}_{inf}.c_{p,a}.(T_i - T_o) \qquad (6.12)$$

Eq. (6.12) assumes that infiltrating air enters the condition space (kept at temperature T_i) has the same temperature as the outdoor temperature T_o. However, various recent studies showed that actually infiltrating air can warm up through the building envelope before entering a condition space. This heat exchange occurs especially when the air leakage occurs in a diffuse manner (though long airflow channels inside the building envelope). As a consequence the actual thermal load due to air infiltration is lower than that determined by Eq.(6.12) by a fraction that depends on the heat exchange rate between air infiltration and heat conduction through the building envelope.

When the building envelope is not air tight, it can be assumed that the airflow by infiltration occurs through direct and short paths from outdoors to the indoors without significant heat recovery and thus Eq. (6.12) can be used to estimate the thermal load due to air infiltration.

6.2.3 Variable Base Degree Days Method

The degree-days method provides an estimation of the heating and cooling loads of a building due to transmission losses through the envelope and any solar and internal heat gains. The degree-days method is based on steady-state analysis of the heat balance across the boundaries of the building. A building is typically subject to several heat flows including conduction, infiltration, solar gains, and internal gains as illustrated in Figure 6.6. The net heat loss or heat gain at any instant is determined by applying a heat balance (i.e., first law of Thermodynamics) to the building. For instance, for heating load calculation, the instantaneous heat balance provides:

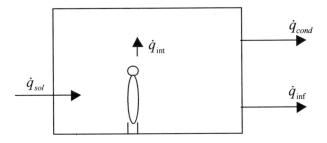

Figure 6.6: A simplified heat balance model for a building

$$\dot{q}_H = BLC \cdot (T_i - T_o) - \dot{q}_g \qquad (6.13)$$

Where:

- BLC is the building load coefficient as defined in Eq.(6.6) but modified to include the effects of both transmission and infiltration losses. Thus, the BLC for any building can be calculated as follows:

$$BLC = \sum_{j=1}^{N_E} U_{T,j} \cdot A_j + \dot{m}_{inf} \cdot c_{p,a} \qquad (6.14)$$

- q_g is the net heat gains due solar radiation, q_{sol}; internal gains (people, lights, and equipment), q_{int}; and in some cases the ground losses, q_{grd}, if they are significant:

$$\dot{q}_g = \dot{q}_{sol} + \dot{q}_{int} - \dot{q}_{grd}$$

This equation can be rearranged to introduce the balance temperature, T_b, for the building

$$q_H = BLC \cdot \left[(T_i - \frac{q_g}{BLC}) - T_o \right] = BLC \cdot (T_b - T_o) \qquad (6.15)$$

Therefore, the balance temperature adjusts the interior temperature set-point by the amount of temperature increase due a reduction in the building heating load resulting from the internal gains. Before the oil crisis, the transmission and the infiltration losses were significant (and thus the BLC value was high relative to the internal gains). It is estimated that the net internal gains contributes to about 3°C (or 5 °F) in most buildings. Therefore, the balance temperature was assumed to be 18 °C (or 65 °F) for all the buildings. However, with the increase in thermal efficiency of the building envelope and the use of more equipment within the buildings, the internal heat gains are more significant and thus can contribute in significantly reducing the heating load of the buildings.

By integrating the instantaneous heating load over the heating season, the total building heating load can be determined. Note only the positive values of q_H are considered in the integration. In practice, the integration is approximated by the sum of the heating loads averaged over short time intervals (one hour or one day). If daily averages are used, the seasonal total building heating load is estimated as:

$$Q_H = 24 \cdot \sum_{i=1}^{N_H} \dot{q}_{H,i}^+ = 24 \cdot BLC \cdot \sum_{i=1}^{N_H} (T_b - T_{o,i})^+ \qquad (6.16)$$

The sum is performed over the number, N_H, of days in the heating season. From Eq. (6.16), a parameter that characterizes the heating load of the building can be defined as the heating degree-days (DD_H) which is function of only the outdoor temperatures and the balance temperature which varies with the building heating set-point temperature and the building internal gains:

$$DD_H(T_b) = \sum_{i=1}^{N_H} (T_b - T_{o,i})^+$$ (6.17)

The total energy use, E_H, to meet the heating load of the building can be estimated by assuming a constant efficiency of the heating equipment over the heating season (for instance several heating equipment manufacturers provide the annual fuel use efficiency rating or AFUE for their boilers or furnaces):

$$E_H = \frac{Q_H}{\eta_H} = \frac{24.BLC.DD_H(T_b)}{\eta_H}$$ (6.18)

The variable base degree days method stated by Eq.(6.18) can also be applied to determine the cooling load by estimating the cooling season degree days (DD) using an equation similar to Eq. (6.16):

$$DD_C(T_b) = \sum_{i=1}^{N_C} (T_{o,i} - T_b)^+$$ (6.19)

Where N_c is the number of days in the cooling season.

It should be noted that the variable base degree days method can provide remarkable accurate estimation of the annual energy use due to heating especially for buildings dominated by losses through the building envelope including infiltration. Unfortunately, the degree-days method is not as accurate for calculating the cooling loads (Claridge et al, 1987) due to several factors including effects of building thermal mass that delays the action of internal gains, mild outdoor temperatures in summer resulting in large errors in the estimation the cooling degree-days, and the large variation in infiltration or ventilation rates as occupants open windows or economizer cycles are used.

6.3 Simplified Calculation Tools for Building Envelope Audit

To determine the cost-effectiveness of any energy conservation measure for the building envelope, the energy use savings has to be estimated. In this section, a general calculation procedure based on the variable base degree days method is provided with some recommendations to determine the values of the parameters required to estimate the energy use savings.

6.3.1   Estimation of the Energy use Savings

When an energy conservation measure is performed to improve the efficiency of the building envelope (for instance by adding thermal insulation to a roof or by reducing the air leakage area for the building envelope), the building load coefficient (BLC) is reduced. Assuming no change in the indoor temperature set-point and in the internal gains within the building, the heating balance temperature actually decrease due to the envelope retrofit as can be concluded from the definition of the heating balance temperature illustrated by Eq. (6.15). Therefore, the envelope retrofit reduces the heating load and thus the energy use since both the BLC and the DD(T_b) are reduced. The energy use savings due the retrofit can be generally calculated as follows:

$$DE_{H,R} = E_{H,E} - E_{H,R} = \frac{24.\left(BLC_E.DD_H(T_{b,E}) - BLC_R.DD_H(T_{b,R})\right)}{h_H} \qquad (6.20)$$

The efficiency of the heating system is assumed to remain the same before and after the retrofit. It is generally the case unless the heating system is replaced or retrofitted. In many applications, the variation caused by the retrofit of the balance temperature is rather small. In these instances, the degree-days (DD$_H$) can be considered constant before and after the retrofit so that the energy use savings can be estimated more easily with the following equation:

$$\Delta E_{H,R} = \frac{24.(BLC_E - BLC_R).DD_H(T_{b,E})}{\eta_H} \qquad (6.21)$$

Note than when only one element of the building envelope is retrofitted (for instance the roof), the difference (BLC$_E$-BLC$_R$) is equivalent to the difference in the roof UA values before and after the retrofit (i.e., UA$_{roof,E}$-UA$_{roof,R}$).

To use either Eq. (6.20) or Eq. (6.21), it is clear that the auditor needs to estimate the heating degree-days and the existing overall building load. Some recommendations on how to calculate these two parameters are summarized below.

6.3.2   Estimation of the BLC for the building

The building load coefficient can be estimated using two approaches as briefly described below. Depending on the data available, the auditor should select the appropriate approach.

1) <u>Direct Calculation</u>: The auditor should have all the data (either through the architectural drawings or from observation during a site walk-through)

needed to estimate the R-value or U-value of all the components of the building envelope and their associated surface areas. Several references are available to provide the R-value of various construction layers commonly used in buildings (ASHRAE 1997). In addition, the auditor should estimate the infiltration/ventilation rates either by rules of thumb or by direct measurement as discussed in section 6.2.2. With these data, the building load coefficient can be calculated using Eq. (6.14).

2) Indirect Estimation: In this method, the auditor can rely on the utility energy use (even monthly data would be sufficient for this purpose) and its correlation with the outdoor temperature to provide an accurate estimation of the BLC. This method is similar to the Princeton Scorekeeping Methods (PRISM) described in details by Fels (1986). Examples of the use of the method to determine the BLC are illustrated by Figures 6.7 and 6.8 to determine the BLC for respectively the heating and the cooling mode. In both Figures, the BLC is determined by the slope of the regression line correlating the building energy use to the outdoor air temperature. It should be noted that the outdoor air temperature should be averaged over the same periods for which the utility data are available.

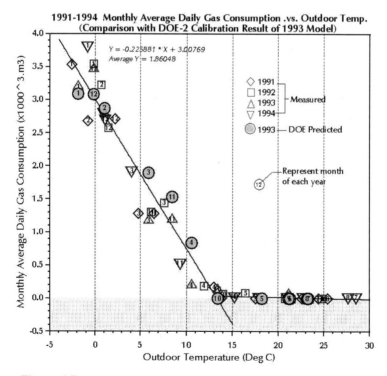

Figure 6.7: Determination of the BLC for the heating season based on the gas consumption (Yoon et al., 1997)

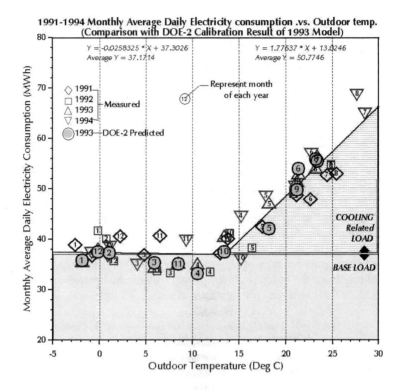

Figure 6.8: Determination of the BLC for the cooling season based on the electricity consumption (Yoon et al., 1997)

6.3.3   Estimation of the Degree Days

Data for heating degree-days can be found in several sources for various values of balance temperature. Table 6.3 provides heating degree-days for selected US cities for various balance temperatures: 65°F (18°C). Additional degree-days data can be found in Appendix B.

Table 6.4 presents the values of DD_H (18°C) calculated over one year and over the heating season of 8 months (period from October 1 through May 21) for 15 locations representatives of various climate zones in France.

Table 6.3: Heating degree days for base temperature 65°F (18°C), 55°F (13°C), and 45°F (7°C) for selected locations in US

Location	DD$_H$ (65°F)	DD$_H$ (55°F)	DD$_H$ (45°F)
Albuquerque, NM	4292	2330	963
Bismarck, ND	9044	6425	4374
Chicago, IL	6127	3912	2219
Dallas/Ft. Worth, TX	2290	949	250
Denver, CO	6016	3601	1852
Los Angeles, CA	1245	158	0
Miami, FL	206	8	0
Nashville, TN	3696	1964	1338
New York, NY	4909	2806	1311
Seattle, WA	4727	2091	602

Table 6.4: Heating degree days for base temperature 18°C for selected locations in France

Location	DD$_H$ (18°C) over one year	DD$_H$(18°C) for 8 months
Embrun	3087	2875
Bourg-St Maurice	3426	3135
Besançon	2995	3093
St Quentin	3085	2777
Le Bourget (Paris)	2758	2549
Lyon	2656	2529
Marignane (Marseille)	1760	1744
Bordeaux	2205	2082
Toulouse	2205	2123
Toulon	1376	1367
La Rochelle	2179	2073
Nantes	2413	2244
Deauville	2961	2604
Ouessant	2314	1954

Rarely the balance temperature for an audited building is adequately estimated to be equal to 65°F (18 °C). A simplified method is described in this section to determine the degree-days for any balance point temperature from limited climatic data. The basic idea of the simplified method first proposed by Erbs et al. (1983) is the assumption that the outdoor ambient temperature for each month follows a probability distribution with a standard deviation σ_m, an average temperature $\overline{T}_{o,m}$, and a frequency distribution F function of the balance temperature T_b:

$$F(T_b) = 1/\left(1 + e^{-2.a.\theta}\right) \tag{6.22}$$

Where, θ is the normalized average outdoor temperature and is defined as:

$$\theta = \frac{T_b - \overline{T}_{o,m}}{\sigma_m.N_m^{1/2}} \tag{6.23}$$

with N_m is the number of days in the month considered.

Using the variation of the frequency distribution F, the heating degree days can be obtained as a function of T_b:

$$DD_H(T_b) = \sigma_m.N_m^{3/2}\left[\frac{\theta}{2} + \frac{Ln(e^{-a\theta} + e^{a\theta})}{2.a}\right] \tag{6.24}$$

For locations spanning most climates in the US and Canada, Erbs et al. (1983) found that the coefficients a and σ_m can be estimated using the following expressions:

$$a = \sqrt{N_m} \tag{6.25}$$

and,

$$\sigma_m = 3.54 - 0.029 * \overline{T}_{o,m} + 0.0664 * \sigma_{yr} \tag{6.26}$$

with σ_{yr} is the standard deviation of the monthly temperatures relative to the annual average temperature, $\overline{T}_{o,yr}$:

$$\sigma_{yr} = \sqrt{\frac{\sum_{m=1}^{12}(\overline{T}_{o,m} - \overline{T}_{o,yr})}{12}} \tag{6.27}$$

For some other countries like France, only daily (rather than hourly) average temperature can be typically obtained. Therefore, the values for a and σ_m to be used for these countries are different from those proposed by Erbs et al. (1983) which are based on hourly average outdoor temperature. For France, Bourges (1987) determined that the following parameters should be used:

- $a = 2.1$ and

- a standard deviation based on the maximum and minimum quintals (F_{max} and F_{min}) of the monthly outdoor temperatures corresponding to the frequency values of 80% and 20%, respectively:

$$\sigma_m = \frac{F_{max} - F_{min}}{1.683} \qquad (6.28)$$

6.3.4 Foundation Heat Transfer Calculations

The practice of insulating building foundations has become more common over the last few decades. However, the vast majority of existing residential buildings are not insulated. It was estimated that in 1985 less than 5% of the existing building stock had an insulated foundation. Globally, earth-contact heat transfer appears to be responsible for 1 to 3 quadrillion kJ of annual energy use in the United States. This energy use is similar to the impact due to infiltration on annual cooling and heating loads in residential buildings (Claridge, 1988). In addition to the energy saving potential, insulating building foundation can improve the thermal comfort especially for occupants of buildings with basements or earth-sheltered foundations.

Typically, the foundation heat transfer is a major part of heating/cooling loads for low-rise buildings including single-family dwellings, small commercial and institutional buildings, refrigerated structures, and large warehouses. A detailed discussion of the insulation configurations for various building types as well as various calculation technique to estimate foundation heat transfer can be found in Krarti (1999). In this section, only simplified calculation method is provided to calculate annual and seasonal foundation heat loss/gain from residential foundations.

It should be noted than in the US, there are three common foundation types for residential buildings: slab-on-grade floors, basements, and crawlspaces. The basement foundations can be either deep or shallow. Typically, shallow basements and crawlspaces are unconditioned spaces. Figure 6.9 shows the three common building foundation types. In some applications, the building foundation can include any combination of the three foundation types such as a basement with a slab-on grade floor. Among the factors that affect the selection of the foundation type include the geographical location and the speculative real estate market.

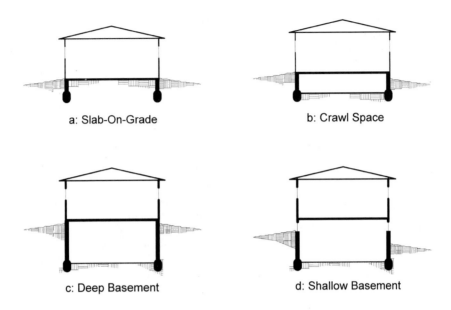

Figure 6.9: Foundation types for the buildings

A recent report from the US Census Bureau indicates that the share of houses built with crawlspaces remained constant at about 20% over the last 7 years (Krarti, 1999). However, the percentage of houses with slab foundations has increased from 38% in 1991 to 45% in 1997. Meanwhile, the share of houses built with basements has declined from a peak of 42% in 1992 to 37% in 1997. In 1993, houses were built with almost an equal number of basement and slab foundations. Moreover, data from US Census Bureau clearly indicates that the foundation type selection depends on the geographical location. In the Northeast and Midwest regions, the basement foundation is the most common with a share of about 80% during the period between 1991 and 1997. Meanwhile, the slab foundation is more dominant in the South and the West.

6.3.5 Simplified Calculation Method for Building Foundation Heat Loss/Gain

A simplified design tool for calculating heat loss for slabs and basements has been developed by Krarti and Chuangchid (1999). This design tool is easy to use and requires straightforward input parameters with continuously variable values including foundation size, insulation R-values, soil thermal properties, and indoor and outdoor temperatures. The simplified method provides a set of equations, which are suitable for estimating the design, and the annual total heat loss for both slab and basement foundations as a function of a wide range of variables.

Specifically, the simplified calculation method can calculate the annual or seasonal foundation heat transfer using two equations to estimate respectively: the mean Q_m and the amplitude Q_a of the annual foundation heat loss.

For the annual mean foundation heat loss:

$$Q_m = U_{eff,m} \cdot A \cdot (T_a - T_r) \qquad (6.29)$$

Where,

$$U_{eff,m} = m \cdot U_o \cdot D$$

For the annual amplitude foundation heat loss:

$$Q_a = U_{eff,a} \cdot A \cdot T_a \qquad (6.30)$$

Where,

$$U_{eff,a} = a \cdot U_o \cdot D^{0.16} \cdot G^{-0.6}$$

The coefficients a and m depend on the insulation placement configurations and are provided in Table 6.1.

Table 6.5: Coefficients m and a to be used in Equations (6.29) and (6.30) for foundation heat gain calculations

Insulation Placement	*m*	*a*
Uniform - Horizontal	0.40	0.25
Partial - Horizontal	0.34	0.20
Partial -Vertical	0.28	0.13

The normalized parameters used by both Eq.(6.29) and Eq.(6.30) are defined below

$$U_o = \frac{k_s}{(A/P)_{eff,b}} ; \qquad G = k_s . R_{eq} \cdot \sqrt{\frac{\omega}{\alpha_s}}$$

$$D = \ln\left[(1+H)\left(1+\frac{1}{H}\right)^{H} \right];$$

$$H = \frac{(A/P)_{eff,b}}{k_s.R_{eq}}$$

For partial insulation configurations (for slab foundations, the partial insulation can be either placed horizontally extending beyond the foundation, or vertically along the foundation walls):

$$R_{eq} = R_f \times \frac{1}{\left[1 - \left(\frac{c}{A/P} \times \frac{R_i}{(R_i + R_f)} \right) \right]}$$

For uniform insulation configurations: $R_{eq} = R_f + R_i$

Where,

$$(A/P)_{eff,b,mean} = \left[1 + b_{eff} \times \left(-0.4 + e^{-H_b} \right) \right] \times (A/P)_b$$

$$(A/P)_{eff,b,amp} = \left[1 + b_{eff} \times e^{-H_b} \right] \times (A/P)_b$$

$$H_b = \frac{(A/P)_b}{k_s.R_{eq}} \quad \text{and} \quad b_{eff} = \frac{B}{(A/P)_b}$$

In the equations above, the following parameters are used:

A   Basement/slab area (total of floor and wall) [m² or ft²]
B   Basement depth [m or ft]
b_{eff} Term defined in Eq (6.29) and Eq (6.30)
C_p Soil specific heat [J/kg°C or Btu/lbm. °F]
c   Insulation length of basement/slab [m or ft]
D   Term defined in Eq (6.29) and (6.30)
G   Term defined in Eq (6.30)
H   Term defined in Eq (6.29) and (6.30)
k_s soil thermal conductivity [Wm⁻¹.°C⁻¹ or Btu/hr.ft.°F]
P   perimeter of basement/slab [m or ft]
Q   Total heat loss [W or Btu/hr]
Q_m The annual mean of the total heat loss [W or Btu/hr]
Q_a The annual amplitude of the total heat loss [W or Btu/hr]
R_{eq} Equivalent thermal resistance R-value of entire foundation [m²K/W or ft².°F.hr/Btu]
R_f Thermal resistance R-value of floor [m²K/W or ft².°F.hr/Btu]

R_i Thermal resistance R-value of insulation [m^2K/W or $ft^2.°F.hr/Btu$]

T_a Ambient or outdoor air temperature [°C or °F]

T_r Room or indoor air temperature [°C or °F]

$U_{eff,m}$ Effective U-value for the annual mean [$Wm^{-2}.°C^{-1}$ or $Btu/hr.ft^2.°F$]defined in Eq (6.29)

$U_{eff,a}$ Effective U-value for the annual amplitude [$Wm^{-2}.°C^{-1}$ or $Btu/hr.ft^2.°F$] defined in Eq (6.30)

U_o U-value [$Wm^{-2}.°C^{-1}$ or $Btu/hr.ft^2.°F$] defined in Eq (6.29) and Eq (6.30)

ρ Soil density [kg/m^3 or lbm/ft^3]

ω Annual angular frequency [rad/s or rad/hr]

α_s Thermal diffusivity [m^2/s or ft^2/hr]

It should be noted that the simplified model provides accurate predictions when A/P is larger than 0.5 meter.

The annual average heat flux (heat loss or gain) from the building foundation is simply Q_m. The highest foundation heat flux under design conditions can be obtained as follows: $Q_{des}=Q_m+Q_a$.

To illustrate the use of the simplified models, two calculation examples is presented for a basement structure insulated with uniform insulation.

Calculation Example No. 1: Basement for a Residential Building:
Determine the annual mean and annual amplitude of total basement heat loss for a house. The basic geometry and construction details of the basement are provided below (see data summary in Step 1). The house is located in Denver, CO.

Solution:

Step 1. Provide the required input data (from ASHRAE Handbook, 1997):

Dimensions
Basement width: = 10.0 m (32.81 ft)
Basement length: = 15.0 m (49.22 ft)
Basement wall height: = 1.5 m (4.92 ft)
Basement total area: = 225.0 m^2 (2422.0 ft^2)
Ratio of basement area to basement perimeter: $(A/P)_b$ = 3.629 m (11.91 ft)
4 inches thick reinforce concrete basement, thermal resistance R-value:
$$= 0.5 \ m^2K/W \ (2.84 \ h.ft^2.°F/Btu)$$
Soil Thermal Properties
Soil thermal conductivity: k_s = 1.21 W/m.K (0.70 Btu/h.ft.°F)
Soil thermal diffusivity: α_s = 4.47 x 10^{-7} m^2/s (48.12 x 10^{-7} ft^2/s)

Insulation
Uniform insulation R-value = 1.152 m^2K/W (6.54 $h.ft^2.°F/Btu$)

Temperatures

Indoor temperature: $T_r = 22$ °C (71.6 °F)
Annual average ambient temperature: $T_a = 10$ °C (50 °F)
Annual amplitude ambient temperature: $T_{amp} = 12.7$ K (23 R)
Annual angular frequency: $\omega = 1.992 \times 10^{-7}$ rad/s

Step 2. Calculate Q_m, and Q_a values:

Using Equations (6.29) and (6.30), the various normalized parameters are first calculated. Then the annual mean and amplitude of the basement heat loss are determined.

$$H_b = \frac{(A/P)_b}{k_s.R_{eq}} = \frac{3.629}{1.21\times(0.5+1.152)} = 1.8155$$

$$b_{eff} = \frac{B}{(A/P)_b} = \frac{1.5}{3.629} = 0.4133$$

$$(A/P)_{eff,b,mean} = \left[1+0.4133\times\left(-0.4+e^{-1.8155}\right)\right]\times3.629 = 3.2731$$

$$(A/P)_{eff,b,amp} = \left[1+0.4133\times e^{-1.8155}\right]\times3.629 = 3.8731$$

$$U_{o,m} = \frac{k_s}{(A/P)_{eff,b,mean}} = \frac{1.21}{3.2731} = 0.3697$$

$$U_{o,a} = \frac{k_s}{(A/P)_{eff,b,mean}} = \frac{1.21}{3.8731} = 0.3124$$

$$H_{mean} = \frac{(A/P)_{eff,b,mean}}{k_s.R_{eq}} = \frac{3.2731}{1.21\times(0.5+1.152)} = 1.6374$$

$$H_{amp} = \frac{(A/P)_{eff,b,amp}}{k_s.R_{eq}} = \frac{3.8731}{1.21\times(0.5+1.152)} = 1.9376$$

$$D_{mean} = \ln\left[(1+H)\left(1+\frac{1}{H}\right)^H\right] = 1.7503$$

$$D_{amp} = \ln\left[(1+H)\left(1+\frac{1}{H}\right)^H\right] = 1.8839$$

$$G = k_s.R_{eq} \cdot \sqrt{\frac{\omega}{\alpha_s}} = 1.21\times(0.5+1.152)\times\sqrt{\frac{1.992\times10^{-7}}{4.47\times10^{-7}}} = 1.3344$$

Therefore,

$$Q_m = U_{eff,m}A(T_a - T_r) = 0.4\times0.3697\times1.7503\times225\times(22.0-10.0) = 698.85 \text{ W } (2384.48 \text{ Btu/h})$$

and,

$$Q_a = U_{eff,a}AT_a = 0.25\times0.3124\times1.8839^{0.16}\times1.3344^{-0.6}\times225\times12.7 = 207.72 \text{ W } (708.74 \text{ Btu/h})$$

Table 6.6: *shows the comparison of the results between the simplified and the ITPE solution. The heat loss per unit area are provided in W/m²*

Method	Mean (Q_m)	Amplitude (Q_a)
Simplified	699	208
ITPE solution*	658	212

*Note: For more details about the ITPE (Interzone Temperature Profile Estimation) solution technique for foundation heat transfer problems, the reader is referred to Krarti (1999).

Calculation Example No. 2: Freezer Slab:

For a freezer warehouse depicted in the sketch below, determine the total freezer heat gain under design conditions. The warehouse is located in Denver, Colorado. Estimate the cost-effectiveness of uniformly insulating the freezer foundation slab using 4 inches of extruded polystyrene insulation.

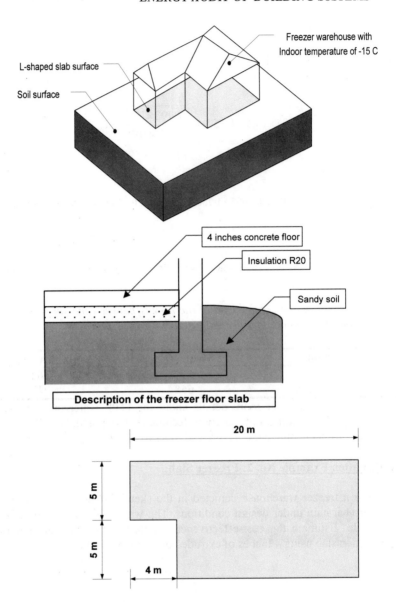

Description of the freezer floor slab

Step 1. Provide the required input data:

Dimensions
Slab width: = 10.0 m (32.81 ft)
Slab length: = 20.0 m (49.22 ft)
Ratio of slab area to slab perimeter: A/P = 3.0 m (9.84 ft)
100 mm (4 inches) lightweight concrete slab [C14] with thermal resistance

R-value: $= 0.587$ m^{2}K/W (3.33 h.ft^{2}F/Btu)*
[*Source: 1997 ASHRAE Fundamentals Handbook; Table 11, p. 28.19]

Soil Thermal Properties
Type of soil : Sandy soils
Soil thermal conductivity: $k_s = 1.51$ W/m.K (0.87 Btu/h.ft.F)*
[*Source: 1997 ASHRAE Fundamentals Handbook; Table 7, p. 24.15]
Soil density: $\rho_s = 2740$ kg/m3 (171.06 lb$_m$/ft^3)[1]
Soil heat capacity: $c_s = 774.0$ J/kg.C (0.18 Btu/lb$_m$.F)[1]
Soil thermal difussivity: $\alpha_s = k_s/\rho_s c_s = 7.12$ x 10^{-7} m^2/s (76.6 x 10^{-7} ft^2/s)
[1Source: M.S. Kersten, 1949]

Insulation
100 mm (4 inches) extruded Polystyrene: R-value $= 3.52$ m^{2}K/W (20.0 h.ft^{2}F/Btu)*
[*Source: 1997 ASHRAE Fundamentals Handbook; Table 4, p. 24.5]

Temperatures
Indoor temperature: $T_r = -15$ °C (5 °F) (for freezer storage)
Annual average ambient temperature: $T_m = 6.3$ °C (43 °F)[2]
Annual amplitude ambient temperature: $T_a = 30$ °C (54 °F)[2]
Annual angular frequency: $\omega = 1.992$ x 10^{-7} rad/s
[2Source: 1997 ASHRAE Fundamentals Handbook; Table 1A, p. 26.8]

Cases of insulation configurations
Case 1: no insulation
Case 2: uniform insulation

Step 2. Calculate Q_{des}:
Using Eqs. (6.29) and (6.30), the various normalized parameters are first calculated. Then the design heat gain for freezer slab is determined.

Case 1: no insulation

$$R_{eq} = 0.587 + 0.0 = 0.587 \text{ m}^2\text{K/W}$$

$$U_o = \frac{1.51}{3.0} = 0.5033$$

$$H = \frac{3.0}{1.51 \times 0.587} = 3.3846$$

$$D = \ln\left[(4.3846)(1.2954)^{3.3846}\right] = 2.3541$$

$$G = (1.51)(0.587) \times \sqrt{\frac{1.992 \times 10^{-7}}{7.12 \times 10^{-7}}} = 0.4688$$

$$U_{eff,m} = 0.4 \times 0.5033 \times 2.3541 = 0.4739$$

$$U_{eff,a} = 0.25 \times 0.5033 \times (2.3541)^{0.16} (0.4688)^{-0.6} = 0.2273$$

Therefore, the design total heat gain for freezer slab with no insulation is:

$$\frac{Q_{des}}{A} = (0.4739)(6.3 - (-15.0)) + (0.2273)(30.0) = 16.91 \text{ W/m}^2 \ (5.36 \text{ Btu/h.ft}^2)$$

Case 2: uniform insulation

$$R_{eq} = 0.587 + 3.52 = 4.107 \text{ m}^2 \text{K/W}$$

$$U_o = \frac{1.51}{3.0} = 0.5033$$

$$H = \frac{3.0}{1.51 \times 4.107} = 0.4837$$

$$D = \ln\left[(1.4837)(3.0674)^{0.4837}\right] = 0.9367$$

$$G = (1.51)(4.107) \times \sqrt{\frac{1.992 \times 10^{-7}}{7.12 \times 10^{-7}}} = 3.2802$$

$$U_{eff,m} = 0.4 \times 0.5033 \times 0.9367 = 0.1886$$

$$U_{eff,a} = 0.25 \times 0.5033 \times (0.9367)^{0.16} (3.2802)^{-0.6} = 0.0610$$

Therefore, the design total heat gain for freezer slab with uniform insulation is:

$$\frac{Q_{des}}{A} = (0.1886)(6.3-(-15.0))+(0.0610)(30.0)=5.85 \text{ W/m}^2 \ (1.85 \text{ Btu/h.ft}^2)$$

Table 6.7: Comparison of the simplified design tool and the ITPE (Krarti, 1999) solution predictions. The heat gain per unit area is provided in W/m².

Method	No insulation	Uniform insulation
Simplified	16.91	5.85
ITPE solution	17.28	5.69

Step 3. Perform an economic analysis:
Assume that warehouse is operated 24 hours a day for whole year.
kW/ton for freezer is: 2.35 kW/ton*
Electricity cost in Denver is: $0.08/kWh
Cost of extruded polystyrene is: $3.00/m².in (Material cost) and $2.50/m² (Labor cost) [Cost data are specific to Denver, CO].

To evaluate the annual performance of the insulation, only annual mean heat gains are considered (for both cases: with and without 4 inches of insulation along the foundation slab) in the economical analysis outlined below.

First, the savings on the annual average heat gain (expressed in Watts) are estimated:

Heat gain Saving = (No insulation heatgain - Uniform insulation heatgain) * Slab Area
$$=(10.07-4.02)\times180=1093.1 \text{ Watt}$$

Thus, the annual total heat gain savings expressed in kWh/yr are calculated as follows:
$$\text{Annual Heat gain Saving} = \frac{Watt\times24\times7\times52}{1000}=9,549 \text{ kWh/yr}$$

The electrical energy savings for the refrigeration equipment are then determined to be:
$$\text{ElectricalEnergy Saving} = \frac{kWh\times2.35\text{kW/ton}}{3.517\text{ kW/ton.refrig}}=6,380.5 \text{ kWh/yr}$$

Therefore, the annual cost savings attributed to the addition of the foundation insulation are:

$$\text{Annual Cost Saving} = \text{kWh/yr}\times\$0.08/\text{kWh} = \$511/\text{yr}$$

The cost of the foundation insulation is estimated to be:

$$\text{Investment Cost} = (4\,\text{inches} \times \$3.00/m^2\text{in} + \$2.50/m^2) \times 180 = \$2,650$$

Therefore, the payback period for the addition of the foundation insulation is:

$$\text{Payback Period} = \frac{\$2,650}{\$511/\,yr} = 5.2\,\text{years}$$

To prevent heaving problems, freezer foundation has to be insulated in any case, unless a floor heating system is installed under the slab foundation.

6.4   Selected Retrofits for Building Envelope

Generally, improvements in the energy efficiency of the building envelope are expensive since labor intensive modifications are typically involved (such as addition of thermal insulation and replacement of windows). As a consequence, the payback periods of most building envelope retrofits are rather long. In these instances, the building envelope retrofits can still be justified for reasons other than energy efficiency such as increase in occupant thermal comfort or reduction of moisture condensation to avoid structural damages. However, there are cases where retrofits of the building envelope can be justified based solely on improvement in energy efficiency. Some of these retrofit measures are discussed in this section with some examples to illustrate how the energy savings and the payback periods are calculated.

6.4.1   Insulation of Poorly Insulated Building Envelope Components

When an element of a building envelope is not insulated or poorly insulated, it may be cost-effective to add insulation in order to reduce transmission losses. While the calculation of the energy savings due to such retrofit may require a detailed simulation tool to account for effects of the building thermal mass and/or the building HVAC systems, Eq. (6.20) or Eq. (6.21) can be used to determine the energy savings during the heating season. If the building is heated and cooled, the total energy savings due to adding insulation to the building envelope can be estimated by summing the energy savings obtained from a reduction in heating loads and those obtained from a decrease (or increase) in cooling loads.

In most cases where the building envelope is already adequately insulated, the addition of thermal insulation is not cost-effective based only on energy cost savings.

Example 6.2: *A machine shop has 500 m^2 metal frame roof that is uninsulated. Determine the payback period of adding insulation (R=2.0 °C.m²/W). The building is electrically heated. The cost of electricity is $0.07/kWh. The machine shop is located in Paris (Le Bourget, France) and operates 24 hours/day, 7 days/Week through throughout the heating season. Assume that the installed cost of the insulation is $15/m²*

Solution: Based on ASHRAE handbook, the existing U-value for a metal frame roof is about 1.44 W/m²·°C. To determine the energy savings due the addition of insulation, we will assume that the annual heating degree days before and after the retrofit remained unchanged and are close to 18 °C. Using Eq. (6.21) with the retrofitted roof U-value to be 0.37 W/m²·°C and heating system efficiency set to be unity (electrical system), the energy savings are calculated to be:

$$\Delta E = 24.500 m^2 * [(1.44 - 0.37)W / m^2 {}^\circ C] * 2758 {}^\circ C.day / yr = 35,413 kWh / yr$$

Thus, the payback period for adding insulation on the roof can be estimated to be:

$$Payback = \frac{500 m^2 * 15\$ / m^2}{35,413 kWh / yr * 0.07\$ / kWh} = 3.0 years$$

Therefore, the addition of insulation seems to be cost-effective. Further analysis is warranted to determine more precisely the cost-effectiveness of this measure.

6.4.2   Window Improvements

Window improvements such as installation of high-performance windows, window films and coatings, or storm windows can save energy through reductions in the building heating and cooling thermal loads. Improvements in windows can impact both the thermal transmission and solar heat gains. In addition, energy-efficient windows create more comfortable environments with evenly distributed temperatures and quality lighting. Energy-efficiency improvements can be made to all the components of a window assembly including:

- Insulating the spacers between glass panes to reduce conduction heat transfer.
- Installing multiple coating or film layers to reduce heat transfer by radiation.

- Inserting argon or krypton gas in the space between the panes can decrease the convection heat transfer.
- Providing exterior shading devices can reduce the solar radiation transmission to the occupied space.

To determine accurately the annual energy performance of window retrofits, dynamic hourly modeling techniques are generally needed since fenestration can impact the building thermal loads through several mechanisms. However, the simplified calculation method based on Eq. (6.20) to account for both heating and cooling savings can be used to provide a preliminary assessment of the cost-effectiveness of window retrofits.

Example 6.3: *A window upgrade is considered for an apartment building from double-pane metal frame windows (U_E=4.61 W/m^2. °C) to double-pane with low-e film and wood frame windows (U_R=2.02 W/m^2. °C). The total window area to be retrofitted is 200 m^2. The building is located in Nantes (France) and is conditioned 24 hours/day, 7 days/Week throughout the heating season. Electric baseboard provides heating while a window AC provides cooling (EER=8.0). Assume that the cost of electricity is $0.10/kWh.*

Solution: To determine the energy savings due the addition of insulation, we will assume that the annual heating and cooling degree days before and after the retrofit remained unchanged (this assumption is justified by the fact that the window contribution to the BLC is relatively small) and are respectively DD_H= 2244 °C-day/yr and DDC= 255 °C-day/yr. The energy savings during heating (assume that the system efficiency is 1.0 for electric heating) are calculated to be:

$$\Delta E = 24.200 m^2 * [(4.61 - 2.02)W / m^2 °C] * 2244°C.day / yr = 27,897 kWh / yr$$

If an EER (energy efficiency ratio) value of 8.0 is assumed for the AC system, the energy savings during cooling are estimated as follow:

$$\Delta E = 24.200 m^2 * [(4.61 - 2.02)W / m^2] * 255°C.day / yr * 1/8.0 = 396 kWh / yr$$

Therefore the total energy se savings due to up-grading the windows is 28,293 kWh which corresponds to about $2,829 when electricity cost is $.10/kWh. The cost of replacing the windows is rather high (it is estimated to be $150/m^2 for this project). The payback period of the window retrofit can be estimated to be:

$$Payback = \frac{200m^2 * 150\$ / m^2}{28,293kWh / yr * 0.10\$ / kWh} = 10.4 years$$

Therefore, the window upgrade is not cost-effectives based solely on thermal performance. The investment on new windows may be however justifiable based on other factors such increase in comfort within the space.

6.4.3   Reduction of Air Infiltration

In several low-rise facilities, the thermal loads due to air infiltration can be significant. It is estimated that for a well-insulated residential building, the infiltration can contribute up to 40% to the total building heating load. Tuluca et al. (1997) reported that measurements in eight US office buildings found average air leakage rates of 0.1 to 0.5 air changes per hour (ACH). This air infiltration accounted for an estimated 10 to 25% of the peak heating load. Shermann and Matson (1993) have shown that the stock of housing in the US is significantly over-ventilated from air infiltration and that there are 2 exa-joules of potential annual savings that could be captured. As described in section 6.2.2, two measurement techniques can be used to evaluate the existing amount of infiltrating air. The blower technique is relatively cheap and quick to set up and can be useful for small buildings to locate air leaks, while the tracer gas technique is more expensive and time-consuming and is appropriate to measure the outdoor air flow rate from ventilation, and infiltration entering large commercial/institutional buildings.

While several studies exist to evaluate the leakage distribution for residential buildings (Dickerhoff et al. 1982 and Harrje and Born 1982), very little work is available for US commercial and industrial buildings. However, some results indicate that the envelope air tightness levels for commercial buildings are similar to those in typical US houses. In particular, it was found that leaks in walls (frames of windows, electrical outlets, plumbing penetrations) constitute the major source of air leakage from both residential and commercial buildings. For instance for office buildings, Tamura and Shaw (1976) found that typical air leakage values per unit wall area at 75 Pa (pressure differential between indoors and outdoors) are 500, and 1500, and 300 $cm^3/(s.m^2)$ for respectively, tight, average, and leaky walls. Other sources of air leakage identified for large commercial buildings are internal partitions (such as elevator and service shafts), and exterior doors (especially for retail stores).

To improve the air tightness of the building envelope several methods and techniques are available including:

i.   Caulking: Several types of caulking (urethane, latex, and polyvinyl) can be applied to seal various leaks such as those around the window and door frames, and any wall penetrations such as holes for water pipes.

ii. Weather Stripping: By applying foam rubber with adhesive backing, windows and doors can be air sealed.

iii. Landscaping: This is a rather long term project and consists of planting shrubs and/or trees around the building to reduce the wind effects and reduce air infiltration.

iv. Air Retarders: These systems consist of one or more air-impermeable components that can be applied around the building exterior shell to form a continuous wrap around the building walls. There are several air retarder (AR) types such as liquid-applied bituminous, liquid-applied rubber, sheet bituminous, and sheet plastic. The AR membranes can be applied to impede the vapor movement through the building envelope and thus act as vapor retarders. Unless they are part of an overall building envelope retrofit, these systems are typically expensive to install for existing buildings.

To assess the energy savings due to a reduction in air infiltration, Eq. (6.20) or Eq; (6.21) can be used. Whenever available, the degree days IDD_H (T_b) determined specifically to calculate infiltration loads can be used instead of the conventional temperature based degree days DD_H (T_b). The infiltration heating degree-days for the balance temperature is defined as follows:

$$IDD_H(T_b) = \sum_{i=1}^{N_H} \frac{\dot{V}}{\dot{V}_{ref}}(T_b - T_{o,i})^+ \tag{6.31}$$

with $\dot{V}$ is calculated as shown in Eq. (6.9) to account for the climatic data and $\dot{V}_{ref}$ is the reference volume rate defined for Eq. (6.8). However, the conventional variable-base degree-days method (which basically ignores the effects of weather on the variation of the infiltration rate) provides generally good estimation of the energy savings incurred from a reduction in air infiltration.

Example 6.3: *Consider a heated manufacturing shop with a total conditioned volume of 1000 m³. A measurement of the air leakage characteristics of the shop showed an infiltration rate of 1.5 ACH. Determine the energy savings due to caulking and weather-stripping improvements of the exterior envelope of the facility to reduce air infiltration by half. Assume the* *shop is located in Seattle, WA and is heated by a gas-fired boiler with a seasonal efficiency of 80%.*

Solution: To determine the energy savings due the addition of insulation, we will assume that the annual heating degree days before and after the retrofit remained unchanged and are close to 18 °C. For Seattle, the degree-days (for base temperature of 65 °F or 18°C) are about 2656 °C-days.

The existing air infiltration has an equivalent UA-value of $UA_{inf}= mc_{p,a}=$ 500 W/°C. From Eq. (6.21) with the new air infiltration equivalent UA-value is 250 W/°C and heating system efficiency set to be 80% (gas-fired boiler), the energy savings are calculated to be:

$$\Delta E = \frac{24}{0.80}.[(500-250)W \ /°C]*2656°C.day \ / \ yr = 19,920kWh \ / \ yr$$

The cost of caulking and weather-stripping is estimated to be about $1,500 (if only material costs are included). For a gas price of $0.05/kWh, the payback period for reducing the infiltration rate can be estimated to be:

$$Payback = \frac{\$1,500}{19,920 \ kWh \ / \ yr * 0.05\$ \ / \ kWh} = 1.5 \ years$$

Therefore, the caulking and weather-stripping can be justified based only on energy cost savings. An additional benefit of reducing infiltration is improved thermal comfort.

6.5   Summary

Energy efficiency improvements of building envelope systems are generally expensive and are not cost-effective especially for large commercial buildings. However, increasing the energy performance of building shell can be justified for low-rise and small buildings based on energy cost savings but also based on improvement in indoor thermal comfort and integrity of the building structure. For residential buildings, weather-stripping to reduce infiltration losses is almost always economically justifiable.

Problems

6.1 Find the monthly heating degree-days for Denver, CO; New York City, NY; and Los Angeles, CA. Consider the balance temperature of 67°F, 65°F, 60°F, and 55°F. Use the following monthly average outdoor temperatures in your calculations:

Month	Denver, CO	New York City, NY	Los Angeles, CA
Jan	29.9	32.2	54.5
Feb	32.8	33.4	55.6
Mar	37.0	41.1	56.5
Apr	47.5	52.1	58.8
May	57.0	62.3	61.9
Jun	66.0	71.6	64.5
Jul	73.0	76.6	68.5
Aug	71.6	74.9	69.6
Sep	62.8	68.4	68.7
Oct	52.0	58.7	65.2
Nov	39.4	47.4	60.5
Dec	32.6	35.5	56.9

For the case of a balance temperature of 65°F, compare your results to the heating degrees days for the three cities directly computed from hourly weather data (these degree-days are widely reported in the existing literature. Appendix B provides the degree-days for selected US locations). Comment on the results of your comparative analysis.

6.2 An energy audit of the roof indicates the following:

Roof area:	10,000 sq.ft.
Existing roof R-value:	5
Degree-days (winter):	6,000
Degrees-hours (summer):	17,000
Fuel cost:	$4/MMbtu
Boiler efficiency:	0.70
Electric rate	$0.07/kWh
Air-conditioning requirement:	0.7 kW/ton

It is proposed to add R-19 insulation in the roof.
  (1) Comment on the potential energy savings of adding insulation.
  (2) If the insulation costs $0.35 per sq.ft, determine the payback of adding insulation.

(3) Is the expenditure on the additional insulation justified based on a minimum rate of return of 12%? Assume a 30-year life cycle. Solve the problem using both the present worth and annual cost methods.

6.3 Do you recommend to add R-20 insulation to a 25,000 ft^2 uninsulated roof (R-5) when you consider a 25-year life cycle with a rate of return of 8%? Assume the following:

Heating degree-days:	5,000
Cooling degrees-hours:	10,000
Fuel cost:	$5/MMBtu
Boiler efficiency:	0.70
Electric rate:	$0.10/kWh
Refrigeration requirement:	0.75kW/ton
Insulation cost:	$0.225 per sq.ft

6.4 A blower door test on a 30 ft × 50 ft × 9 ft house located in Denver, CO, revealed that under 4 Pa pressure differential between indoors and outdoors, the leakage area is about 200 in^2.

(a) Determine in air changes per hour, the annual average infiltration rate for the house.

(b) It was decided to weather-strip the house so that the infiltration is reduced to just 0.25 ACH (for 4 Pa pressure differential). Determine the payback period of weather-stripping the house given the following parameters:

Heating degree-days:	6,000
Cooling degrees-hours:	5,000
Fuel cost:	$5/MMBtu
Boiler efficiency:	0.70
Electric rate:	$0.10/kWh
Refrigeration requirement:	0.75kW/ton
Cost of weather-stripping:	$150

6.5 Estimate the annual energy loss from a 100 ft X 50 ft slab-on-grade floor of a building located in Denver, CO. The floor is made up of 4-in concrete slab. Estimate the annual heating savings if the slab is insulated uniformly with R-10. Assume that the boiler efficiency is 80%.

7

SECONDARY HVAC SYSTEMS RETROFIT

Abstract

A summary of measures to improve the energy efficiency of the secondary HVAC systems is provided in this chapter. In particular, calculation methods are provided to estimate energy savings due to ventilation control, conversion of constant volume to variable air volume systems, and the use of variable speed drives for variable air volume fans.

7.1   Introduction

The heating ventilating and air conditioning (HVAC) system maintains and controls temperature and humidity levels to provide an adequate indoor environment for people activity or for processing goods. The cost of operating a HVAC system can be significant in commercial buildings and in some industrial facilities. In the US, it is estimated that the energy used to operate the HVAC systems can represent about 50% of the total electrical energy use in a typical commercial building (EIA, 1994). It is therefore important that the auditor recognizes some of the characteristics of the HVAC systems and determines if any retrofits can be recommended to improve the energy of HVAC systems.

7.2   Types of Secondary HVAC Systems

A basic HVAC air distribution system consists of an air-handling unit with the following components as shown in Figure 7.1:

- Dampers to control the amount of air to be distributed by the HVAC system including: outside air (OA) damper, return air (RA) damper, exhaust air (EA) damper, and supply air (SA) damper.
- Preheat coil in case the outside air is too cold, to avoid any freezing problems.
- Filter to clear any dirt from the air.

- Cooling coils to condition the air supply to meet the cooling load of the conditioned spaces.
- Humidifiers to add moisture to the air supply in case a humidity control is provided to the conditioned spaces.
- A distribution system (i.e. ducts) where the air is channeled to various locations and spaces.

Each of the above listed components can come in several types and styles. The integration of all the components constitutes the secondary HVAC system for the sole purpose of conditioned air distribution. Two main categories of secondary HVAC central air systems can be distinguished:

1) Constant Air Volume Systems: These systems provide a constant amount of supply air conditioned at proper temperature to meet the thermal loads in each space based on a thermostat setting. Typically, the supply air temperature is controlled by either mixing cooled air with heated or bypassed air or by directly reheating cooled air. Therefore, these systems waste energy because of the mixing and/or reheat especially under partial thermal load conditions. Among the constant air volume systems commonly used to condition existing buildings are:

    (i) Constant air volume with terminal reheat systems. These systems require that circulated air to be cooled to meet design thermal loads. If partial thermal load conditions occur, reheat of pre-cooled air is required.
    (ii) Constant air volume systems with terminal reheat in interior spaces and perimeter induction or fan-coil units. For these systems, the energy waste is not significant for the perimeter spaces. Indeed, a large portion of the air supplied to the perimeter spaces is re-circulated within each perimeter space by either induction or fan-coil units.
    (iii) All-air induction systems with perimeter reheat. The system attempts to capture heat from lights by returning air through the light fixtures. The induction units accept varying amounts of warm return air to mix with primary air for temperature control. The energy waste due to reheat is small for these systems. However, extensive static pressure control is required at the terminals.
    (iv) Constant air volume dual duct systems. These systems have a cold air duct and hot air duct. The supply air temperature is controlled by mixing cold air with hot air proportionally to meet the thermal load of the space. Energy waste occurs during partial thermal load conditions when mixing is needed.

2) Variable Air Volume Systems: These systems provide a variable amount of supply air conditioned at a constant temperature to meet thermal loads in all spaces based on thermostat settings. The supply air volume can be

controlled and modulated using various techniques such as outlet dampers, inlet vanes, and variable speed drives. Typically, only cooled air is supplied at central air handling unit. In each space, reheat is provided depending on the space thermal load. These systems waste significantly less energy than the constant air volume. Retrofitting existing constant volume systems to variable air volume systems constitutes a common and generally a cost-effective energy conservation measure for secondary HVAC systems. Among the variable air volume systems commonly used are:

(i) Variable air volume with terminal reheat systems. These systems reduce the amount of air supplied as the cooling load reduces until a preset minimum volume is reached. At this minimum volume, reheat is provided to the supply air to meet the thermal load. Because of this volume reduction, reheat energy waste is significantly reduced relative to the constant air volume systems with reheat terminal.

(ii) Variable air volume systems with perimeter heating systems. These systems provide cooling only and heating is performed by other auxiliary systems such as hot water baseboard units. The baseboard heating units are often controlled by outside air temperature since the perimeter heating load is function of the transmission losses.

(iii) Variable air volume dual duct systems. These systems have a cold air duct and hot air duct and operate in a similar manner than the variable air volume systems with terminal reheat. As the cooling load decreases, only the cold air is supplied until a preset minimum volume is reached. At this minimum volume, the hot air is mixed with the cold air stream.

To summarize, variable air volume (VAV) systems are more energy efficient than the constant air volume (CAV or CV) since they minimize reheat energy waste. However, several energy conservation opportunities can be considered even if the existing HVAC system is a VAV. The potential for energy savings in the secondary HVAC systems depends on several factors including the design of systems, the method of operation, and the maintenance of the systems. Generally, energy can be conserved in the HVAC systems by following one or several principles listed below:

♦ Operate the HVAC systems only when needed. For instance, there is no need to provide ventilation during unoccupied periods.

♦ Eliminate overcooling and overheating of the conditioned spaces to improve comfort levels and avoid energy waste.

♦ Reduce reheat since it wastes energy.

- ◆ Provide free cooling and heating whenever possible by using economizer cycles or heat recovery systems to eliminate the need for mechanical air conditioning.
- ◆ Reduce the amount of air delivered by the HVAC systems by reducing the supply air and especially make up and exhaust air.

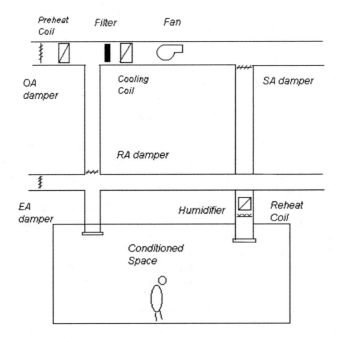

Figure 7.1: Typical Air Handling Unit for an air HVAC system

In the following sections, selected energy conservation measures are described starting from measures specific to each component of an air handling unit to other measures related to conversion of a constant volume system to a variable air volume system.

7.3   Ventilation

The energy required to condition ventilation air can be significant in both commercial buildings and industrial facilities especially in locations with extreme weather conditions. While the ventilation is used to provide fresh air to occupants in commercial buildings, it is used to control the level of dust, gases, fumes, or vapors in several industrial applications. The auditor should estimate the existing volume of fresh air and compare this estimated amount of the ventilation air with that required by the appropriate standards and codes. Excess in air ventilation should be reduced if it can lead to increases in heating and/or cooling loads. Some energy conservation measures related to ventilation is described in this section. However, in some climates and periods of the year or the day, providing more air ventilation can be beneficial and may actually reduce cooling and heating loads through the use of air-side economizer cycles.

7.3.1   Reduced Ventilation Air

The auditor should first estimate the existing level of ventilation air brought by the mechanical system (rather than by natural means such as infiltration through the building envelope). As described in Chapter 6, the tracer gas technique can be used to determine the amount of fresh air entering a facility. However, this technique does not differentiate between the outside air coming from the mechanical ventilation system to that from infiltration. Currently, several measurement techniques are available to measure the flow of air through duct. Some of these techniques are summarized in Table 7.1 which provide the range and the accuracy of each listed technique.

Table 7.1: Accuracy and Range of Airflow Measurement Techniques (*Source: Krarti et al. 1999b*).

Technique	range	Accuracy	Comments
Pitot-Tube	1 m/s – 45 m/s	1% - 5%	For low flows (1-3 m/s), high accuracy DP is needed.
Thermal Anemometer	> 0.005 m/s	2 % - 5%	Sensitive to turbulence. Need frequent calibration
Rotating Vanes Anemometer	0.5 m/s – 15 m/s	2% - 5%	Susceptible to changes in flow rates. Need periodic calibration.
Swinging Vanes Anemometer	0.25 m/s – 50 m/s	10%	Not sufficiently accurate for OA measurements.
Vortex Shedding Meter	> 2.5 m/s	1% - 5%	Not accurate for low flow rates.
Integrated damper/measuring device	1 m/s – 45 m/s	1% - 5%	Same errors and limitation that Pitot-Tube.
Laser Doppler Anemometer	0.005 m/s– 25 m/s	1% - 3%	Accurate at low rates. Too costly for field applications.
Orifice Meter	> 0.1 m/s	1% - 5%	Accuracy is affected by installation conditions.

It should be noted that all the techniques listed in Table 7.1 provide direct measurement of ventilation air. However, these techniques are relatively expensive and are generally difficult to set up in existing systems. To have an estimation of ventilation air provided by the mechanical system, an enthalpy balance technique can be used. In this technique, the temperature is measured at three locations in the duct system as shown in Figure 7.2: before the outdoor air damper (to measure the outdoor air temperature, T_{oa}), in the return duct (to measure the return air temperature, T_{ra}), and in the mixing plenum area (to measure the mixing air temperature, T_{ma}). The outside air fraction X_{oa} (fraction of the ventilation air over the total supply air) is then determined using the following equation (based on the first law of thermodynamics):

$$T_{ma} = X_{oa}.T_{oa} + (1 - X_{oa}).T_{ra} \qquad (7.1)$$

Thus, the amount of ventilation, V_{oa}, can be determined under design conditions using the capacity of the air-handling unit, $V_{des,}$ as indicated in Eq. (7.2):

$$\dot{V}_{oa} = X_{oa}.\dot{V}_{des} = \left(\frac{T_{ra} - T_{ma}}{T_{ra} - T_{oa}} \right) \dot{V}_{des} \qquad (7.2)$$

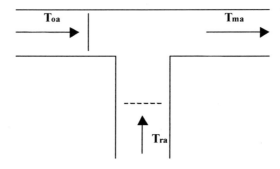

Figure 7.2: Location of the temperature sensors to measure the ventilation air fraction in an air-handling unit.

It should be noted that the accuracy of the estimation for the ventilation air using Eq. (7.2) is reduced, as the difference between the return air and outside air temperature is small. Figure 7.3 indicates the accuracy of the temperature balance for two outside air fractions 20% and 40% as function of the temperature difference between return air and outside air (Krarti et al. 1999b). Thus, it is recommended that the auditor performs temperature measurements when the outdoor temperatures are extremes (i.e.; during heating or cooling seasons).

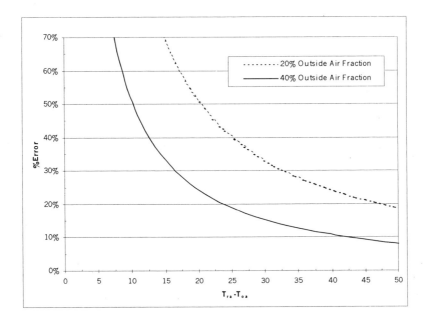

Figure 7.3: Predicted errors of temperature balance.

Once the existing ventilation air is estimated, it has to be compared to the ventilation requirements by the applicable standards. Table 7.2 summarizes some of the minimum outdoor air requirements for selected spaces in commercial buildings.

Table 7.2: Minimum ventilation rate requirements for selected spaces in commercial buildings.

SPACE AND OR APPLICATION	MINIMUM OUTSIDE AIR REQUIREMENTS	REFERENCE
Office Space	9.5 L/s (20 cfm) per person	ASHRAE Standard 62-89
Corridor	0.25 L/s per m^2 (0.05 cfm/ft^2)	
Restroom	24 L/s (50 cfm) per toilet	
Smoking Lounge	28.5 L/s (60 cfm) per person	
Parking Garage	7.5 L/s (1.5 cfm/ft^2)	

If excess ventilation air is found, the outside air damper setting can be adjusted to supply the ventilation that meets the minimum outside requirements as listed in Table 7.2. Further reductions in outdoor air can be obtained by using demand ventilation controls by supplying outside air only during periods when there is need for fresh air. A popular approach for demand ventilation is the monitoring of CO_2 concentration level within the spaces. CO_2 is considered as a good indicator of pollutants generated by occupants and other construction materials. The outside air damper position is controlled to maintain a CO_2 set-point within the space.

The energy savings due to reduction in the ventilation air can be attributed to lower heating and cooling loads required to condition outdoor air. The instantaneous heating and cooling savings can be estimated using respectively, Eq. (7.3), and Eq. (7.4):

$$\Delta e_H = \rho_a . c_{P,a} . (\dot{V}_{oa,E} - \dot{V}_{oa,R}).(T_i - T_o) \tag{7.3}$$

and,

$$\Delta e_C = \rho_a . c_{P,a} . (\dot{V}_{oa,E} - \dot{V}_{oa,R}).(h_o - h_i) \tag{7.4}$$

Where:

- $\dot{V}_{oa,E}$, and $\dot{V}_{oa,R}$ are respectively the ventilation air rate before and after retrofit.
- ρ_a, and $c_{p,a}$ are respectively the density and the specific heat of ventilation air.
- T_i and T_o are the air temperatures of respectively the indoor space and outdoor ambient during winter.
- h_i and h_o are the air enthalpies of respectively the indoor space and the outdoor ambient during summer.

It should be noted that the humidity control is not typically performed during the winter and thus the latent energy is neglected as indicated in Eq. (7.3).

To determine the total energy use savings attributed to ventilation air reduction, the annual savings in heating and cooling loads has to be estimated. Without using detailed energy simulation, the savings in heating and cooling loads can be calculated using Eq. (7.3) and Eq. (7.4) for various bin temperatures and summing the changes in the thermal loads over all the bins. By taking into account the energy efficiency of the heating and cooling equipment, the energy use savings due to a reduction in the ventilation air can be estimated for both winter and summer as illustrated in Eq. (7.5) and Eq. (7.6) respectively:

$$\Delta kWh_H = \frac{3.6 * \sum_{k=1}^{N_{bin}} N_{h,k} . \Delta e_{H,k}}{\eta_H} \tag{7.5}$$

and,

$$\Delta kWh_C = \frac{3.6 * \sum_{k=1}^{N_{bin}} N_{h,k} . \Delta e_{C,k}}{EER_C} \quad (7.6)$$

Where:

- $N_{h,k}$ is the number of hours in bin k
- EER_C is the average seasonal efficiency ratio for the cooling system
- η_H is the average seasonal efficiency of the heating system

When an air-side economizer is present, the summation in both Eq. (7.5) and Eq. (7.6) should be performed for only the bin temperatures corresponding to time periods when the outside air damper is set at its minimum position.

The calculations of the energy use savings can be further simplified if the air density is assumed to be constant (i.e., independent of temperature) and if the HVAC system has a year-round operation. Under these conditions, the energy savings due to heating can be estimated as follows:

$$\Delta kWh_H = \frac{3.6 * \rho_a . c_{p,a} . N_h . (\dot{V}_{oa,E} - \dot{V}_{oa,R}).(T_i - \bar{T}_o)}{\eta_H} \quad (7.7)$$

Where:

- N_h is the total number of hours in the heating season. When the ventilation air is not provided during all hours, N_h can be adjusted to include only occupied hours (i.e., periods when ventilation is provided) assuming that the average outdoor air temperature does not vary significantly with this adjustment.

- $\bar{T}_o$ is the average outdoor air temperature during the heating season.

For the energy use savings due to cooling, a simplified form of Eq. (7.5) can be obtained with the introduction of a seasonal cooling load to condition a reference amount (such as 1000 m³/hr) of outdoor air, ΔH_C which depends on the climate and the indoor temperature setting:

$$\Delta kWh_C = \frac{3.6 * \rho_a . N_h . (\dot{V}_{oa,E} - \dot{V}_{oa,R}) \Delta H_C}{EER_C} \quad (7.8)$$

Other measures that reduce the ventilation air through the HVAC system include the following actions:

1. Reduce leakage through the outside air damper especially when this damper is set to be closed. The change in the ventilation air can be calculated using

the leakage percentage. The low leakage dampers can restrict leakage to less than 1% while the standards can allow 5% up to 10% leakage when closed.
2.   Eliminate the ventilation during unoccupied periods or when ventilation is not needed.

The calculations of the energy savings for the above listed measures follow the same methods illustrated by Eq. (7.3) through Eq. (7.8). Some examples are provided at the end of this Chapter to show how the energy use savings can be calculated for these measures.

It should be noted that when the ventilation air is reduced, the amount of exhaust air should also be adjusted. Otherwise, a negative static pressure can be obtained in the building (since more air is exhausted than introduced to the building). Several problems can occur because of the negative pressure within the building including:

•   Difficulty in opening exterior doors and windows.
•   Draft can be felt at the perimeter of the building since outside cold air is drawn near the windows and doors.
•   Accumulation of fumes, odors, dirt, and dust is increased since exhaust fans cannot operate at rated capacity under negative pressure.
•   Combustion efficiency of boilers and ovens can decrease if the equipment depends on natural draft to operate properly.

7.3.2   Air-Side Economizers

When the outdoor air conditions are favorable, excess ventilation air can actually be used to condition the building and thus reduce the cooling energy use for the HVAC system.   There are typically two control strategies to determine the switchover point and decide when it is better to use more than the minimum required amount of the outdoor air to cool a building: one strategy is based on dry-bulb temperatures and the other on enthalpies. These control strategies are known respectively as temperature and enthalpy air-side economizers. For both strategies, the operation of the HVAC system is said to be on an economizer cycle. Several existing HVAC systems do not have an economizer cycle and thus do not take advantage of its potential energy use savings.   The two economizer cycles are briefly described below.

Temperature Economizer Cycle:
For this cycle, the outside air intake damper is opened beyond the minimum position whenever the outside air temperature is colder than the return air temperature. However, when the outdoor air temperature is either too cold or too hot, the outside air intake damper is set back to its minimum position. Therefore, there are outdoor air temperature limits beyond which the economizer cycle should not operate. These temperature limits are called respectively economizer

low temperature limit, T_{el}, and economizer high temperature limit, T_{eh}, as illustrated in Figure 7.4. Beyond the two economizer temperature limits, the outside air intake damper is set to its minimum position.

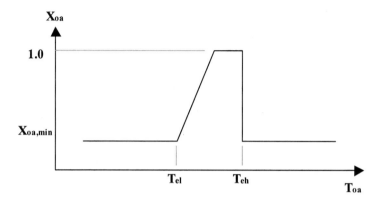

Figure 7.4: Economizer temperature limits.

While the economizer high temperature limit is typically difficult to define and is often set to be same as the return air temperature, the economizer low limit temperature can be determined as function of the conditions of both the return air and supply air using a basic principle. There is no need to introduce more than the required amount (i.e., for ventilation purpose) of outdoor air if it has more heat content than the return air.

Enthalpy Economizer Cycle:

This cycle is similar to the temperature-based economizer cycle except that enthalpy of the air streams are used instead of the temperature. Therefore, two parameters are typically measured for each air stream in order to estimate its enthalpy (dry and wet bulb temperatures, for instance). Because of this requirement, the enthalpy economizers are less common since they are more expensive to implement and less robust to use even though it may achieve greater savings if properly operated.

7.4   Ventilation of Parking Garages

Automobile parking garages can be partially open or fully enclosed. Partially open garages are typically above-grade with open sides and usually do not need mechanical ventilation. However, fully enclosed parking garages are usually underground and require mechanical ventilation. Indeed, in absence of ventilation, enclosed parking facilities present several indoor air quality problems. The most serious is the emission of high levels of carbon monoxide (CO) by cars within the parking garages. Other concerns related to enclosed garages are the presence of oil and gasoline fumes, and other contaminants such as oxides of nitrogen (NOx) and smoke haze from diesel engines.

To determine the adequate ventilation rate for garages, two factors are typically considered: the number of cars in operation and the emission quantities. The number of cars in operation depends on the type of the facility served by the parking garage and may vary from 3% (in shopping areas) up to 20% (in sports stadium) of the total vehicle capacity (ASHRAE, 1999). The emission of carbon monoxide depends on individual cars including such factors as the age of the car, the engine power, and the level of car maintenance.

For enclosed parking facilities, ASHRAE standard 62-1989 specifies a fixed ventilation rate of below 7.62 L/s.m^2 (1.5 cfm/ft^2) of gross floor area (ASHRAE, 1989). Therefore, a ventilation flow of about 11.25 air changes per hour is required for garages with 2.5-m ceiling height. However, some of the model code authorities specify an air change rate of 4 to 6 air changes per hour. Some of the model code authorities allow ventilation rate to vary and be reduced to save fan energy if CO-demand controlled ventilation is implemented, that is, a continuous monitoring of CO concentrations is conducted, with the monitoring system being interlocked with the mechanical exhaust equipment. The acceptable level of contaminant concentrations varies significantly from code to code. A consensus on acceptable contaminant levels for enclosed parking garages is needed. Unfortunately, ASHRAE standard 62-1989 does not address the issue of ventilation control through contaminant monitoring for enclosed garages. Thus, ASHRAE commissioned a research project 945-RP (Krarti et al. 1999a) to evaluate current ventilation standards and recommend rates appropriate to current vehicle emissions/usage. Based on this project, a general methodology has been developed to determine the ventilation requirements for parking garages.

7.4.1   Existing Codes and Standards

Table 7.3 provides a summary of existing codes and standards for ventilating enclosed parking garages in the Unites States, and other selected countries.

As shown in Table 7.3, the recommendations for the CO exposure limits are not consistent between various regulations within the US and between countries. However, the recommendations offer an indication of risks from exposure to CO in parking garages. A limit level of 25 ppm for long-term CO exposure would meet almost all the codes and standards listed in Table 7.3.

Table 7.3: Summary of U.S. and International Standards for Ventilation Requirements of Enclosed Parking Garages.

	Time [hrs]	ppm	Ventilation
ASHRAE	8	9	7.6 L/s.m^2
	1	35	(1.5 cfm/ft^2)
ICBO	8	50	7.6 L/s.m^2
	1	200	(1.5 cfm/ft^2)
NIOSH/OSHA	8	35	
	ceiling	200	
BOCA			6 ACH
SBCCI			6-7 ACH
NFPA			6 ACH
ACGIH	8	25	
Canada	8	11/13	
	1	25/30	
Finland	8	30	2.7 L/s.m^2
	15 minutes	75	(0.53 cfm/ft^2)
France	ceiling	200	165 L/s.car
	20 minutes	100	(350 cfm/car)
Germany			3.3 L/s.m^2
			(0.66 cfm/ft^2)
Japan/S. Korea			6.35-7.62 L/s.m^2
			(1.25-1.5 cfm/ft^2)
Netherlands	0.5	200	
Sweden			0.91 L/s.m^2
			(0.18 cfm/ft^2)
U.K.	8	50	6-10 ACH
	15 minutes	300	

As part of an ASHRAE sponsored project (945-RP), field measurements for the seven tested parking facilities were performed. For more detailed description of the field measurements, refer to Krarti et al. (1999a). In particular, the following selected results were obtained from the field study (Ayari et al. 2000):

(a) All the tested enclosed parking garages had contaminant levels that are significantly lower than those required by even the most stringent regulations (i.e, 25 ppm of 8-hr weighted average of CO concentration).
(b) The actual ventilation rates supplied to the tested garages were generally well below those recommended by ASHRAE standard 62-1989 [i.e., below 7.6 L/s.m^2 (1.5 cfm/ft^2)].
(c) When it was used, demand controlled ventilation was able to maintain acceptable indoor air quality within the tested enclosed parking facilities.
(d) The location of the supply and exhaust vents, the traffic flow pattern, the number of moving cars, and the travel time were important factors that

affect the effectiveness of the ventilation system in maintaining acceptable CO (or NOx) levels within enclosed parking garages. Any design guidelines should account for these factors to determine the ventilation requirements for enclosed parking facilities.

It is clear from the results of the field study that the current ventilation rate specified in the ASHRAE standard 62-1989 is out-dated for enclosed parking garages and can lead to significant energy waste. To prevent waste, ventilation requirements for enclosed parking garages can be estimated using a new design method developed by Krarti et al. (1999a). The new method allows the estimation of the minimum ventilation rate required to maintain contaminant concentrations within parking facilities at the acceptable levels set by the relevant health authorities without large penalties in fan energy use. Moreover, the new method accounts for variability in the parking garage traffic flow, car emissions, travel time, and number of moving cars.

7.4.2   General Methodology for Estimating the Ventilation Requirements for Parking Garages

Based on the results of several parametric analyses (Krarti et al., 1999a), a simple design method was developed to determine the ventilation flow rate required to maintain acceptable CO level within enclosed parking facilities. Ventilation rates for enclosed parking garages can be expressed in terms of either flow rate per unit floor area (L/s.m^2 or cfm/ft^2) or air volume changes per unit time (ACH). The design ventilation rate required for an enclosed parking facility depends on four factors:

- Contaminant level acceptable within the parking facility;
- Number of cars in operation during peak conditions;
- Length of travel and operation time of cars in the parking garage; and,
- Emission rate of a typical car under various conditions.

Data for the above listed factors should be available to determine accurately the design ventilation rate for enclosed parking garages. A simple design approach is presented in the following section to determine the required ventilation rate for both existing and newly constructed enclosed parking garages.

To determine the required design flow rate to ventilate an enclosed parking garage, the following procedure can be followed (Ayari and Krarti, 2000):

Step 1. Collect the following data:

(i)   Number of cars in operation during the hour of peak use, N (# of cars). The ITE Trip Generation Handbook (ITE, 1995) is a good source to estimate the value of N.

(ii)   Average CO emission rate for a typical car per hr, ER, (gr/hr). The CO emission rate for a car depends on several factors such as vehicle characteristics, fuel types, vehicle operation conditions, and environment conditions. Data provided in the ASHRAE Handbook (ASHRAE, 1999) and reproduced in Table 7.4 below can be used to estimate CO emission rates for a typical car. Typically, hot starts are common in facilities where cars are parked for short periods such as shopping malls. In the other hand, cold starts characterize facilities where cars park during long periods such as office buildings.

(iii)   Average length of operation and travel time for a typical car, T (seconds). The ASHRAE Handbook gives average entrance/exit times for vehicles. However, higher values may be used for worst case scenarios such as during rush hours or special events.

(iv)   The level of CO concentration acceptable within the garage, CO_{max} (ppm). This level can be defined based on the recommendations of the applicable standards (typically, $CO_{max}= 25$ ppm).

(v)   Total floor area of the parking area, A_f (ft^2 or m^2).

Table 7.4: Typical CO emissions within parking garages (*Source*: ASHRAE, Applications Handbook, 1999. With permission.)

Season	HOT EMISSIONS (STABILIZED), GRAMS/MIN		COLD EMISSIONS, GRAMS/MIN	
	1991	1996	1991	1996
Summer [32°C (90°F)]	2.54	1.89	4.27	3.66
Winter [0°C (32°F)]	3.61	3.38	20.74	18.96

Step 2. (a) Determine the peak generation rate, GR [gr/hr.m^2 (gr/hr.ft^2)], for the parking garage per unit floor area using Eq. (7.9):

$$GR = \frac{N^* \, ER}{A_f} \qquad (7.9)$$

(b) Normalize the value of generation rate using a reference value $GR_o=26.8$ gr/hr.m^2 ($GR_o=2.48$ gr/hr.ft^2). This reference value was obtained using the worst emission conditions (cold emissions in winter season) for an actual enclosed parking facility (Krarti et al., 1999a):

$$f = \frac{GR}{GR_o} * 100 \qquad (7.10)$$

Step 3. Determine the required ventilation rate per unit floor area ($L/s.m^2$ or $cfm.ft^2$) the correlation presented by Eq. (7.11) depending on the maximum level of CO concentration CO_{max}:

$$L/s.m^2 = C. \text{ f. } T \qquad (7.11)$$

Where, the correlation coefficient, C is given below:

$$C = \begin{cases} 1.204 \times 10^{-3} \text{ } L/m^2.s^2 \text{ } (2.370 \times 10^{-4} \text{ } cfm/ft^2.s) \text{ for } CO_{max} = 15 \text{ ppm} \\ 0.692 \times 10^{-3} \text{ } L/m^2.s^2 \text{ } (1.363 \times 10^{-4} \text{ } cfm/ft^2.s) \text{ for } CO_{max} = 25 \text{ ppm} \\ 0.482 \times 10^{-3} \text{ } L/m^2.s^2 \text{ } (0.948 \times 10^{-4} \text{ } cfm/ft^2.s) \text{ for } CO_{max} = 35 \text{ ppm} \end{cases}$$

and T is the average travel time of cars within the garage in seconds.

Example 7.1: *An office building has a two-level enclosed parking garage with a total capacity of 450 cars, a total floor area of 8300 m^2 (89,290 ft^2), and an average height of 2.75 m (9.0 ft). The total length of time for a typical car operation within the garage is 2 minutes (120 s).*

(a) Determine the required ventilation rate for the enclosed parking garage in $L/s.m^2$ (or cfm/ ft^2) and in ACH so that CO levels never exceeds 25 ppm. Assume that the number of cars in operation is 40% of the total vehicle capacity (a shopping mall facility).

(b) Determine the annual fan energy savings if the ventilation found in (a) is used instead of 1.5 cfm/ ft^2 recommend by ASHRAE 62-1989. The ventilation system is operated 16 hours/day over 365 hours/year. The cost of electricity is $0.07/kWh. Assume the size of the initial ventilation fan is 30 hp (22.9 kW).

Solution:

(a) The procedure for the new design methodology can be applied to estimate the minimum ventilation requirements for the garage defined in this example:

Step 1. Garage data: N = 450* 0.4 = 180 cars, ER = 11.66 gr/min (average emission rate for a winter day using the data from Table 7.4), T = 120 s, CO_{max} = 25 ppm.

Step 2. Calculate CO generation rate:

(a) $GR = \dfrac{180*11.66 \text{ gr/min} *60 \text{ min/hr}}{8300 \text{ m}^2} = 15.17 \text{ gr/hr.m}^2$

(b) $f = \dfrac{15.17}{26.8} *100 = 56.6$

Step 3. Determine the ventilation requirement:

Using the correlation of Eq. (7.11) for CO_{max} = 25 ppm, the design ventilation rate in $L/s.m^2$ can be calculated:

$$L/s.m^2 = 0.692 \text{ x } 10^{-3}* 56.6 * 120 \text{ s} = 4.7$$

This ventilation rate corresponds to 0.91 cfm/ft^2. In terms of air change per hour, the ventilation rate required for the enclosed parking garage is:

$$ACH = \dfrac{4.7 \ \ L/m.s^2 \text{ x } 10^{-3} \text{ L/m}^3 \text{ x } 3600 \ \ s/hr}{2.75 \text{ m}} = 6.1$$

Due to the reduced ventilation rate (from 1.5 cfm/ft^2 to 0.91 cfm/ft^2), the fan energy use can be reduced either by installing smaller fans or by reducing the fan speeds (for fans equipped with variable speed drives). In either cases, the theoretical reduction in fan energy use can be estimated using the fan laws (in particular, these laws state that the fan power consumption is proportional to the cube of the ventilation flow rate):

$$\Delta kW_{e, fan} = \dfrac{kW_{exist}}{\eta_m} . (\dfrac{\dot{m}_{new}}{\dot{m}_{exist}})^3 . N_h$$

Since N_h = 16 hrs/day * 365 days/yr = 5,840 hrs/yr and kW_{exist} =22.9 kW with $\dot{m}_{new} \big/ \dot{m}_{exist}$ = 0.91/1.5 =0.61, the annual fan energy savings can be estimated to be (assuming 90% motor efficiency):

$$\Delta kW_{e, fan} = \dfrac{22.9 kW}{0.90} .(0.61)^3 .5,840 \ \ hrs \ / \ yr = 33,728 \ kWh \ / \ yr$$

Based on an electricity rate of $0.07/kWh, the cost savings attributed to savings in fan energy use is:

$$\Delta C_{e, fan} = \$0.07 \ / \ kWh * 33,728 \ \ kWh \ / \ yr = \$2,361 \ / \ yr$$

To further conserve energy, fan systems can be controlled by CO meters to vary the amount of air supplied, if permitted by local codes. For example, fan systems could consist of multiple fans with single-or variable- speed motors or variable pitch blades. In multi-level parking garages or single-level structures of extensive area, independent fan systems, each under individual control are preferred. Figure 7.5 provides the maximum CO level in a tested garage (Krarti et al., 1999a) for three car movement profiles (as illustrated in Figure 7.6) and three ventilation control strategies:

(i)   CV where the ventilation is on during the entire occupancy period,
(ii)   On-Off with the fans are operated based on CO-sensors, (either On or Off)
(iii)   VAV with variable volume fans that are adjusted depending on the CO level within the garage.

Figure 7.5 also indicates the fan energy savings achieved by the On-Off and VAV systems (relative to the fan energy use by the CV system). As illustrated in Figure 7.5, significant fan energy savings can be obtained when demand CO-ventilation control strategy is used to operate the ventilation system while maintaining acceptable CO levels within the enclosed parking facility.

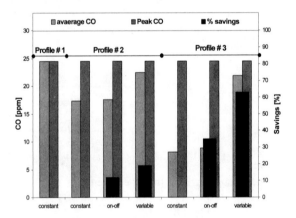

Figure 7.5: Typical energy savings and maximum CO level obtained for demand CO-ventilation controls

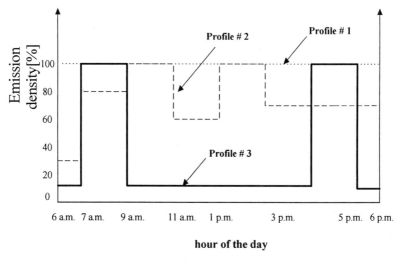

Figure 7.6: Car movement profiles used in the analysis conducted by Krarti et al. (1999).

7.5   Indoor Temperature Controls

The indoor temperature settings during both heating and cooling seasons has significant impacts on the thermal comfort within occupied spaces and on the energy use of the HVAC systems. It is therefore important for the auditor to assess the existing indoor air temperature controls within the facility to evaluate the potential for reducing energy use and/or improving indoor thermal comfort without any substantial initial investment. There are four options for adjustments of the indoor temperature setting that can save heating and cooling energy:

i.   Eliminating overcooling by increasing the cooling set-point during the summer.

ii. Eliminating overheating by reducing the heating set-point during the winter.

iii. Preventing simultaneous heating and cooling operation for the HVAC system by separating heating and cooling set-points.

iv. Reducing heating and/or cooling requirements during unoccupied hours by setting back the set-point temperature during heating and setting up the set-point temperature (or letting the indoor temperature float) during cooling.

214   *ENERGY AUDIT OF BUILDING SYSTEMS*

The calculations of the energy use savings for these measures can be estimated based on degree-days methods as outlined in Chapter 6. However, even more simplified calculation methods can be considered to estimate the magnitude of the energy savings. Some examples are provided to illustrate the energy use savings due to adjustments in indoor temperature settings.

It should be noted that some of the above listed measures could actually increase the energy use if they are not adequately implemented. For instance when the indoor temperature is set lower during the winter, the interior spaces may require more energy since they need to be cooled rather than heated. Similarly, the setting of the indoor temperature higher can lead to an increase in the reheat energy use for the zones with reheat systems.

7.6   UPGRADE OF FAN SYSTEMS

7.6.1   Introduction

Fans are used in several HVAC systems to distribute air throughout the buildings. In particular, fans are used to move conditioned air from central air handling units to heat or cool various zones within a building. According to a survey reported by the US Energy Information Administration (EIA, 1997), the energy use for fans represents about 25% of the total electrical energy use of a typical building. Thus, improvements in the operation of fan systems can provide significant energy savings.

In a typical air handling unit, fans create the pressure required to move air through ducts, heating or cooling coils, filters, and any other obstacles within the duct system. Two types of fans are used in HVAC systems: centrifugal and vane-axial fans. The centrifugal fan consists of rotating wheel, generally referred to as an impeller, mounted in the center of a round housing. The impeller is driven by an electric motor through a belt drive. The vane-axial fan includes a cylindrical housing with the impeller mounted inside along the axis of the cylindrical housing. The impeller of an axial fan has blades mounted around a central hub similar to an airplane propeller. Typically, axial fans are more efficient than centrifugal fans but are more expensive since they are difficult to construct. Currently, the centrifugal fans are significantly more common in existing HVAC systems.

There are several energy conservation measures that help reduce the energy use of fan systems. Some of these measures are described in this section. A brief review of basic laws that characterize the fan operation is first provided.

7.6.2   Basic Principles of Fan Operation

To characterize the operation of a fan, several parameters need to be determined including the electrical energy input required in kW (or Hp), the maximum amount of air it can move in L/s (or cfm) for a total pressure differential (ΔP_T) or a static pressure differential (ΔP_s), and the fan efficiency. A simple relationship exists that allows the calculation of the electrical energy input required for a fan as a function of the amount air flow, the pressure differential and its fan efficiency. If the total pressure is used, Eq. (7.9) provides the electrical energy input using metric units.

$$kW_{fan} = \frac{\dot{V}_f.\Delta P_T}{\eta_{f,t}} \qquad (7.12)$$

Using English units, the horsepower of the fan can be determined as follows:

$$Hp_{fan} = \frac{\dot{V}_f.\Delta P_T}{6,356 * \eta_{f,t}}$$

If the static pressure is used, Eq. (7.13) should be used instead. Note that the static fan efficiency is needed.

$$kW_{fan} = \frac{\dot{V}_f.\Delta P_s}{\eta_{f,s}} \qquad (7.13)$$

To measure the total pressure of the fan, a Pitot Tube can be used in two locations within the duct that houses the fan as illustrated in Figure 7.7

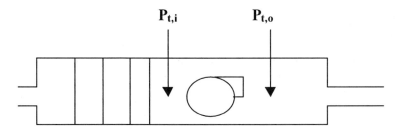

Figure 7.7: Measurement of Total pressure of the fan

As shown in Fig. 7.7, the total pressures $P_{t,i}$ and $P_{t,o}$ at respectively the inlet and outlet of the fan are first measured. Then, the fan total pressure is simply found by taking the difference:

$$\Delta P_T = P_{t,o} - P_{t,i} \qquad (7.14)$$

When any of the fan parameters (such as total pressure ΔP_T, static pressure ΔP_s, or air flow rate $\dot{V}_f$) vary, they are still related through certain laws, often referred to as the fan laws. Fan laws can be very useful to estimate the energy savings from any change in fan operation. These fan laws are summarized in the following three equations:

Equation (7.15) states that the air flow rate $\dot{V}_f$, is proportional to the fan speed ω :

$$\frac{\dot{V}_{f,1}}{\dot{V}_{f,2}} = \frac{\omega_1}{\omega_2} \tag{7.15}$$

Equation (7.16) indicates that the fan static pressure ΔP_s, varies as the fan air flow rate $\dot{V}_f$, or the square of the fan speed ω :

$$\frac{\Delta P_{s,1}}{\Delta P_{s,2}} = \left(\frac{\dot{V}_{f,1}}{\dot{V}_{f,2}}\right)^2 = \left(\frac{\omega_1}{\omega_2}\right)^2 \tag{7.16}$$

Finally Equation (7.17) states that the power used by the fan kW_{fan} varies as the cube of the fan air flow rate $\dot{V}_f$, or the square of the fan speed ω :

$$\frac{kW_{fan,1}}{kW_{fan,2}} = \left(\frac{\dot{V}_{f,1}}{\dot{V}_{f,2}}\right)^3 = \left(\frac{\omega_1}{\omega_2}\right)^3 \tag{7.17}$$

Generally, fans are rated based on standard air density. However, when the air density changes, the fan static pressure and the power required to drive the fan will change. Both fan static pressure and fan power varies in direct proportion to the change in air density.

There are several applications to the fan laws. In particular, when volume of air needs to be reduced (as in the case of CV to VAV retrofits) two options can be considered:

- Static Pressure can be increased by closing dampers (for instance, by using outlet dampers or variable inlet vanes)
- Speed can be reduced by using variable speed drives (through the use of variable frequency drives)

It can be noted from the fan laws that by reducing the amount of air to be moved by the fan, the electrical energy input required is reduced significantly. For instance, a 50% reduction in the volume of air results in 87.5% reduction in fan energy use. This fact hints clearly to the advantage of using variable air volume fan systems compared to constant volume fans.

In HVAC systems, there are three common approaches to reduce the air volume including

- Outlet dampers: located at the outlet of the fan. To reduce the air volume, the dampers are controlled to the desired position. Effectively, the outlet dampers increase air flow resistance.

- Variable Inlet Vanes: to change the characteristics of the fan. The position of the fan inlet vanes can be changed to allow variable volume of air.

- Variable Speed Drives: to change the speed of the fan. Using typically variable frequency drives (see Chapter 5), it is possible to control fan speed and thus supplied air volume. Theoretically and according to the fan laws, fan power consumption can be reduced substantially when fan speed is reduced. However, due to back pressure effects, the actual reduction in fan power consumption is less significant (see Example 7.6 for more details).

Table 7.5 compares the reduction of fan power as a function of the reduction in air volume for the three control options. For the variable speed drives, the fan performance curve is given for both actual and theoretical conditions.

Table 7.5: Percent reduction of fan power as a function of the percent reduction in air volume for the three control options.

Fan Control Option	Fan Air Volume (in % of Maximum Flow)			
	100%	80%	60%	40%
Outlet Dampers	100%	96%	88%	77%
Variable Inlet Vanes	100%	78%	61%	51%
Variable Speed Drives:				
Actual Performance	100%	64%	36%	16%
Theoretical Performance	100%	51%	22%	6 %

To determine the best control approach to vary the air volume (for VAV systems, for instance) an economic analysis is recommended. However, the use of variable frequency drives is often more cost-effective than the other control options. Example 7.3 illustrates a simplified energy and economic analysis to determine the best control option for VAV fan system.

Example 7.3: *A 50,000-cfm CV air handling unit is to be converted to a VAV system. The motor for the existing supply fan is rated at 35 Hp with 90% efficiency. The fan is operated for 3,500 hours/year according to the following load profile:*

➤ *100% load, 10% of the time,*
➤ *80% load, 40% of the time,*
➤ *60% load, 40% of the time, and*
➤ *40% load, 10% of the time.*

Three options for variable air volume controls have been considered:
(a) Outlet dampers at a cost of $3,500; (b) variable inlet vanes at a cost of $8,000; and (c) variable speed drive at a cost of $ 9,000.

Determine the best option for the fan control using simple payback period analysis. Assume that electricity costs $0.05/kWh.

Solution: The electrical energy input of the supply fan motor is first determined for each control option by calculating a weighted average fan horse power rating using the fan performance curves given in Table 7.5:
For the outlet dampers:

$$\overline{kW}_{e,fan}^{OD}=35 \ Hp \ *(1.0*0.1+0.96*0.4+0.88*0.40+0.77*0.1)=31.95 \ Hp$$

For the variable inlet vanes:
$$\overline{kW}_{e,fan}^{VIV}=35 \ Hp \ *(1.0*0.1+0.78*0.4+0.61*0.40+0.51*0.1)=24.76 \ Hp$$

For the variable speed drives, both the actual and the theoretical fan performance curves will be considered:

The actual fan performance leads to:

$$\overline{kW}_{e,fan}^{VSD}=35 \ Hp \ *(1.0*0.1+0.64*0.4+0.36*0.40+0.16*0.1)=18.06 \ Hp$$

While, the theoretical fan performance gives:

$$\overline{kW}_{e,fan}^{VSD}=35 \ Hp \ *(1.0*0.1+0.51*0.4+0.22*0.40+0.064*0.1)=13.99 \ Hp$$

The total number of hours for operating the fan during one year is $N_{f,h}$=3,500 hours for all the fan types including the constant volume fan (used as a base case for the payback analysis). First, the electrical energy use for each fan control option is estimated:

For the existing constant volume fan:

$$kWh_{e,fan}^{CV} = \left(\frac{35Hp * 0.746kW / Hp}{0.90}\right) * 3,500 hrs = 101,539 \quad kWh / yr$$

For the outlet dampers (OD):

$$kWh_{e,fan}^{OD} = \left(\frac{31.95Hp * 0.746kW / Hp}{0.90}\right) * 3,500 hrs = 92,690 \quad kWh / yr$$

For the variable inlet vanes (VIV):

$$kWh_{e,fan}^{OD} = \left(\frac{24.76Hp * 0.746kW / Hp}{0.90}\right) * 3,500 hrs = 71,832 \quad kWh / yr$$

For the actual performance of the variable speed drives (VSD):

$$kWh_{e,fan}^{OD} = \left(\frac{18.06Hp * 0.746kW / Hp}{0.90}\right) * 3,500 hrs = 52,394 \quad kWh / yr$$

For the theoretical performance of the variable speed drives (VSD):

$$kWh_{e,fan}^{OD} = \left(\frac{13.99Hp * 0.746kW / Hp}{0.90}\right) * 3,500 hrs = 40,587 \quad kWh / yr$$

The fan energy use cost and the payback period for each fan control option are summarized below:

Fan Control Option	Annual Energy Use (kWh/yr)	Annual Energy Cost ($/yr)	Annual Cost Savings ($/yr)	Simple Payback Period (yrs)
CV	101,539	5,077	0	----
OD	92,690	4,635	442	7.9
VIV	71,832	3,592	1,485	5.4
VSD (act.)	52,394	2,620	2,457	3.7
VSD (th.)	40,587	2,029	3,048	3.0

It is clear from the results of the economic analysis presented above, that the fan equipped with a variable speed drive is the most cost-effective option.

7.6.3   Size Adjustment

A recent EPA study found that 60% of building fan systems were oversized by at least 10%. By reducing the size of these fans to the required capacity, it is estimated that average savings of 50% can be achieved in energy use of fan systems. Even more savings can be expected if the fan size adjustment is implemented with other measures related to fan systems such energy-efficient motors, energy-efficient belts, and variable speed drives.

The size adjustment of fans can be implemented for both constant volume and variable air volume systems. In addition to energy savings, the benefits of using the proper fan size include:

- Better comfort: If the fan system is oversized, more air than needed may be supplied to the zones which may be reduce the comfort of the occupants in addition to wasting energy.
- Longer equipment life: an oversized fan equipped with variable speed drive operates at low capacities. This mode of operation can reduce the useful life of motors and other equipment.

To determine if a fan system is oversized, some in-situ measurements can be made depending on the type of the HVAC system: constant volume or variable air volume.

For constant volume systems, the measurement of supply fan static pressure is generally sufficient to assess whether or not the fan is properly sized. To ensure that static pressure for the main supply fan is measured when the HVAC system is operating close to its design capacity, the testing should be made during a hot and humid day. Moreover, all dampers and fan vanes should be fully open during the tests. If the measured static pressure is larger than the design pressure which is typically provided in the building mechanical drawings (otherwise, design value should be determined), the fan is supplying too much air and is most probably oversized.

For variable air volume systems, three methods can be used to determine if the fan system is oversized including:

(a) Measurement of the electrical current drawn by the fan motor. If this current is lower than 75% of the nameplate full-load amperage rating (this value can be obtained directly on the motor's nameplate or from the operations and maintenance manual), then the fan is oversized.

(b) Check of the position of fan control vanes and dampers. If the vanes or dampers are closed more than 20%, the fan is oversized.

(c) Measurement of the static pressure for the main supply fan. If the measured static pressure is larger than the set-point, the fan system is oversized.

It should be noted that all the diagnostic methods need to be performed when the HVAC system is operating under peak load such as during a hot and humid day.

When it is clear that the fan system is oversized, the adjustment of its size can be achieved by one or a combination of these measures:

(i)   Installation of larger pulleys to reduce the speed of the existing fan. By reducing the speed of the fan, not only is the flow of air moved through the duct proportionally reduced, but also the energy use of the fan system is reduced significantly. For instance a 20% reduction in a fan's speed will reduce its energy consumption by 50%.

(ii)   Replacement of the existing oversized motor by a smaller energy-efficient motor that matches the peak load. The use of a smaller motor will obviously reduce the energy use of the fan system. For instance, replacing a 50 kW standard motor with a 35 kW energy-efficient motor will reduce the energy use of the fan system by about one third (33%).

(iii)   Adjustment of the static pressure setting (for the VAV systems, only). By reducing the static pressure set-point to a level low enough to maintain indoor comfort, the energy use by the fan system will be reduced. For instance, a VAV system operating at a static pressure of 6 inches of water can be reduced to 4 inches without loss of thermal comfort. This 33% reduction of static pressure will achieve about 45% of energy savings in the fan system operation.

It is important to analyze the performance of the entire HVAC system with any of the measures described above to ensure effective adjustment of the fan size. Indeed, changes to the fan system can affect the operation and/or control of other components of the HVAC system.

7.7   Common HVAC Retrofit Measures

In this section, simplified analysis methods are provided to estimate the energy savings potential for retrofit measures related to the operation of HVAC systems.

7.7.1   Reduction of Outdoor Air Volume:

This measure can be implemented by replacing or appropriately controlling the outside air dampers (as depicted in Figure 7.8) using several approaches including:

(a) Use of low leakage dampers to reduce the amount of unwanted ventilation air when outdoor air dampers are closed. Standards dampers can allow up to 15% air leakage when closed. Low-leakage dampers restrict air leakage to less than 1%.

(b) Reduce the ventilation air requirements to the levels recommended by ASHRAE Standard 62-1999. In addition, ventilation air requirements may be reduced by continuously monitoring indoor contaminant levels (for instance CO_2 for offices and schools and CO for parking garages).

(c) Eliminate or reduce the ventilation air during unoccupied periods. Some HVAC systems are operated during unoccupied hours to maintain specified temperature set-points (even for systems operated with night temperature set back) or to start condition the space before occupancy (optimum start controls). In these cases, the ventilation requirements can be safely reduced as long as there is no occupant.

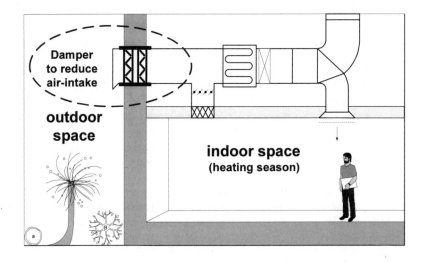

Figure 7.8: Outdoor air dampers can be a source of significant energy waste.

To estimate the energy savings obtained from a reduction of outdoor air volume, Eqs. (7.3) through (7.6) can be used. Example 7.4 illustrates the simplified calculation procedure for a retrofit of outside air duct with low-leakage dampers.

Example 7.4: *Estimate the fuel savings obtained by using low-leakage dampers for a 40,000 cfm air handling unit (AHU) which served an office building located in Chicago, IL. The office is conditioned during the entire heating season to 72 °F (even during unoccupied hours). It was found that the existing dampers have 15% leakage when closed. The outdoor air damper system is scheduled to be closed 3,500 hours during the heating season. For the analysis, assume that he cost of fuel is $6/MMBtu and the boiler efficiency is 80%. The average heating season outdoor temperature in Chicago is 34.2 °F. The low-leakage dampers have a 1% leakage.*

Solution: The annual fuel savings is estimated based on the energy saved due to a reduction in the amount of outdoor air volume to be heated from $T_{out} = 34.2° F$ to $T_{in} = 72° F$:

$$\Delta FU_H = \frac{\Delta \dot{m}_a . c_p . (T_{out} - T_{in})}{\eta_b} . N_{h,H}$$

Therefore:

$$\Delta FU_H = \frac{40,000 cfm * (0.15 - 0.01) * (1.08 Btu / cfm.hr.° F) * (72° F - 34.2° F)}{(0.80) * (10^6 Btu / MMBtu)} * 3,500 hrs / yr$$

Or

$$\Delta FU_H = 1,000 \, MMBtu / yr$$

In terms of cost savings, the use of low-leakage dampers can reduce fuel costs by $6,000/yr.

7.7.2   Reset Hot or Cold Deck temperatures

In several HVAC systems with dual ducts, fixed temperature set-point is used throughout the year for the hot deck and/or the cold deck. This operation strategy can waste significant heating and cooling energy by mixing air that can be too hot with air that can be too cold. Simple controls can be set to allow the temperature set-point for both the hot and cold decks to be changed according to the zone loads.

Again, Eqs. (7.3) through (7.6) can be used to estimate the energy savings for hot/cold deck temperature setting. Example 7.5 illustrates the simplified calculation procedure for hot-deck temperature reset.

Example 7.5: *Estimate the fuel savings obtained by resetting the hot deck of a 60,000 cfm dual duct system by 2 °F during the winter and 1 °F during the summer. The HVAC system is operated 80 hours per week and serves a building in Chicago, IL. For the analysis, assume that the cost of fuel is $6/MMBtu and the boiler efficiency is 80%. The average heating season outdoor temperature in Chicago is 34.2 °F . It should be noted that for Chicago, the number of weeks is 30 for the heating season (winter) and 21 for the cooling season (summer).*

Solution: The annual fuel savings is estimated based on the energy saved through a change in the temperature difference: $\Delta T_H = 2^\circ F$ in the winter and $\Delta T_C = 1^\circ F$ in the summer:

$$\Delta FU_H = \frac{\dot{m}_a.c_p.(\Delta T_H.N_{h,H} - \Delta T_C.N_{h,C})}{\eta_b}$$

Therefore:

$$\Delta FU_H = \frac{60,000cfm*(1.08Btu/cfm.hr.^\circ F)*[2^\circ F*(30*80hrs)+1^\circ F*(21*80hrs)]}{(0.80)*(10^6\,Btu/MMBtu)}$$

Or

$$\Delta FU_H = 525\,MMBtu/yr$$

Thus, the hot deck temperature reset can reduce the fuel cost by \$3,150/yr.

7.7.3   CV to VAV system Retrofit

The conversion of constant volume (CV) system to variable air volume (VAV) system can provide significant energy savings for three reasons: (i) reduction in fuel use for heating requirements, (ii) reduction in electrical energy use for cooling requirements, and (iii) reduction in fan electrical power use for air circulation.

Two types of constant volume systems are good candidates for VAV conversion. These systems are:

(a) Single duct reheat systems which provide constant volume of cold air continuously to satisfy the peak cooling load requirements for critical zones. For other zones or under other operating conditions, the cold air is reheated using hot water or electrical terminal heaters. This reheat of a cold air represents waste of energy. The conversion from CV to VAV systems can be easily implemented for single duct reheat systems by adding a variable air volume sensor/controller to the terminal units. Reheat may be retained in case heating and cooling occur simultaneously.
(b) Dual duct systems that have both hot and cold air streams which are continually mixed to meet the zone load. In peak heating or cooling conditions, mixing may not occur and only hot air or cold air is provided to meet zone loads.

The energy analysis of CV to VAV retrofit measures requires detailed simulation programs. Simplified methods can be used but provide only an order of magnitude of the total energy savings. Example 7.6 presents a simplified analysis to estimate the energy and cost savings attributed to the conversion of a single duct reheat system to VAV system. Example 7.3 compares the fan energy savings for various fan systems that can be used to deliver variable air flow volumes.

Example 7.6: *Estimate the heating and cooling energy and cost savings incurred by converting a 20,000-cfm single duct reheat system to a VAV system in an office building located in Chicago, IL. For the CV system, the return air is measured to be 76 °F and the cold duct air temperature is set to be 58°F. The HVAC system is operated 80 hours per week. For the analysis, consider only the sensible loads. The cost of fuel is $5/MMBtu and the boiler efficiency is 80%. The cost of electricity is $0.05/kWh and the chiller efficiency is 0.9 kW/ton. For Chicago, the number of weeks is 30 for the heating season (winter) and 21 for the cooling season (summer).*

Solution:
(a) First, fuel savings are determined. These fuel savings are the results of the heating load reduction incurred from the CV to VAV conversion of the 20,000-cfm air handling unit. Assuming that the average heating load X is 50% over the entire year, the annual fuel savings are attributed to the energy saved through the reduction of the air volume (by a fraction of 1-X). This air volume has to be heated from: the cold air temperature $T_{cold} = 58^o F$ to the return air temperature $T_{return} = 76^o F$ during the entire year (52 weeks):

$$\Delta FU_H = \frac{\dot{m}_a.c_p.(1-X)(T_{return} - T_{coldC}).N_{h,yr}}{\eta_b}$$

Therefore:

$$\Delta FU_H = \frac{20,000 cfm*(1.08 Btu/cfmhr.^o F)*(1-0.5)*(76^o F-58^o F)*(52*80 hrs)]}{(0.80)*(10^6 Btu/MMBtu)}$$

Or
$$\Delta FU_H = 1,010.9\, MMBtu/yr$$

Thus, the fuel cost saving from the CV to VAV conversion is $5,054/yr.

(b) The electricity savings are incurred from the cooling load reduction. Again, assuming that the average cooling load X is 50% over the cooling season, the annual cooling energy savings is estimated based on the energy saved through the reduction of the supplied air volume (by a fraction of 1-X). This air volume has to be cooled from: the return air temperature $T_{return} = 76^{\circ} F$ to the cold air temperature $T_{cold} = 58^{\circ} F$ during the cooling season (21 weeks):

$$\Delta kWh_C = \frac{\dot{m}_a.c_p.(1 - X)(T_{return} - T_{coldC}).N_{h,C}}{12,000 \, Btu \, / \, ton..hr}.kW \, / \, ton$$

Thus:

$$\Delta kWh_C = \frac{20,000 cfm * (1.08 Btu / cfm.hr.^{\circ} F) * (1 - 0.5) * (76^{\circ} F - 58^{\circ} F) * (21 * 80 hrs)]}{12,000 \, Btu / ton.hr)} * 0.90 kW \, / \, ton$$

Or

$$\Delta kWh_C = 24,494 \, kWh \, / \, yr$$

The electricity cost savings amounts to $1,225/yr. Therefore, savings of $6,279 are achieved for both heating and cooling energy costs as a result of the CV to VAV conversion. Additional savings can be obtained from the reduction in fan power as illustrated in Example 7.3.

7.8   Summary

In commercial buildings, significant energy savings can be obtained by improving the energy efficiency of secondary HVAC systems. These improvements can be achieved with simple operating and maintenance measures with no or little investment. Better operation and control of HVAC systems provide not only energy and cost savings but also improved thermal comfort. Finally and in most cases, conversion of constant volume systems to variable air volume systems is cost effective and should be considered for existing commercial and institutional buildings.

Problems

7.1 It is proposed to retrofit the supply fan of a 60,000 cfm Air Handling Unit (AHU). The fan has a 25-HP motor with a total efficiency of 85.5%. If the fan is operating 5,500 hrs per year and if the cost of electricity is $0.08/kWh, determine the simple payback period when the constant volume fan is retrofitted to:

(a) A temperature measurement indicated that when the outdoor air is at 50°F, the mixed air is at 70°F and the return air is at 75°F. Determine the heating fuel savings by reducing the outdoor air to the minimum ventilation requirements (i.e, 20 cfm/person). Outdoor air is provided only during occupied period.

(b) Determine the heating energy and cost savings due to a temperate setback to 60°F during unoccupied period.

(c) If the system is converted into a VAV system, determine the energy and cost savings due to the fan. The average load on the fan is 60% of the peak.

Note: For Denver, the average outdoor DB winter temperature is 35.2°F and the length of the heating season is 29.4 weeks.

7.2 Consider a 200,000 sqft enclosed parking garage that can hold 1,000 cars.

(a) Determine the size of the supply/exhaust fans to ventilate this garage if 1.5 cfm/sqft is required. Assume a fan pressure of 1.5 in of water and a fan efficiency of 0.75.

(b) Using the design method, determine the needed cfm/sqft to ventilate the garage if at peak hour, 30% of the cars are moving (assume the worse case scenario for emissions – winter and cold start.). Assume that the acceptable CO level to maintain is 25 ppm. Size the supply/exhaust fans. Discuss the energy saving if the garage is operated 12 hours/day, 7days/week, and 50 weeks/year. Is it worth it to retrofit the ventilation system if the energy price is $0.08.kWh? (The cost of fans and their drives can be estimated as the cost of motors multiplied by 2.5).

(c) Discuss any other means to reduce the energy cost in ventilating the garage.

7.3 Consider a 80,000 sqft two story building. A 60,000 cfm air handling unit services the building. The building is occupied 50 hrs per week. The fuel cost is $5.00 per 10^6 Btu's and electricity cost is $0.09 per kWh. Determine the energy and cost savings for the following ECOs for three locations: Denver, CO; Chicage, IL; Miami, FL.

(a) Reset hot Deck Temperature: 3°F in summer and 4°F in winter. Assume 50% of airflow is in hot deck.

(b) Reset Cold Deck Temperature: 2.5 Btu/lb if enthalpy. Assume 50% of airflow is in cold deck.

(c) Reduced Minimum Outdoor Air: from 25% to the minimum required. Assume 20 cfm/person and 300 people in the building. Average indoor temperature is 70°F.

(d) Low Leakage Dampers: reduce damper leakage from 10% to 1%.

(e) Install a dry-bulb temperature economizer or an enthalpy economizer.

7.4 An 80,000 sqft office building located in Chicago has a dual duct air conditioning system with a 60,000 cfm air handling capacity (60HP). Determine the annual energy and cost savings that could be achieved with a VAV conversion. The following data apply:

	Winter	Summer
Mixed air temp.	65°F	80°F
Diffuser discharge temp.	60°F	67°F
Room temp.	70°F	74°F
Cold deck temp.	57°F	61°F
Hot deck temp.	110°F	95°F
Operating hours	72 hrs/wk	72 hrs/wk
Length of season	30wks	22wks
Cost of electricity	$0.06/kWh	$0.09/kWh
Cost of fuel	$5/MMBtu	$5/MMBtu

7.5 Consider a building in Denver with a 70,000 cfm single duct CV air handling unit. The building is occupied 60 hours per week by 500 people. The indoor temperature is always maintained at 73°F. Fuel cost is $4.50 per 10^6 Btu. Electricity cost is $0.09 per kWh. The fan pressure differential inside the AHU is 4.0 in. The fan has efficiency of 0.78 and is operating 80 hours per week. Determine the cost savings incurred by converting this CV to a VAV system.

8

CENTRAL HEATING SYSTEMS

Abstract

This chapter provides some recommendations to upgrade and improve the energy efficiency of central heating systems in commercial buildings and industrial facilities. First, a basic overview of combustion principles is described to pinpoint the basic parameters that affect the energy efficiency of a heating system. Then, various energy conservation measures for central heating plant are suggested.

8.1   Introduction

According to a survey reported by the US Energy Information Administration (EIA, 1997), four types of heating systems are used extensively in commercial buildings including:

- Boilers
- Packaged heating units
- Individual space heaters
- Furnaces

As Indicated in Table 8.1, boilers provide heating to almost 33% of the total heated floor-space of US commercial buildings. However, boilers are used in only 15% of heated commercial buildings, significantly less than furnaces which are used in more than 42% of US commercial buildings. This difference in usage stems from the fact that boilers are more commonly used in larger buildings while furnaces are the heating system of choice for smaller buildings. The same EIA survey (EIA, 1997) indicated that small buildings [less than 500 m^2 (5,000 ft^2)] constitute more than half (52%) of the total US commercial building stock.

Table 8.1: Heating equipment used for main and other use in US commercial buildings in 1995 (*Source:* EIA, 1997)

Heating System Type	Percentage of Heated Floor-space	Percentage of Heated Buildings
Boilers	33 %	15 %
Package heating units	31 %	26 %
Individual space heaters	30 %	29 %
Furnaces	26 %	42 %
District heating	13 %	3 %
Heat pumps	11 %	10 %
Other	12 %	4 %

Of the existing boilers used in US commercial buildings, 65 percent are gas-fired, 28 percent are oil-fired, and only 7 percent are electric. The average combustion efficiency of the existing boilers is in the range of 65 to 75 percent. New energy-efficient gas or oil fired boilers can be in the range of 85 to 95 percent.

8.2   Basic Combustion Principles

8.2.1   Fuel Types

Fuels used in boilers consist of hydrocarbons including alkynes (C_nH_{2n-2}) such as the acetylene (n=2), alkenes (C_nH_{2n}) such as the ethylene (n=2), alkanes (C_nH_{2n+2}) such as the octane (n=8). A typical combustion reaction involves an atom of carbon with two atoms of oxygen with a generation of heat according to the following generic combustion reaction:

$$C + O_2 \rightarrow CO_2 + Heat \qquad (8.1)$$

The heat generated in the combustion reaction is referred to as the heating value, HV, of a fuel. Typically, the heating value is given when the fuel is dry. The moisture actually reduces the heating value of fuels according to the following simplified equation:

$$HV = HV_{dry}.(1 - M) \qquad (8.2)$$

Where, M is the moisture content of the fuel.

In addition, the heating value of fuels decreases with the altitude. As a rule of thumb, the heating value reduces by 4% for every 300 m (1,000 ft) increase in the altitude.

Two analyses are typically used to determine the basic components of a fuel. The first analysis is called proximate analysis and determines the fuel content in percentage by weight of moisture, the volatile matter, fixed carbon, ash, and sulfur. The second analysis is referred to as the ultimate analysis and determines the fuel content in percentage by weight of carbon, hydrogen, nitrogen, and oxygen. It should be noted that the heating value of a fuel increases with its carbon content. Tables 8.2 and 8.3 provide the results of respectively proximate and ultimate analyses for coal extracted from two sites in the US.

Table 8.2: Results of proximate analysis of coal extracted from two sites in the US

Coal Type	Moisture	Volatile Matter	Fixed Carbon	Ash	Sulfur
Lackawana, PA	2.0	6.3	79.7	12	0.6
Weld, CO	24.0	30.2	40.8	5	0.3

Table 8.3: Results of ultimate analysis of coal extracted from two sites in the US

Coal Type	Carbon	Hydrogen	Oxygen	Nitrogran	Heating Value
Lackawana, PA	93.5	2.6	2.3	0.9	13000
Weld, CO	75.0	5.1	17.9	1.5	9200

Table 8.4 illustrates the results of the ultimate analysis of another solid fuel: wood. It is clear that pine has higher carbon content (by weight) and thus higher heating value.

Table 8.4: Results of ultimate analysis of selected wood types

Wood Type	Carbon	Hydrogen	Oxygen	Nitrogran	Heating Value
Oak	49.5	6.6	43.7	0.2	7980
Pine	59.0	7.2	32.7	1.1	10400
Ash	49.7	6.9	43.0	0.3	8200

Liquid or distillate fuels are generally graded in different categories depending on its properties. For fuel oils, there are six different grades depending on the viscosity level. Table 8.5 provides the heating values and common usage of five oil fuels commonly sold in the US. Fuel oil No. 3 has now been incorporated as part of fuel oil No. 2.

Table 8.5: Heating value and specific gravity of oil fuels used in the US.

Oil Grade	Specific Gravity	Heating Value KWh/L (MBtu/gal)	Applications
No.1	0.805	9.7 (134)	For vaporizing pot-type burners
No. 2	0.850	10.4 (139)	For general purpose domestic heating
No. 3	0.903	10.9 (145)	For burners without preheating
No. 5	0.933	11.1 (148)	Requires preheating to 75-95 °C
No. 6	0.965	11.3 (151)	Requires preheating to 95-115 °C

Similar grades are used for diesel fuels with diesel No. 1 used for high speed engines and diesel No. 2 used for industrial applications and heavy cars.

The liquid petroleum gas (LPG) is a mixture of propane and butane while natural gas is a mixture of methane and ethane.

8.2.2   Boiler Configurations and Components

Typically, boilers have several parts including an insulated jacket, a burner, a mechanical draft system, tubes and chambers for combustion gas, tubes and chambers for water or steam circulation, and controls.

There are several factors that influence the design of boilers including fuel characteristics, firing method, steam pressure, and heating capacity. However, commercial and industrial boilers can be divided into two basic groups: fire-tube or water-tube depending on the relative location of the hot combustion gases and the fluid being heated within the boiler. In the following sections, brief description of common boiler and burner types is provided.

Boiler Types

Most commercial boilers are manufactured of steel. Some smaller-size boilers are made up of cast iron. The steel boilers transfer combustion heat to the fluid using an assembly of tubes which can be either watertubes or firetubes.

Fire-tube Boilers:

In these boilers, the hot combustion products flow through tubes submerged in the boiler water as depicted in Figure 8.1. To increase the contact surface area between the hot gases and the water, 2 to 4 passes are used for the tubes. The multi-pass tubes increase the efficiency of the boiler but require greater fan power. Due to economics, the highest capacity of fire-tube boilers is currently in the 10,000 kg of steam per hour with an operating pressure of 16 atm (250 psi).

The fire-tube boilers are generally simple to install and maintain. Moreover, they have the ability to meet sudden and wide load fluctuations with only small pressure changes.

Water-tube Boilers:

In these boilers, the water flows inside tubes surrounded by flue combustion gases as shown in Figure 8.2. The water flow is generally maintained by the density variation between cold feed water and the hot water/steam mixture in the riser. The water-tube boilers are classified in several groups depending on the shape, and drum location, capacity, and number. The size of water-tube boilers can be as small as 400 kg of steam per hour and as large as 1000 MW unit. The largest industrial boilers are generally about 250,000 kg of steam per hour.

Cast Iron Boilers:

These boilers are used in small installations (below 1 MW) where long service life is important. These boilers are made up of precast sections and thus are more readily field assembled than steel boilers. At similar capacities, the cast-iron boilers are typically more expensive than fire-tube or water-tube boilers. Figure 8.3 illustrates selected types of cast iron boilers.

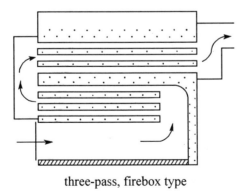

three-pass, firebox type

Figure 8.1: Typical configuration for fire-tube boilers (*Source*: ASHRAE, Systems and Equipment Handbook, 2000. With permission.)

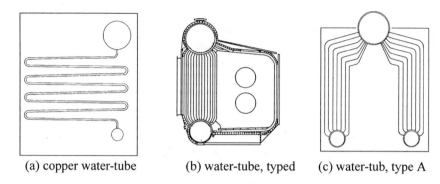

(a) copper water-tube   (b) water-tube, typed   (c) water-tub, type A

Figure 8.2: Common types of water-tube boilers (*Source*: ASHRAE,
Systems and Equipment Handbook, 2000. With permission.)

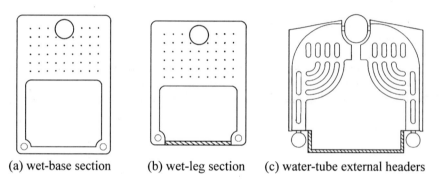

(a) wet-base section   (b) wet-leg section   (c) water-tube external headers

Figure 8.3: Selected Types of cast iron boilers (*Source*: ASHRAE, Systems
and Equipment Handbook, 2000. With permission.)

Firing Systems

The firing system of a boiler depends on the fuel used. The characteristics
of the firing system for each fuel type are summarized below:

Gas Fired Units:
Natural gas is the simplest fuel to burn since it mixes easily with
combustion air supply. Gas is generally introduced at the burner through several
orifices that provide jets of natural gas that mix rapidly with the combustion air
supply. There is a wide range of burners designs depending on the orientation,
the number, and the location of the orifices.

As part of routine tune-up and maintenance of gas fired boilers, it is important to inspect gas injection orifices to check that all the passages are unobstructed. In addition, it is important to identify and replace any burned off or missing burner parts.

Oil Fired Units:
Oil fuels need some form of preparation and treatment before final delivery to the burner. The preparation of the oil fuels may include the following:

- Use of strainers and filters to clean the oil fuel and remove any deposit or solid foreign material.

- Addition of flow line preheaters to deliver the fuel oil with a proper viscosity.

- Use of atomizers to deliver the fuel oil in small droplets before mixing with the combustion air supply. The atomization of the fuel oil can be carried out by a gun fitted with a tip that has several orifices which can produce fine spray. Moreover, oil cups that spin the oil into a fine mist are also used on small boiler units.

During tune-up of central heating systems, it is important to check that the burner is adequate for the boiler unit. In particular, it is important to verify that the atomizer has the proper design, size, and location. In addition, the oil-tip orifices should be cleaned and inspected for any damages to ensure proper oil-spray pattern.

Coal Fired Units:
In some central heating systems, coal can be used as the primary fuel to fire the burner. There are two main coal firing systems:

i.   Pulverized coal fired systems that pulverize, dry, classify, and transport the coal to the burner's incoming air supply. The pulverized coal fired systems are generally considered to be economical for units with large capacities (more than 100,000 kg of steam per hour).

ii. Coal stoker units that have a bed combustion on the boiler grate through which the combustion air is supplied. There are currently several stoker firing methods used in industrial applications such underfed, overfed, and spreader. Both underfed and overfed firing methods require that the coal be transported directly to the bed combustion and usually respond slowly to sudden load variations. The spreader stokers partially burn the coal in suspension before transporting it to the grate. The spreader stokers can burn a wide range of fuels including waste products.

The efficiency of coal firing systems depends on the firing system, the type of boiler or furnace, and the ash characteristics of the coal. Some units are equipped with ash re-injection systems that allow collected ash that contains some unburned carbon to be redelivered into the burner.

8.2.3   Boiler Thermal Efficiency

As illustrated by Eq. (8.1), fuel combustion involves a chemical reaction of carbon and oxygen atoms to produce heat. The oxygen comes from the air supplied to the burner that fires the boiler. A specific amount of air is needed to ideally complete the combustion of the fuel. This amount of air is typically referred to as the stoichiometric air. However, in actual combustion reactions, more air than the ideal (or stoichiometric) amount is needed to totally complete the combustion of fuel. The main challenge to ensure optimal operating conditions for boilers is to provide the proper excess air for the fuel combustion. It is generally agreed that 10% excess air provides the optimum air to fuel ratio for complete combustion. Too much excess air causes higher stack losses and requires more fuel to heat ambient air to stack temperatures. On the other hand, if insufficient air is supplied, incomplete combustion occurs and the flame temperature is reduced.

The general definition of overall boiler thermal efficiency is the ratio of the heat ouput, E_{out}, over the heat input, E_{in}:

$$\eta_b = \frac{E_{out}}{E_{in}} \tag{8.3}$$

The overall efficiency accounts for combustion efficiency, the stack heat loss, and the heat losses from the outside surfaces of the boiler. The combustion efficiency refers to the effectiveness of the burner in providing the optimum fuel/air ratio for complete fuel combustion.

To determine the overall boiler thermal efficiency, some measurements are required. The most common test used for boilers is the flue gas analysis using an Orsat apparatus to determine the percentage by volume the amount of CO_2, CO, O_2 and N_2 in the combustion gas leaving the stack. Based on the flue gas composition and temperature, some adjustments can be made to tune-up the boiler and to determine the best air-to-fuel ratio in order to improve the boiler efficiency. The following general rules of thumb can be used to adjust the operation of the boiler:

- *Stack temperature*: The lower the stack temperature, the more efficient is the combustion. High flue gas temperatures indicate that there is no good heat transfer between the hot combustion gas and the water. The tubes and the chambers within the boiler should be cleaned to remove any soot, deposit, and fouling that may reduce the heat transfer. However, the stack temperature should not be too low, to avoid water condensation along the stack. The water from the condensation mixes with sulfur and can cause corrosion of the stack. Table 8.6 provides the minimum stack exit temperature for common fuel types to avoid corrosion problems.

- CO_2 level: The higher the CO_2 level, the more efficient is the combustion. The low limits acceptable for CO_2 level is 10% for gas fired boilers and 14% for oil fired boilers. If the CO_2 levels are lower than these limits, the combustion is most likely incomplete. The air-to-fuel ratio should be adjusted to provide more excess air.

- CO level: No CO should be present in the flue gas. Indeed, any presence of CO indicates that the combustion reaction is incomplete and thus that there is not enough excess air. The presence of CO in the flue gas can be detected by the presence of smoke which leads to soot deposit in the boiler tubes and chambers.

- O_2 level: The lower the O_2 level, the more efficient is the combustion. Indeed, high level of O_2 is an indication of too much excess air. The high limit acceptable for O_2 level is 10%. When O_2 levels greater than 10% are found, the excess air should be reduced.

Table 8.6: Minimum exit flue gas temperatures to avoid stack corrosion

Fuel type used by the Boiler	Temperature Limit (°C)
Fuel Oil	200
Bituminous Coal	150
Natural gas	105

When the excess air is not adequate, the following boiler adjustment procedure can be used:

1. Operate the boiler for a specific firing rate and put the combustion controls on manual.

2. After stable operation, take a complete set of measurements (decomposition and temperature of the stack flue gas).

3. Increase the excess air by 1 to 2% and take a new set of measurements (after reaching stable boiler operating conditions).

4. Decrease the excess air by small steps until a minimum excess O_2 condition is reached (i.e., when the combustion becomes incomplete and a noticeable

CO level - above 400 ppm - can be detected in the flue gas). Take measurements following each change (allow the boiler to reach stable operating conditions).

5. Plot the measured data to determine the variation of CO level as a function of the percent of O_2 in the flue gas. A margin in excess O_2 above the minimum value can be established. Typically, a margin ranging from 0.5 to 2% O_2 above the minimum value is used.

6. Reset the burner controls to maintain the excess O_2 within the margin established in step 5.

7. Repeat steps 1 through 6 for various firing rates to be considered in the operation of the boiler. It is recommended that the tests be performed from higher to lower firing rates.

The new operating controls should be closely monitored for a sufficient length of time (one to two months) to ensure proper operation of the boiler.

Monographs are available to determine the overall boiler efficiency based on measurement of flue gas composition and temperature. One of these monographs applies to both gas fired and oil fired boilers and is reproduced in Figure 8.4 and Example 8.1 illustrate how the monograph can be used to determine the boiler efficiency.

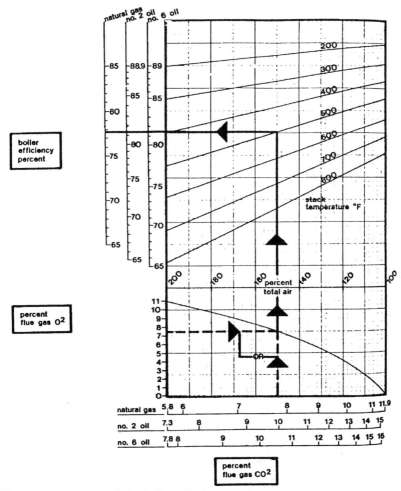

Figure 8.4: Monograph for boiler efficiency estimation (*Source*: Thumann and Mehta, 1997. With Permission)

Example 8.1: *A flue gas analysis of a oil fuel fired boiler indicates that the CO_2 content is 11% with a gas flue temperature of 343 °C (650 °F). Determine the overall thermal efficiency of the boiler. Use oil No. 2.*

Solution: By reading the monograph of Fig. 8.4, the combustion occurs with an excess air of 38% and excess O_2 of 6%. The overall boiler thermal efficiency is about 78%.

240   *ENERGY AUDIT OF BUILDING SYSTEMS*

8.3   Boiler Efficiency Improvements

There are several measures by which the boiler efficiency of an existing heating plant can be improved. Among these measures:

♦   Tune-up the existing boiler.
♦   Replace the existing boiler with a high efficiency boiler.
♦   Use modular boilers.

The net effect of all these measures is some savings in the fuel use by the heating plant. To calculate the savings in fuel use, ΔFU related to the change in the boiler efficiency, the following equation can be used:

$$\Delta FU = \frac{\eta_{eff} - \eta_{std}}{\eta_{eff}} . FU_{std}$$ (8.3)

Where,

▪ η_{std}, η_{eff} are respectively the old and the new efficiency of the boiler.

▪ FU_{std} is the fuel consumption before any retrofit of the boiler system.

It is therefore important to obtain both the old and the new overall thermal efficiency of the boiler to estimate the energy savings. The following sections provide more detailed description of the various boiler improvement measures.

8.3.1   Existing Boiler Tune-up

By analyzing the flue gas composition and temperature, the boiler thermal efficiency can be estimated using the monograph provided in Fig. 8.4. If it was found that the efficiency is low due to inappropriate excess air, the boiler can be adjusted and its efficiency improved as described in the step-by-step procedure described in the previous section. To perform this adjustment, some instrumentation is needed (gas flue analyzer and a temperature measurement device). Example 8.2 illustrates how the cost-effectiveness of a boiler tune-up can be evaluated.

Example 8.2: *The boiler of Example 8.1 users 1,500,000 L of fuel oil per year. An instrumentation is purchased at a cost of $20,000 and is used to adjust the boiler operation so its excess of O_2 is only 3%. Determine the payback of the instrumentation if the cost of fuel oil is $0.20/L.*

Solution: From Example 8.1, the existing boiler has an excess O_2 of 6% and an overall boiler thermal efficiency of about 78%(i.e, $\eta_{std} = 0.78$).

After the boiler tune-up, the excess O_2 is 3%. Using the monograph of Fig. 8.1, the new boiler efficiency can be determined from the flue gas temperature (*343 °C or 650 °F*) and the excess O_2 (3%). It is found to be $\eta_{eff} = 84\% = 0.84$. Using Eq. (8.4), the fuel savings can be calculated:

$$\Delta FU = \frac{0.84 - 0.78}{0.84}.1,500,000 = 107,140 \quad L/yr$$

Therefore, the simple payback period for the instrumentation is:

$$SBP = \frac{\$20,000}{10,714\,L/yr * \$0.20/L}. \approx 1.0 \quad year$$

Other measures that can be considered to increase the overall efficiency of the boiler are summarized below:

➤ Install turbulators in the fire-tubes to create more turbulence and thus increase the heat transfer between the hot combustion gas and the water. The improvement in boiler efficiency can be determined by measuring the stack flue gas temperature. The stack gas temperature should decrease when the turbulators are installed. As a rule of thumb, a 2.5% increase in the boiler efficiency is expected for each 50 °C decrease in the stack flue gas temperature.

➤ Insulate the jacket of the boiler to reduce heat losses. The improvement of the boiler efficiency depends on the surface temperature.

➤ Install soot-blowers to remove boiler tube deposits that reduce heat transfer between the hot combustion gas and the water. The improvement in the boiler efficiency depends on the flue gas temperature.

➤ Use economizers to transfer energy from stack flue gases to incoming feedwater. The stack temperature should not be lowered below the limits provided in Table 8.6 to avoid corrosion problems. As a rule of thumb, a 1% increase in the boiler efficiency is expected for each 5 °C increase in the feedwater temperature.

> ➢ Use of air preheaters to transfer energy from stack flue gases to combustion air. Again, the stack temperature should not be lower than the values provided in Table 8.6.

The stack flue gas heat recovery equipment (i.e., air preheaters and economizers) is typically the most cost-effective auxiliary equipment that can be added to improve the overall thermal efficiency of the boiler system.

8.3.2   High Efficiency Boilers

Manufacturers continue to improve both the combustion and the overall efficiency of boilers. Currently, commercial sized units can achieve over 95% combustion efficiency. For conventional boilers, anything over 85% is traditionally considered efficient. One of the most innovative combustion technologies currently available in the market is the gas-fired pulse-combustion boilers. This technology is first introduced in the early 1980s for residential water heaters, and is now available in several commercial-size boilers for both space heating and hot water heating.

Pulse-combustion boilers operate essentially like automotive internal combustion engines. First, air and gas are introduced in a sealed combustion chamber in carefully measured amounts. This gas/air mixture is then ignited by a spark plug. Almost all the heat from the combustion is used to heat the water in the boiler. Indeed, the exhaust gases have only a relatively low temperature of about 50 °C. Once the combustion chamber is fully heated, successive air/fuel mixtures or "pulses" ignite spontaneously (without the need for an electrical spark). Thus, no fuel-consuming burner or standing pilot light is required. When pulse boilers operate, they extract latent heat from the products of combustion by condensing the flue gas. Therefore, the boiler efficiency is increased and the flue gas is left with low water vapor content. The corrosion problems at the stack are then avoided.

The combustion efficiency of the pulse-combustion boilers can reach 95% to 99%. When combined with other high-performance elements for heat transfer, the overall thermal efficiency of the combustion-pulse boilers can attain 90%. In addition to savings energy, the pulse-combustion boilers can reach operating temperatures in as little as one-half the time of conventional boilers. Moreover, pulse-combustion burners produce lower emissions than conventional gas burners.

8.3.3   Modular Boilers

Almost all the heating systems are most efficient when they operate at full capacity. Improvements in peak-load efficiency result in lower energy use. However, the reduction in the fuel use is not necessarily proportional to the improvement in the heating system efficiency. Indeed, peak loads occur rarely in most heating installations. Therefore, the boiler is most often operating under part-load conditions. Some boilers may be forced to operate in an on/off cycling mode. This on/off cycling is an inefficient mode of operating the boiler. Indeed, the boiler loses heat through the flue and to the ambient space when it cycles off. Moreover, the water in the distribution pipes cools down. All these losses have to be made up when the boiler restarts. If the boiler capacity is much higher than the load, the cycling can be frequent and the losses can increase and thus, significantly reducing the seasonal efficiency of the heating system.

Instead of operating the boiler in an on/off mode when the load is lower than its capacity, controls using step-firing rates (high/low/off) or modulating firing rates (from 100% to 15%) can be specified. Another effective measure to avoid cycling the boilers is to install a group of smaller boilers or modular boilers. In a modular heating plant, one boiler is first operated to meet small heating loads. Then, and as the heating load increases, new boilers are fired and enter on-line to increase gradually the capacity of the heating system. Similarly, as the heating load decreases, the boilers are taken off-line one by one.

Some manufacturers offer pre-assembled modular boiler packages of various sizes, ranging from approximately 50 kW to 1 MW. However, individual units can be piped and wired together in the field to form an efficient modular heating plant system. In addition to energy savings, modular boilers allow more flexibility in the use of space since they can be transported through doors that cannot accommodate a large boiler. Thus, modular boiler can be located in confined spaces.

Modular boiler plants are suitable for applications with widely varying heating, steam, and hot water loads, such as hotels, schools, or high-rise buildings. The modular boilers can increase the overall seasonal efficiency of the heating system by 15 to 30%. For instance, a 5000 m^2 shopping mall in Iowa with more than 16 stores and food service area was retrofitted with 12 modular boilers (each with 40 kW capacity). According to the system manufacturer, the heating cost savings is about 33% relative to a conventional gas fueled boiler (Tuluca, 1997).

8.4   Summary

Basic types of central heating systems are discussed in this chapter. In addition, cost-effective measures to improve the energy efficiency of boilers are presented. Simplified energy analysis methods are illustrated using examples of calculation to estimate energy savings incurred from higher efficiency central heating plants. As a rule of thumb, the auditor should consider simple operating and maintenance measures to increase the energy efficiency of existing heating systems before recommending new more efficient boilers.

Problems

8.1: A recent analysis of your gas-fired boiler showed that you have 30% excess combustion air. Discussion with the local gas company has revealed that you could use 10% excess air if the burner's controls were better adjusted. This represents calculated efficiency improvements of 8%.

   (a) What would be the change in the flue gas temperature – due to better controls – if the current temperature is measured to be 600 °F?
   (b) How large an annual gas bill is needed before adding a maintenance person for the boiler alone is justified if this person would cost $35,000/yr?

8.2: A crude but effective approach to estimate steam leak (in lb/hr) from an orifice is to use the following expression:

$$lb/hr = C_d.k.\Delta t.A.P_i^{0.97}$$

where:
- C_d is the coefficient of discharge (for perfectly round orifice $C_d=1$. In most cases, $C_d=0.7$).
- k is a constant (k=0.0165).
- Δt is the numb of seconds in one hour (i.e., 3600).
- A is the area of the orifice (in in^2).
- P_i is the pressure inside the steam pipe (in psia).

Estimate the hourly, monthly, and annual costs of steam leaks from 300-psig pipe. The steam is generated from a gas-boiler with an efficiency of 80%. The cost of natural gas is $0.90/therm (1 therm is equivalent to 100,000 Btu/hr). The heating season spans 7 months per year. Provide the costs for the following orifice diameters: 1/16; 1/8; 1/4; 3/8; 1/2; and 1 in.

8.3 An efficiency test of a boiler fired by fuel No.2 indicated a flue gas temperature of 700°F and excess air of 40%. The annual fuel consumption of the boiler is 85,000 gal/year. The cost of fuel No.2 is $1.15/gal.
   (a) Estimate the efficiency of the boiler as well as the percent in CO_2 and O_2 is the flue gas.
   (b) Determine the new boiler efficiency and the annual cost savings in fuel use if the percent O_2 in the flue gas is reduced to 3% and the stack temperature is set to 500°F.
   (c) Determine the payback period of installing an automatic control system – at a cost of $8,500 – to maintain the same efficiency found in (b) throughout the life of the boiler.

8.4 A combustion efficiency test performed on gas-fired boiler indicated that the stack temperature is 700°F and the CO_2 content is 7%.

(a) Estimate the excess air and the boiler efficiency.
(b) The boiler efficiency is increased to 84% by adjusting the fuel-air ratio so that the excess air is limited to 20%. Determine the new stack temperature and the CO2 content.
(c) Estimate the annual natural gas consumption (in therms) if a combustion gas analyzer kit – installed at a cost of $8,500 – is paid for after 3 years. The kit helps maintains the boiler efficiency at 84%. The cost of natural gas is $1.10/therm.

9

COOLING EQUIPMENT

Abstract

This chapter outlines some measures to improve the energy efficiency of the cooling systems in commercial buildings. First, a basic overview of refrigeration principles is described to explain the basic parameters that affect the energy efficiency of a cooling system. Then, a brief description of commonly installed cooling systems for space air conditioning in the US is presented. In particular, the evolution of energy efficiency of unitary equipment and chillers is discussed with some analysis of their market penetration. Finally, some energy conservation measures for central cooling systems are suggested.

9.1   Introduction

According to a survey reported by the US Energy Information Administration (EIA, 1997), several types of cooling systems are used in commercial buildings including:

- Packaged Air Conditioner Units
- Central Chillers
- Individual Air Conditioners
- Heat Pumps
- Residential-Type Central Air Conditioners
- District Chilled Water
- Swamp (or Evaporative) Coolers

Table 9.1 summarizes the results of an EIA study to assess the percentage of respectively floor space and number of commercial buildings that are conditioned by each type of cooling equipment. In particular, the packaged air conditioner (AC) units are the main equipment used to condition buildings in the US, both in terms of percentage of total cooled buildings (42%) and percentage of the total cooled floor-space (55 %). On the other hand, central chillers are used in only 4% of conditioned commercial buildings but cool 27% of the floor area. Indeed, central chillers are typically used in larger buildings while packaged AC units are installed in smaller buildings. Specifically, the same EIA

247

study indicates that for buildings over 20,000 m^2 (or 200,000 ft^2) of floor area, central chillers are used to cool almost 78% of the floor area. Packaged AC units condition 58% of the total floor area in the commercial sector in the US.

Table 9.1: Cooling equipment used for main space conditioning and other uses in US commercial buildings in 1995 (*Source*: EIA, 1997)

Cooling System Type	Percentage of Cooled Floor-space	Percentage of Cooled Buildings
Packaged AC Units	55 %	42 %
Central Chillers	27 %	4 %
Individual Air Conditioners	19 %	26 %
Residential-Type AC Units	18 %	27 %
Heat Pumps	16 %	12 %
District Chiller Water	7 %	2 %
Swamp Coolers	5 %	6 %
Other	2 %	1 %

9.2   Basic Cooling Principles

A typical cooling system consists of several components including a compressor, a condenser, an expansion device, an evaporator, and other auxiliary equipment. Figure 9.1 illustrates a simple chilling system where the compressor is driven by a motor.

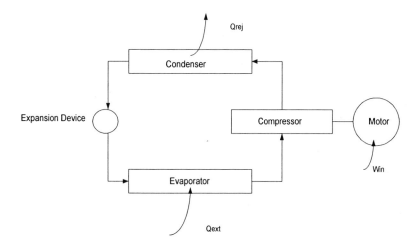

Figure 9.1: Typical Cooling System driven by an electrical motor

Note that chilling (or heat extraction) occurs at the evaporator while heat rejection is done by the condenser. Both the evaporator and the condenser are heat exchangers. At the evaporator, heat is extracted by the refrigerant from water that is circulated through cooling coils of an air handling unit. At the condenser, heat is extracted from the refrigerant and rejected to the ambient air (for air cooled condensers) or water (for water cooled condensers connected to cooling towers).

Generally, the energy efficiency of a cooling system is characterized by its Coefficient of Performance (COP). The COP is defined as the ratio of the heat extracted divided by the energy input required. In case of an electrically driven cooling system as represented in Figure 9.1, the COP can be expressed as:

$$COP = \frac{Q_{ext}}{W_{in}} \qquad (9.1)$$

Both Q_{ext} and W_{in} should be expressed in the same unit (i.e., W or kW), so that the COP has no dimension.

The maximum theoretical value for COP can be estimated using the ideal Carnot cycle COP. The Carnot cycle consists of isentropic compression and expansion and isothermal evaporation and condensation. In real cooling systems, the energy efficiency of the Carnot cycle cannot be attained because of irreversible losses. Among these losses are the irreversible losses in the compression and expansion of the refrigerant, the pressure losses in the lines, and the heat losses in the rejection and absorption processes due to the non-uniform temperature through the heat exchangers. However, it is useful to compare the COP of an actual cooling system to that of the Carnot cycle operating between the same temperatures to determine the potential of any added energy efficiency improvements in the design of cooling units. The COP of an ideal Carnot cycle can be expressed in terms of the absolute temperature of the evaporator, T_C (the lowest temperature in the cycle), and the condenser, T_H (the highest temperature in the cycle) as follows:

$$COP_{Carnot} = \frac{T_C}{T_H - T_C}$$ (9.2)

For instance, using the ARI standard 550/590 (1998) rated conditions for water chillers, T_H is 308 K and T_C is 280 K, the COP of Carnot cycle can be estimated by Eq (9.2) to be 9.88. Currently, the most energy efficient centrifugal water chiller has a COP of about 7.0 or about 70% of the ideal Carnot cycle.

Most manufacturers provide typically the COP of their cooling systems for full load conditions. The capacity of cooling systems is expressed in kW and is defined in terms of the maximum amount of heat that can be extracted. In the US, manufacturers and HVAC engineers use refrigeration tons to rate the capacity of the cooling systems (1 ton is about 3.516 kW), and kW/ton to express their energy efficiency. In addition, the energy efficiency of the electrically powered cooling systems can be expressed in terms of the energy efficiency ratio (EER) which is defined as the ratio of the heat extracted (expressed in Btu/hr) over the energy input required (expressed in Watts). Therefore, the relationship between the EER and the COP is given as follows:

$$EER = 3.413 * COP$$ (9.3)

The definition of the EER provided above is specific to the US HVAC industry. In Europe, the EER is defined to be exactly the same as the COP. However, the adopted European standard EN 814 (Cenelec, 1997) specifies that the term COP is to be used for only heating mode operation of heat pumps. Otherwise, the standard requires the use of the term EER to rate the energy efficiency of air conditioners and heat pumps.

ARI standard 550/590 allows the rating the energy efficiency of water chilling systems using the vapor compression cycle by one of the three parameters: COP, EER, or kW/ton. These three parameters are related as indicated below:

$$kW/ton = \frac{3.516}{COP} = \frac{12}{EER}$$

The rating tests for chillers need typically to be performed under specific conditions for leaving chilled water temperatures and entering condenser water temperatures or air dry-bulb and wet-bulb temperatures.

Generally, the energy efficiency of a cooling system varies under part load conditions. Since cooling systems operate often under part-load conditions throughout the year, other energy efficiency coefficients have been proposed in an attempt to provide a better estimation of the energy performance of the cooling units over a wide range of operating conditions. Currently, two parameters are commonly used in the HVAC industry: the seasonal energy efficiency ratio (SEER) and the integrated part load value (IPLV). ARI 550/590 standard defines the IPLV based on the COP (or EER) values at 100%, 75%, 50%, and 25% using the following equation:

$$IPLV = 0.01 * A + 0.42 * B + 0.45 * C + 0.12 * D \qquad (9.4)$$

Where,
- A is the COP (or EER) at 100% load
- B is the COP (or EER) at 75% load
- C is the COP (or EER) at 50% load
- D is the COP (or EER) at 25% load

The COP and the EER at part loads are determined using specific conditions. Table 9.2 summarizes the conditions for both the entering water temperatures (EWT) for water-cooled condensers and the entering air dry-bulb temperatures (EDB) for air-cooled condensers. For all conditions and types, the evaporator leaving chilled water temperature is specified to be 6.7 °C.

Table 9.2: Part load conditions for ARI 550/590 standard rating

Cooling Load (% of Capacity)	Water-Cooled Condensers EWT (°C) [°F]	Air-Cooled Condensers EDB (°C) [°F]
100 %	29.4 [85]	35.0 [95]
75%	23.9 [75]	26.7 [80]
50%	18.3 [65]	18.3 [65]
25%	18.3 [65]	12.8 [55]
0%	18.3 [65]	12.8 [55]

The following section provides a brief description of commonly used cooling systems and their typical energy efficiencies. In later sections, some common improvement measures to increase the energy efficiency of cooling systems are discussed.

9.3   Types of Cooling Systems

As mentioned in the introduction, several types of cooling systems are currently available for space air conditioning. The most common cooling systems used for space air conditioning can be grouped into major categories: unitary AC systems and chillers. The unitary AC systems include packaged AC units, individual air conditioners, residential type AC units, and heat pumps.

9.3.1   Unitary AC Systems

Unitary AC systems are typically factory assembled units that provide either cooling only or both cooling and heating. Compared to chillers, the unitary AC systems have a lower life span and lower energy efficiency. They are typically installed in small commercial buildings (with less than three floors) including small office buildings, retail spaces, and classrooms.

9.3.2   Packaged AC Units

The packaged AC units are compact cooling systems encased in cabinets. There are various types of packaged AC units including:

- **Rooftop Systems** are typically located in the roof (thus, the name rooftop AC units). For commercial buildings, the rooftop AC units are available in the range of 17 to 70 kW (or 5 to 20 tons) even though custom-built units can have larger capacities (up to 350 kW or 100 tons). For residential buildings, capacities between 3 to 7 kW (or 0.75 to 2 tons) are common. Most units are equipped with a heating system (a built-in gas furnace, an electric resistance, or a heat pump) to provide both cooling and heating.

- **Vertical Packaged Systems** are typically designed for indoor installation. Most systems have water-cooled condensers.

- **Split Packaged Systems** have typically air cooled condenser and compressor installed outdoors and the evaporator installed in an indoor air-handling unit.

9.3.3   Heat Pumps

Heat pump can be used for both cooling and heating by simply reversing the refrigeration flow through the unit. The heat sink (or source) for the heat pump can be air, water, or ground. For commercial and industrial applications, air to air heat pumps can have capacities up to 90 kW (or 25 tons); hydronic heat pumps can have higher cooling capacities. Ground-coupled heat pumps are still small and are mostly suitable for residential applications.

9.3.4   Central Chillers

In large buildings, central chillers are used to cool water for space air conditioning. Central chillers are powered by electric motors, fossil fuel engines, or turbines. Some chillers use hot water to steam to generate chilled water. A description of various types of central chillers is provided below.

Electric Chillers

The are currently three major types of electric chillers available in the market using centrifugal, reciprocating, or rotary compressors. All these chillers use mechanical vapor compression cycle.

- **Centrifugal compressors** use rotating impellers to increase refrigerant gas pressure and temperature. Chillers with centrifugal compressors have capacities in the range of 300 kW to 25,000 kW (or 85 to 7,000 tons). For capacities above 4,500 kW (or 1,250 tons), the centrifugal compressors are typically field erected.

- **Reciprocating compressors** use pistons to raise the pressure and the temperature of refrigerant gases. Two or more compressors can be used under part-load conditions to achieve higher operating efficiencies. Capacities of 35 kW to 700 kW (or 10 to 200 tons) are typical for chillers with reciprocating compressors.

- **Rotary compressors** use revolving motions to increase refrigerant gas pressure. One of the most ingenious rotary compressor is the scroll compressor. The most conventional rotary compressors are the screw compressors that can have several configurations. The capacity of the rotary chillers can range from 3 kW to 1750 kW (1 to 500 tons).

Absorption Chillers

Absorption chillers operate using a concentration-dilution cycle to change the energy level of refrigerant (water) by using lithium bromide to alternately absorb heat at low temperatures and reject heat at high temperatures. The absorption chillers can be direct fired (using natural gas or oil fuel), or indirect fired. Indirect fired units may use as heat source steam or hot water (from a boiler, a district heating network, an industrial process, or a waste heat). A typical absorption chiller includes an evaporator, a concentrator, a condenser, and an absorber.

- **Direct-fired absorption chillers** can be cost-effective when the price of natural-gas is favorable. Some of the direct-fired chillers can be used to produce both chilled and hot water. Thus, they can provide cooling and heating and are sometimes referred to as chillers/heaters. These chillers/heaters can be cost-effective especially when heating needs exist during the cooling season (for instance, buildings with large service hot water requirements). Two types of absorption chillers are available in the market: single-effect and double-effect chillers. Some prototypes of triple-effect absorption chillers have been built in the US. Capacities ranging from 100 kW to 5,000 kW (or 30 to 1,500 tons) are available for direct-fired chillers.

- **Indirect-fired absorption chillers** operate with steam (with pressures as low as 15 psig) or hot water (with temperatures as low as 140 °C). Some small absorption chillers using solar energy (to generate hot water) have been proposed and some prototypes have been developed and evaluated. Cooling capacities from 15 kW to 425 kW (4 to 120 tons) are available even though typical sizes range from 200 kW to 5,000 kW (55 to 1,450 tons). Double-effect chillers can be considered only for high temperature hot water and steam or with hot industrial waste gases.

Engine-Driven Chillers

Like electrically driven chillers, the engine-driven chillers can use reciprocating, rotary, or centrifugal compressors to provide mechanical refrigeration. The compressors can be powered by turbines or gas-fired engines. The engine-driven chillers can have large capacities up to 15,000 kW (4,250 tons) but have usually high first costs.

9.4   Market Analysis and Energy Efficiency of Cooling Systems

The air-conditioning industry has made significant progress over the last twenty years to improve the energy efficiency of its products. For unitary products (including packaged AC units and heat pumps), the average efficiencies have increased about 50% in the last 20 years. For large chillers, the improvement in the energy efficiency has also been significant. For instance, the average COP of centrifugal chillers has increased by 34% over the past 20 years. More detailed analyses of the market share and the energy efficiency evolution for both the unitary and large chillers are presented in the following sections.

9.4.1 Unitary Products

Market Analysis

According to a study reported by the United Nations Environment Program (UNEP, 1999), the installed cooling capacity of unitary products worldwide is about 1,450 10^6 kW (or 410 10^6 tons). The number of units sold worldwide is estimated to be in the range of 30,000,000 units including 9,100,000 units from USA (30.3%); 8,000,000 units from Japan (26.7%); 5,000,000 units from China (16.7%); and 3,500,000 units from other Asian countries (11.7%). The US and Japan provide almost 57% of the unitary air conditioning market while Europe sells only 8%. In particular, the major manufacturers of unitary air conditioners include Carrier (11% of the worldwide market), Matsushito (8%), Hitachi (7%), Melco (7%), Toshiba (7%), Daikin (7%), Sanyo (5%), and Mitsubishi (4%).

The US manufacturers ship almost 6,000,000 units per year with sizes ranging from under 5 kW (or 1.5 tons) to over 150 kW (or 40 tons). Based on the ARI statistical releases (ARI, 1999), US shipments of unitary air conditioners and heat pumps (air-source only) was over 6,200,000 units in 1998 up 16% from the shipments registered in 1997. Among these shipments 1,250,000 (i.e., 20%) units are air-source heat pumps.

In the European Union (EU) countries, the estimated number of installed unitary cooling equipment is 7,400,000 units during 1996 (Orphelin, 1999) with 74% located in the Mediterranean countries (specifically, 17% in France, 10% in Greece, 29% in Italy, and 18% in Spain). However, the use of air-conditioning in Europe remains generally low compared to other developed countries. In particular, space air conditioning is provided only to 0.25% (in United Kingdom with the lowest use rate in EU) and to 4.8% (in Spain with the highest use rate in EU) of the residential buildings. These penetration rates are low compared to 55% for the US and 75% for Japan. In the commercial sector, space air conditioning is more common but its use remains low in the European countries. Indeed, air conditioning equipment is installed in 5% (Germany) and 20% (Greece and Portugal) of office buildings. Meanwhile, air conditioning is used in the majority of existing office buildings for both Japan and US. Specifically, the penetration rate of air conditioning in office buildings is 100% in Japan and over 80% in the US.

Energy Efficiency

Over the past 25 years, the HVAC industry has made significant improvements in the energy efficiency of unitary products. Table 9.3 provides the US shipment-weighted average efficiency of unitary air-to-air heat pumps from 1976 to 1995. It should be noted that prior to 1981, the efficiency rating criterion was the EER (measured only for full load conditions), while after 1981, the seasonal efficiency was used to account for part load conditions. Table 9.3 provides the evolution of the energy efficiency of air-source heat pumps using the US shipment weighted average SEER values [using the US definition with a unit of Btu/Wh] and EER values [based on the European definition with no unit].

Table 9.3: Evolution of US shipment-weighted average efficiencies of unitary heat pumps

Year of Shipment	US EER/SEER (Btu/W·hr)	European EER/SEER (W/W)
1976	6.9	2.0
1978	7.3	2.1
1980	7.5	2.2
1981*	7.7	2.3
1983	8.0	2.4
1985	8.5	2.5
1987	8.8	2.6
1989	9.1	2.7
1991	9.6	2.8
1993	10.6	3.1
1995	11.0	3.2

* From 1981, the SEER was used instead of the EER as a measure of efficiency.

The significant increase in the SEER of the heat pump occurred after 1991 due to the US government regulations that mandate that all new equipment must have a SEER of at least 10. The year-to-year steady increase is due to the technology improvements attributed mostly to motors and compressors. It should be noted that the energy efficiency values provided in Table 9.3 are based on a shipment-weighted average. Units with higher energy efficiency were available during the respective time periods even though they were usually more expensive than the standard units. Currently, air-to-air unitary equipment can have SEER values of 18.

A study based on a selected number of installed unitary air conditioning units in various European Union countries during 1997 (Orphelin, 1999) revealed that the energy efficiency of the units is not directly correlated to their capacity. In addition, the study indicated that the weighted-average EER of installed unitary units in the European Union is about 2.55 (which corresponds to EER=8.7 using the US definition). However, the spread in the energy efficiencies was found to be significant. Indeed, some units have very low energy efficiency with EER of 1.45 and other units have relatively high energy efficiency with EER of 5.4. Thus, it was found that there are several unitary AC models present in the European market that do not meet the standard of other developed countries. In particular, the study reported that only 35% of the European units have energy efficiencies higher that those mandated by Japanese standards.

9.4.2 Chillers

Market Analysis

For large chillers using mechanical vapor compression, the US manufacturers lead the world market in both production and shipment with companies like Carrier, Trane, York, and McQuay. In 1995, US companies produced 10,000 units while in 1996 about 9,200 units. On other hand, Japanese manufacturers prevail in the market for absorption chillers, with 6,600 units sold in 1996 – well ahead of their close competitors, China with 2,500 units and South Korea with 1,800. The interest in absorption cooling by the Asian countries stems from their energy policies that promote direct use of natural gas in favor of savings electrical energy consumption.

In Europe, a study by ICARMA (1997) indicated that of the 37,100 chiller installations, 80% have capacities less than 100 kW (or 30 tons). However, the remaining 20% units that are over 100 kW (or 30 tons) represent more than 75% of the market value. In addition to the US and Japanese manufacturers, several small Italian manufacturers are active in supplying the European market especially with small chilling units (e.g. below 50 kW). In France, the majority of the units sold are small and generally manufactured by either Carrier or Trane. Of the 5,500 chillers sold in 1996, 80% have a capacity that is less than 100 kW (or 30 tons) and 40 % less than 40 kW (12 tons). Only 50 large chiller units are sold in France in 1996 (mostly centrifugal chillers). The chillers with rotary compressors represent only 20% of the total French market. The chillers with reciprocating compressors prevail especially for smaller units. Moreover, the market penetration of absorption chillers remains very low in France. A recent survey by Gas de France (GDF, 1999) indicated that there are about 100 installations that use gas absorption cooling systems in 1998 within France's territories. Almost all of these installations have been implemented after 1995.

Energy Efficiency

In the US commercial air conditioning sector, centrifugal chillers account for 70% of the total installed cooling capacity which is estimated at $211 \ 10^6$ kW (i.e., $60 \ 10^6$ tons). In the last 20 years, the improvement in the energy efficiency has been significant for centrifugal chillers. Table 9.4 summarizes the average full load efficiencies for one US manufacturers (Menzer, 1997).

Table 9.4: Average design efficiencies of centrifugal chillers for one US manufacturer (Menzer, 1997)

Year of Production	Average COP
1976	4.24
1978	4.71
1980	5.15
1982	5.25
1984	5.37
1986	5.44
1988	5.50
1990	5.54
1992	5.59
1994	5.63
1996	5.67

From Table 9.4, it is clear that the progress of the energy efficiency in centrifugal chillers has been relatively rapid in the early years (from 1976 to 1980) but slower during the later years (especially after 1990). This reduction of energy efficiency improvement may indicate that the maximum achievable limit for COP has been almost reached. Indeed, one manufacturer estimated that for large centrifugal chillers, energy efficiency has the potential to be improved by only another 4% while for small chillers energy efficiency can be increased by as high as 7.5%. The potential improvements for the chiller design include subcooling, motor efficiency, motor cooling, spray evaporators, and reduction in piping losses.

Table 9.5 provides some indication of typical energy efficiency for commonly available chiller units.

Table 9.5: Typical COP for different types of chillers.

Type of Chillers	Range for COP
Small Electric Chillers	
Air Cooled	2.2 - 3.2
Large Electric Chillers	
Air Cooled	3.7 – 4.1
Water Cooled	4.6 – 5.3
Absorption Chillers	
Single-Effect	0.4 – 0.6
Double-Effect	0.8 – 1.1
Engine-Driven Chillers	1.2 – 2.0

9.5   Energy Conservation Measures

To reduce the energy use of cooling systems, the energy efficiency of the equipment has to be improved under both full load and part load conditions. In general, the improvement of the energy efficiency of cooling systems can be achieved by one of the following measures:

◆ Replace the existing cooling systems by others that are more energy efficient.
◆ Improve the existing operating controls of the cooling systems.
◆ Use alternative cooling systems.

The energy savings calculation from increased energy efficiency of cooling systems can be estimated using the simplified but general expression provided by Eq. (9.5):

$$\Delta E_C = \left(\frac{\dot{Q}_C . N_{h,C} . LF_C}{SEER} \right)_e - \left(\frac{\dot{Q}_C . N_{h,C} . LF_C}{SEER} \right)_r \qquad (9.5)$$

Where, the indices e and r indicate the values of the parameters respectively before and after retrofitting the cooling unit, and

➢ SEER is the seasonal efficiency ratio of the cooling unit. When available, the average seasonal COP can be used instead of the SEER.

➢ $\dot{Q}_C$ is the rated capacity of the cooling system

➢ $N_{h,c}$ is the number of equivalent cooling full-load cooling hours

➢ LF_C is the rated load factor and is defined as the ratio of the peak cooling load experienced by the building over the rated capacity of the cooling equipment. This load factor compensates for over-sizing of the cooling unit.

It should be noted that the units for both Q_C and SEER have to be consistent, that is, if SEER has no dimension (using the European definition of EER), Q_C has to be expressed in kW.

When the only effect of the retrofit is improved energy efficiency of the cooling system so that only the SEER is changed due to the retrofit, the calculation of the energy savings can be performed using the following equation:

$$\Delta E_C = \dot{Q}_C . N_{h,C} . LF_C . \left(\frac{1}{SEER_e} - \frac{1}{SEER_r} \right) \qquad (9.6)$$

In the following sections, some common energy efficiency measures applicable to cooling systems are described with some calculation examples to estimate energy use and cost savings.

9.5.1   Chiller Replacement

It can be cost-effective to replace an existing chiller with a new and more energy efficient chiller. In recent years, significant improvements in the overall efficiency of mechanical chillers have been achieved by the introduction of two-compressor reciprocating and centrifugal chillers, variable-speed centrifugal chillers, and scroll compressor chillers. A brief description of each of these chiller configurations is presented below with some estimation of their energy efficiency.

- Multiple compressor chillers can be reciprocating, screw, or centrifugal with capacities in the range of 100 kW to 7,000 kW (i.e., 30 tons to 2,000 tons). They are energy efficient to operate, especially under part load conditions. Some studies indicate that chillers equipped with multiple compressors can save up to 25% of the cooling energy use compared to single-compressor chillers (Tuluca et al., 1997).

- Variable-speed compressor chillers are in general centrifugal and operate with variable head pressure using variable speed motors. Therefore, the variable-speed compressor chillers work best when their cooling load is most of the time below the peak. The typical capacity of a variable-speed compressor chiller is in the range of 500 kW to 2,500 kW (i.e., 150 tons to 700 tons). It is reported that chillers with variable speed compressor can reduce the cooling energy use by almost 50% (Tuluca et al., 1997).

- The scroll compressor is a rotary compression device with two primary components, a fixed scroll and an orbiting scroll, both needed to compress and increase the pressure of the refrigerant. The scroll compressors are more energy efficient than the centrifugal compressor since the heat loss between the discharge and the suction gases is reduced. Manufacturers of scroll compressors report that the COP of the scroll chillers exceeds 3.2.

Example 9.1 illustrates a sample of calculation to determine the cost-effectiveness of replacing an existing chiller with a high energy efficiency chiller.

Example 9.1: *An existing chiller with a capacity of 800 kW and with an average seasonal COP of 3.5 is to be replaced by a new chiller with the same capacity but with an average seasonal COP of 4.5. Determine the simple payback period of the chiller replacement if the cost of electricity is $0.07/kWh and the cost differential of the new chiller is $15,000. Assume that the number of* *equivalent full-load hours for the chiller is 1,000 per year both before and after the replacement.*

Solution: In this example, the energy use savings can be calculated using Eq. 9.6 with $SEER_e = 3.5$; $SEER_r = 4.5$; $N_{h,C} = 1000$; and $Q_C = 800$ kW; $LF_C = 1.0$ (it is assumed that the chiller is sized correctly):

$$\Delta E_C = 800kW * 1000hrs / yr * 1.0 * \left(\frac{1}{3.5} - \frac{1}{4.5} \right) = 50,800kWh / yr$$

Therefore, the simple payback period for investing in a high efficiency chiller rather than a standard chiller can be estimated as follows:

$$SPB = \frac{\$15,000}{50,800kWh / yr * \$0.07 / kWh} = 4.2 years$$

A Life Cycle Cost analysis may be required to determine if the investment in a high energy efficiency chiller is really warranted.

In some cases, only some parts of the cooling system may need to be replaced. Indeed, regulations have been enacted that phased out the production and the use of chlorofluorocarbons (CFCs) including R-11 and R-12 by the end of 1995, after their implication in the depletion of the earth's ozone layer. Since the CFCs have been extensively used as refrigerants in air conditioning and refrigeration equipment, the existing stocks of CFCs have been significantly reduced and are becoming expensive. Therefore, replacement and conversion of CFCs chillers to operate with non-CFC refrigerants are becoming attractive options. If the existing chiller is relatively new (less than 10 years old), it may not be cost-effective to replace the entire chiller with a new non-CFC chiller. Just the conversion of the chiller to operate with non-CFC refrigerants may be the most economical option. However, the non-CFC refrigerants (such as R-134a and R-717) may reduce the energy efficiency of the chiller by reducing its cooling capacity due to their inherent properties. Fortunately, this loss in energy efficiency can be limited by upgrading some components of the cooling system including the impellers, orifice plates, gaskets, and even compressors. The specifics of a chiller upgrade and/or conversion are now available from several manufacturers. In some instances, the conversion with equipment upgrade may actually improve the chiller performance.

Some of the strategies that can be used to improve the efficiency of existing chillers by using an upgrade are listed below:

- Increase the evaporator and the condenser surface area for more effective heat transfer

- Improve the compressor efficiency and control
- Enlarge internal refrigerant pipes for lower friction
- Ozonate the condenser water to avoid scaling and biological contamination.

Over-sizing is another problem that may warrant the replacement of cooling systems. Indeed, several existing chillers have a capacity that is significantly higher than their peak cooling load. These chillers operate exclusively under part-load conditions with reduced energy efficiency and thus increased operating and maintenance costs. When the oversized chillers are more than 10 years old, it may be cost-effective to replace them with smaller and more energy efficient chillers operating with non-CFC refrigerants.

9.5.2   Chiller Control Improvement

Before replacing an existing chiller, it is recommended to consider alternative cooling systems or simple operating and control strategies to improve its energy performance. Some common and proven alternative cooling systems such as evaporative cooling and water-side economizers are discussed later on in this Chapter. In this section, measures involving the use of improved controls are discussed. Among these controls are those that are based on two basic strategies:

- Supply chilled water at the highest temperature that meets the cooling load.
- Decrease the condenser water supply temperature (for water-cooled condensers) when the outside air wet bulb temperature is reduced.

Indeed, chiller performance depends not only on the cooling load but also on the chilled water supply temperature and the condenser water temperature. The Carnot efficiency expressed by Eq. (9.2) can be used to illustrate that the COP increases when the condenser temperature (i.e. T_H) is reduced and/or when the evaporator temperature (i.e., T_C with $T_C < T_H$) is increased. For typical water-cooled chillers, Figure 9.2 can be used to evaluate the improvement in the COP of a chiller when the leaving water temperature is increased from 4.5 °C (40 °F). Similarly Figure 9.3 can be used to estimate the effect of reducing the condenser water temperature on the COP of the cooling system. Both Figure 9.2 and 9.3 represent typical chiller performance based on manufacturers data (LBL, 1980).

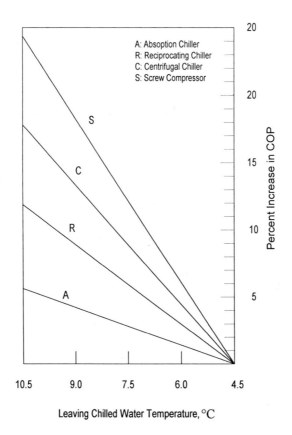

Figure 9.2: Effect of leaving chilled water temperature on the chiller COP.

Example 9.2: *A centrifugal chiller with a capacity of 500 kW and with an average seasonal COP of 4.0 operates with a leaving chilled temperature of 4.5 °C; Determine the cost savings incurred by installing an automatic controller that allows the leaving water temperature to be set on average to be 2.5 °C higher. Assume that the number of equivalent full-load hours for the chiller is 1,500 per year and that the electricity cost is $0.07/kWh.*

Solution: Using Figure 9.2, the increase in the COP for a centrifugal chiller due to increasing the leaving chilled water temperature from 4.5 °C to 7.0 °C is about 8%. The energy use savings can be calculated using Eq. (9.6) with $SEER_e$ = 4.0; $SEER_r$ = 4.0*1.08 = 4.32; $N_{h,C}$ = 1500; and Q_C = 500 kW; LF_C = 1.0 (assume that the chiller is sized correctly):

$$\Delta E_C = 500kW * 1500hrs / yr * 1.0 * \left(\frac{1}{4.0} - \frac{1}{4.32} \right) = 13,890kWh / yr$$

Therefore, the energy cost savings is $970.

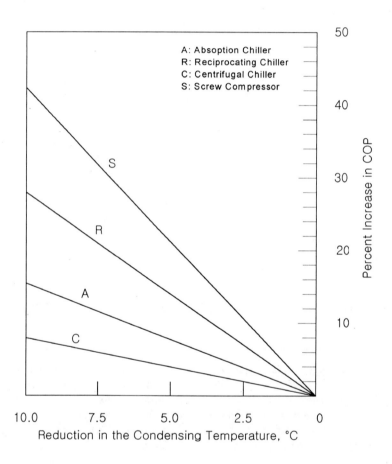

Figure 9.3: Effect of a reduction in the condensing temperature on the chiller COP.

9.5.3 Alternative Cooling Systems

There are a number of alternative systems and technologies that can be used to reduce and even eliminate the cooling loads on the existing cooling systems. Among the alternative systems and technologies are:

- **Water-side economizers** can be used when the outdoor conditions are favorable. Instead of operating the chillers to provide air conditioning, water can be cooled by using only cooling towers and circulated directly to the cooling coils either through the normal chilled water circuit or through heat exchangers.

- **Evaporative Cooling** is a well established technique that uses water sprays or wetted media to cool supply air either directly or indirectly allowing temperatures to approach the wet-bulb temperature of the ambient air. Direct evaporative cooling humidifies the air supply when its temperature is reduced while indirect evaporative cooling is performed through air-to-air heat exchanger with no humidity addition. Typically, indirect evaporative cooling is less effective and more expensive than direct evaporative cooling. In addition to some energy use (mostly electric energy to power fans), both evaporative cooling methods consume a significant amount of water. Evaporative cooling can be used to reduce the cooling load for a conventional mechanical air conditioning systems in climates characterized by dry conditions either throughout the year or during limited periods. The average COPs of evaporative cooling systems can be in the range of 10 to 20 depending on the climate (Huang, 1991).

- **Desiccant Cooling** is basically evaporative cooling in reverse since the air temperature is increased but its humidity is reduced. The dried air is then cooled using heat exchangers in contact with ambient air. Finally, the air is further cooled using evaporative cooling. A source of heat is needed to regenerate the desiccant after it has absorbed water from the air. Desiccant cooling has been used mostly in industrial applications and is less commonly used in the commercial sector. However, future developments of gas-fired desiccant dehumidification systems are expected especially to condition outside air required for ventilating office buildings (Tuluca, 1997).

- **Subcooling** of the refrigerant increases typically the cooling capacity and can decrease the compressor power and thus increases the overall energy efficiency of the cooling system. Subcooling requires the addition of a device such as a heat exchanger to decrease the enthalpy of the refrigerant entering the evaporator, resulting in an increase in the cooling capacity. There are currently three common subcooling technologies. The first technology uses suction-line heat exchanger of the vapor compression system as a heat sink. The second technology involves a second mechanically driven vapor compression cycle coupled with the main cycle using a subcooling heat exchanger located downstream from the condenser. The third technology requires an external heat sink such as a small cooling tower or ground source water loop. Refrigerant subcooling has long been used in low and medium temperature refrigeration systems (Couvillion et al., 1988). Currently, some manufacturers of packaged and split-system for air conditioners and heat pumps are integrated subcooling devices with their systems using alternative refrigerants (such as R-134a).

9.6   Summary

This chapter provided a brief analysis of the type and the energy efficiency of cooling equipment currently available in the US and other countries. In addition, cost-effective energy conservation measures have been proposed with some specific examples to illustrate the energy savings potential for some of the cooling equipment retrofits. Currently, energy retrofits of cooling systems include improvement in controls, replacement with more energy-efficient systems, and use of alternative cooling systems. In the future, it is expected that higher efficiency cooling equipment will be available as well as other innovative air conditioning alternatives such as desiccant cooling systems.

Problems

9.1 Consider a 800-tons chiller operating for 1,600 equivalent full load hours per year. The chiller is a centrifugal chiller that is rated at an average seasonal efficiency of 0.72 kW/ton. Determine the energy and cost savings when the following operating changes have been made:

(a) The condensing temperature is reduced from its current setting by 5°F, 10°F, or 15°F.
(b) The leaving chilled water temperature is increased from 40 °F to 45 °F and from 40°F to 50°F.

Assume the electricity cost is $0.08/kWh.

9.2 Redo Problem 9.1 for a 800-tons absorption chiller using 10 lbs of steam per ton. The cost of steam is $10/1000 lbs.

9.3 A 300-ton chiller, with an average seasonal efficiency of 1.2 kW/ton, is operated 3,500 hours per year with an average load factor of 70%. This chiller needs to be replaced by either a chiller A or chiller B. The manufacturers indicated that the IPLV for chiller A is 4.14 and for chiller B is 4.69.

(a) Estimate the energy cost savings due to replacing the existing chiller with chiller A or with chiller B. The electricity cost is $0.09/kWh.
(b) The cost differential between chiller B and A is $25,000. Determine is it is cost-effective to replace the exiting chiller with chiller B rather than chiller A. For this question, a simple payback analysis can be used.
(c) If the discount rate is 6%. Determine the electricity price for which it is more cost-effective to replace the existing chiller with chiller A rather than chiller B. Assume a life cycle of 10 years for both chillers.

10

ENERGY MANAGEMENT CONTROL SYSTEMS

Abstract

This chapter describes control strategies which can provide significant energy-savings opportunities to various energy consuming equipment such boilers, chillers, and motors. First, basic concepts of HVAC system controls are presented. Then, typical energy management and control systems (EMCS) are described. Finally, some applications of EMCS are discussed with some examples to illustrate the cost benefits of using more energy efficient control strategies to operate building energy systems.

10.1 Introduction

Currently, almost all new buildings have some control systems to manage the operation of various building equipment including HVAC systems. More elaborate control systems can operate simultaneously several mechanical and electrical equipment dispersed throughout the facility. In particular, these energy management control systems can be used to reduce and limit the energy demand of the entire facility. In the last decade, most of the advances in the HVAC equipment are due to the modern electronic controls which are now cheap, flexible, and reliable.

The development of energy management and control systems (EMCS) is mostly attributed to the introduction of computerized building automation systems. In fact, energy management represents one of several tasks performed by an integrated building automation system (IBAS). Among other tasks of the IBAS include fire safety, vertical transportation control, and security regulation. Advanced IBAS include logic for interaction between lighting, HVAC, and security systems. For instance, if an automated occupancy sensor detects the presence of people in specific spaces during late hours (during night or week-ends), the information can be used to adjust indoor temperature (for comfort) and to reinstate elevator service (to ensure that people can leave the building). Moreover, EMCS can provide facility operators with recommendations on maintenance needs (such as lighting fixtures replacement) and alarms for equipment failures (such as motors when they burn out).

The use of energy-efficient equipment does not always guarantee energy savings. Indeed, good management of the operation of this equipment is a significant factor to reduce whole building energy use. Generally, building energy loads are continuously changing with time due to fluctuations in weather and changes in equipment use and occupancy. Thus, effective energy management requires knowledge of the facility loads. Two approaches are typically applied:

- Load Tracking: the operation of equipment is modulated to respond to the actual needs in the facility. As an example, the compressor in a centrifugal chiller may change speed to match the cooling demand. The actual needs of a facility can be determined by a continuous monitoring. As an example, the load on the chiller can be estimated if the chiller water flow, and chilled water supply and return are monitored.

- Load Anticipation: In some applications, the needs of a facility have to be predicted to be able to modulate adequately the operation of equipment. For instance, in cooling plants with a thermal energy storage system, it is beneficial to anticipate the future cooling loads to be able to decide when and how much to charge and discharge the storage tank. Load prediction can be achieved by analyzing the historical pattern variations of the loads.

Using monitored data and other parameters characterizing the building, energy control systems enable operators and managers to operate efficiently HVAC and lighting systems to maintain comfort level. In the following sections, building energy control systems and some of their applications are presented and discussed.

10.2   Basic Control Principles

10.2.1   Control Modes

Control systems are used to match equipment operation to load requirements by changing system variables. A typical control system includes four elements as briefly described below:

i.   Controlled variable is the characteristic of the system to be controlled (for instance, the indoor temperature is often the controlled variable in HVAC systems).

ii. Sensors which measure the controlled variable (for instance thermocouple can be used to measured indoor temperatures).

iii. Controllers that determines the needed actions to achieve the proper setting for the controlled variable (for instance, the damper position of the VAV box terminal can be modulated to increase the air supply in order to increase the indoor temperature of the zone if it falls below a set-point).

iv. Actuators are the controlled devices which need to be activated in order to complete the actions set by the controllers (to vary the air supplied by a VAV box, the position of the damper is changed by an actuator through direct linkages to the damper blades)

Generally, two categories of control systems can be distinguished including closed loop and open loop systems. In a closed loop system (also known as a feedback control system), the sensors are directly affected by (and thus sense) the actions of the actuators. A typical control of a heating coil is an example of a closed loop system. However, in an open loop system (also called a feedforward control system), the sensors do not directly sense the actions of the controllers. The use of timer to set the temperature of the heating coils would be an example of an open loop system since the time may not have a direct connection with the thermal load on the heating coils.

Figure 10.1 shows the various components and terms discussed above as well as an equivalent control diagram for a closed-loop control system for a heating coil.

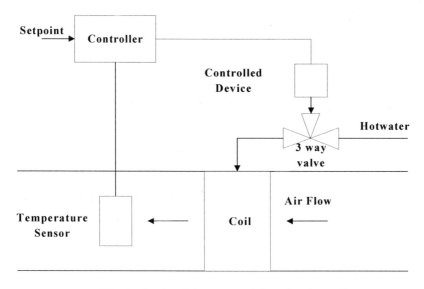

(a) Basic closed-loop control for a heating coil

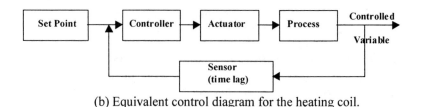

(b) Equivalent control diagram for the heating coil.

Figure 10.1: Typical Representations for a Heating Coil Control System

Each control system can use different control modes to achieve the required objectives of the control actions. Four control modes are commonly used in operating HVAC systems. These four control modes are:

1-Two-position: This control mode allows only two values (on-off or open-closed) for the controlled variable and is best suited for slow-reacting systems. Figure 10.2 (a) shows the effect of two-position control on the time variation of the controlled variable (such as the air temperature due to on-off valve position in a heating coil). In order to avoid rapid cycling, a control differential can be used. Due to the inherent time lag in the sensor response and to the thermal mass of the HVAC system, the controlled variable fluctuates with an operating range (called operating differential) with higher amplitude than the control differential. Thus, the operating differential is always higher than the control differential as illustrated in Figure 10.2 (b).

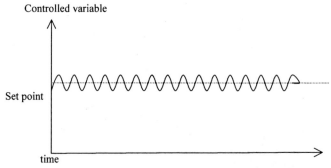

(a) Two-position action when no control differential is used (rapid cycles)

Controlled variable

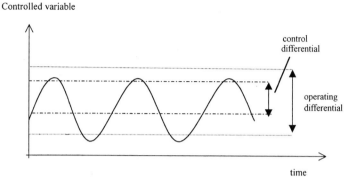

(b) Two-position action with a control differential

Figure 10.2: Effect of two-position control on the time variation of a controlled variable

Examples of two-position controls are domestic hot-water heating, residential space-temperature controls, and HVAC system electric preheat elements.

2-Proportional: This mode has a linear relationship between the incoming sensor signal and the controller's output. The relationship is established within an operating range for the sensor signal. The set point of a proportional controller is the sensor input which results in the controller output to be at the midpoint of its range. Mathematically, the controller output, u, is given by the following equation for a proportional control:

$$u = K_p e + u_0 \qquad (10.1)$$

The offset or error, e, is the difference between the set point and the value of the controlled variable. The proportionality constant, K_p, is called the proportional gain constant. The controller bias, u_0, is the value of the controller output when no error exists.

As depicted by Eq (10.1), the proportional control is not capable of reducing the error since an error is required to produce any controller action. Therefore, the controlled variable fluctuates within a throttling range as depicted in Figure 10.3.

Controlled variable

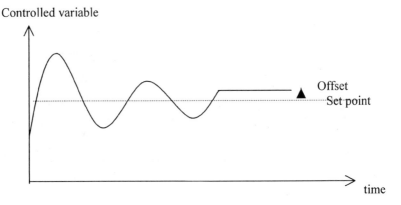

Figure 10.3: Proportional control effect on the time variation of a controlled variable

It should be noted that when the gain constant is very large, an unstable system can be obtained. Example 10.1 shows how the proportional gain constant can be determined.

Example 10.1: *A hot water heating coil has a set-point of 35 °C with a throttling range of 10°C. The heat output of the coil varies from 0 to 50 kW. Assuming that a proportional controller is used to maintain the air temperature set-point, determine the proportional gain for the controller and the relationship between the output air temperature and the heat rate provided by the coil. Assume steady-state operation.*

Solution: Using Eq. (10.1), the relationship between the heat rate Q, and the error in the air temperature at the coil outlet can be put in the form of :

$$Q = K_p \left(T_{setpoint} - T_{air}\right) + Q_o$$

Since:

(i) when the heat rate $Q = Q_{min} = 0$ kW, the coil outlet air temperature is $T_{air} = T_{min} = 35°C - 5°C = 30°C$, and
(ii) when the heat rate $Q = Q_{max} = 50$ kW, the coil outlet air temperature is $T_{air} = T_{max} = 35°C + 5°C = 40°C$

The proportional gain K_p can be determined as follows:

$$Q_{max} - Q_{min} = K_p(T_{min} - T_{max})$$

Or

$$K_p = [Q_{max}-Q_{min}]/[T_{min}-T_{max}] = -50 \text{ kW}/10 \text{ }^{\circ}C = -5 \text{ kW}/^{\circ}C$$

Similarly, the constant Q_o can be determined from

$$Q_{min} = K_p(T_{sepoint}-T_{min})+Q_o$$

Or

$$Q_o = Q_{min}-K_p(T_{setpoint}-T_{min}) = 0+5 \text{ kW}/^{\circ}C*(35^{\circ}C-30^{\circ}C)=+25 \text{ kW}$$

Therefore, the relationship between the heat rate output and the air temperature for the heating coil is:

$$Q = -5 (T_{setpoint} - T_{air}) +25$$

Thus, as long as the heat rate is different from $Q_o=25$ kW, the quantity $(T_{setpoint}-T_{air})$ which is the error in the proportional control equation cannot be equal to zero.

Generally, proportional controllers are used with slow stable systems that have small offset.

3-Integral: This control mode is typically incorporated with a proportional control mode to provide an automatic means to reset the set point in order to eliminate the offset. The combination of the proportional and integral actions is called "proportional-plus-integral" or simply PI control. Mathematically, the PI control can be expressed as follows:

$$u = K_i \int e.dt + K_p e + u_0 \qquad (10.2)$$

where K_i is the integral gain constant (also known as the reset rate) and has the effect of adding a correction to the controller output whenever an error exists. For HVAC systems, typical K_p/K_i ratio is less than 60 minutes.

The PI control can be applied to fast-acting systems that require large proportional bands for stability. Typical applications include mixed-air controls, heating or cooling coil controls, and chiller-discharge controls.

4-Derivative: This control action is used to speed up the response of the system in case of sudden changes. The derivative control mode is included in a combination of proportional-plus-integral-plus derivative (PID) control modes for fast-acting systems that tend to be unstable such as duct static-pressure controls. The mathematical model for the PID control is given by Eq. (10.3):

$$u = K_d \frac{de}{dt} + K_i \int e.dt + K_p e + u_0 \qquad (10.3)$$

where K_d is the derivative gain constant. The derivative term generates a corrective action proportional to the time rate of change of the error. The ratio K_d/K_p is typically less than 15 minutes for most HVAC applications. If the system has a uniform offset, the derivative term has little effect. The use of PID controls are typically less common than the PI controls for HVAC systems since no rapid control responses are needed.

To illustrate the action of P, PI, and PID control modes, Figure 10.4 compares the response of the system to an input step change. As expected the proportional control results in an offset and the controlled variable does not reach the set point. The correction term due to the PI control slowly forces the controlled variable to reach the set point value. Finally, the derivative term of the PID control provides a faster action to allow the controlled variable to attain the set point.

In addition to these conventional control modes, other intelligent controllers have been investigated in various engineering fields including HVAC equipment controls as will be discussed in the following section.

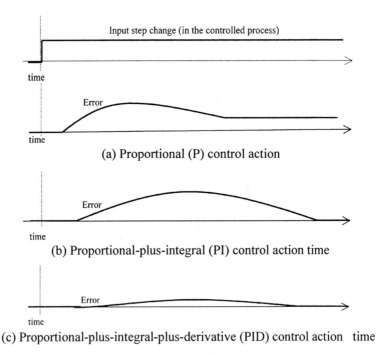

Figure 10.4: Comparison of the reaction of three control modes to an input step change

10.2.2 Intelligent Control Systems

In the future, it is expected that intelligent control systems will be commonly available to operate HVAC systems. A number of research studies are applying intelligent system methodologies so that control systems have human-like capabilities such as pattern recognition, adaptation, learning, reasoning, and associative memory to operate complex systems that convention control techniques cannot handle. Various intelligent control environments have been considered for HVAC systems including:

- **Expert controls**: Rule-based expert systems are based on human-like reasoning and can be powerful to solve practical engineering problems. Indeed, expert control systems have been applied to self-tuning, structured knowledge based adaptive control, fault diagnosis, and scheduling and planning. However, rule-based expert systems are not yet suitable to process numerical knowledge and thus to provide precise solutions. The development of an expert system environment that can deal with both qualitative and quantitative knowledge, and can automatically explore knowledge, is still in the realm of future expert control research.

- **Fuzzy Controls**: Since the development of fuzzy logic in the 1960's, fuzzy control has been one of the most attractive strategies in controlling complex systems with imprecise and/or uncertain knowledge of system information and behavior. Fuzzy logic can be applied to system modeling, estimation, optimal control, and adaptive control with the requirements of only system fuzzy knowledge and input/output data. Applications of fuzzy logic in expert systems are becoming attractive to establish intelligent expert control systems with fuzzy knowledge representation and fuzzy reasoning.

- **Artificial Neural Networks**: Similar to biological neural networks, artificial neural networks consist of large number of simple non-linear processing elements, typically called nodes or neurons, which are interconnected with adjustable weights. Well-trained neural network models can provide both qualitative and quantitative knowledge and have powerful functions in learning and self-organization. These features make neural networks more suitable in dealing with numerical data than expert systems (Kreider, 1997).

10.2.3   Types of Control Systems

To achieve the actions of the control systems discussed above, several types of energy sources are used. In particular, the following types of control systems are used in HVAC applications:

- Pneumatic devices are used with low-pressure compressed air at 0 to 20 psig. Pneumatic systems are common in older installations.
- Electric devices using 24 to 120 volts or even higher voltage sources.
- Electronic devices with low direct current voltages varying from 0 to 10V. These devices are being installed in new commercial buildings especially with direct digital control (DDC) systems.
- Hydraulic systems when large forces are required with pressure larger than 100 psi.
- Self-generated energy derived from the change of state of the controlled variable or from the energy available in the process plant.

For HVAC retrofit applications, it is recommended that direct digital control (DDC) systems be considered. Indeed, currently developed DDC systems use the latest digital technology including features such as intelligent controllers, high speed communication networks, and sophisticated control algorithms. All these features allow more energy-efficient control strategies to be implemented. Moreover, digital devices present additional advantages compared to pneumatic or electric devices as outlined below:

- Little or no maintenance is required for digital devices.
- Calibration of digital devices can be performed through remote instructions issued over a network. Some digital devices have the advantage to be continuously self-calibrating.
- Better accuracy is obtained from digital controls compared to pneumatic or electric devices.

While DDC devices have been used since the 1980's, it is only recently that manufacturers have developed systems that house interposing devices (such as relays, transducers, and hard-wired logic) in the same package with the electronic devices. These new DDC systems are suitable for retrofitting applications since it is now economical to convert existing pneumatic and electric analog controls to DCC systems. The best candidates for DDC retrofitting are air handling units (AHUs), heat exchangers, distribution pumps, and cooling towers. In general, the larger the equipment size, the faster the payback period. Replacing the controls for small HVAC equipment such as package unitary systems (including unit ventilators, heat pumps, and fan coils) may not be cost-effective.

Retrofit of pneumatic controllers can be made using electronic-to-pneumatic (E/P) transducers to convert signals so electronic and pneumatic control components can be combined in the same control loop. For instance, the controller and the sensor in an existing pneumatic HVAC control system can be converted into electronic devices while the actuator can remain pneumatic. The pressure output of the E/P transducer should match the electric signal. The use of E/P transducers allows retrofit of control systems with minor interruption in the operation of the controlled system.

10.3   Energy Management Systems

10.3.1   Basic Components of an EMCS

To control and operate equipment for heating, ventilating, and air conditioning, or for lighting and process equipment, an energy monitoring (or management) and control system (EMCS) can be used. A typical EMCS is configured into a network that includes sensors and actuators at the bottom level, microprocessor controllers in the middle, and a computer at the top with a modem to allow remote monitoring and control of the building energy systems. For a typical commercial building, an EMCS system can be cost-effective in reducing the energy use for HVAC and lighting systems.

Energy management is often just one element of an integrated building automation system (IBAS) which regulates security, fire safety, lighting, HVAC systems, and elevators. Advanced IBAS systems include logic for interaction between various systems such as HVAC, lighting, and security systems. Indeed, the automated occupancy count information obtained for different spaces in a facility can be used to adjust indoor temperature settings, reduce or turn-off lights, and ensure elevator operation.

The size of an EMCS system is typically classified based on the total number of points connected to a system. Five size categories are generally considered for EMCS systems:

- Large EMCS systems with more than 2000 points.
- Medium EMCS systems with 500 to 2000 connected points.
- Small EMCS systems with 200 to 1000 points.
- Small centralized EMCS systems with 50 to 500 points.
- Micro EMCS systems with less than 100 points.

A typical EMCS system is depicted in Figure 10.5 and includes a central control unit (CCU), a processing memory, storage devices, input-output devices, a central communications controller (CCC), data transmission medium (DTM), a field interface devices (FID), multiplexers, instruments, and controls. A brief description of the major components of an EMCS system is presented in the following sections.

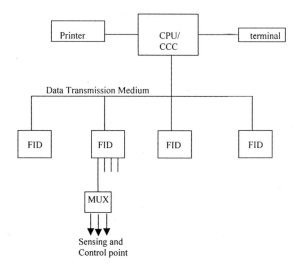

Figure 10.5: Typical EMCS components and configuration

- The central control unit (CCU) consists generally of a computer with memory for the operating system software, command software, and implementation of application algorithms. In particular, computations and logical decision functions for central supervisory control and monitoring are performed by the CCU. However, data and programs are stored in and retrieved from the memory or storage devices such as magnetic tape and disk systems. Typically, the CCU has input-output (I/O) ports for specific equipment such as printers and terminals.

- The central communications controller (CCC) is typically a computer with enough memory to execute specific programs required to reformat, transfer, and perform error checks on data coming from the CCU or the field interface devices (FIDs). The CCC may have a back-up capabilities in case of CCU failure.

- Field interface devices (FIDs) consist generally of computer devices with memory, I/O, communications, and power supply. The FIDs record the monitored and controlled data, perform calculations and logical operations, accept and process system commands. They should be capable of operating in case of CCU failure.

- A multiplexer (MUX) is a device that communicates between the data environment and its associated FID. The MUX is functionally part of the FID and can thus be in the same enclosure.

- The data transmission medium (DTM) is the communications link that allows to transfer data between the CCU and its associated FIDs such as telephone lines, optical fibers, or coaxial cables.

10.3.2   Typical Functions of EMCS

Several control operations and functions can be performed by an EMCS including but not limited to:

(i)   Equipment operation such start-stop, on-off, and open-close controls.

(ii)   Alarm functions such as abnormal equipment status, high or low parameter values (temperature, pressure, refrigerant level, etc.).

(iii)   Computer programming and table look-up functions for energy management and equipment use optimization including enthalpy economizer controls, chiller plant optimization, load shedding based on demand monitoring, lighting control by zones.

(iv)   Monitoring of operation conditions such as temperatures, pressures, and energy end-uses.

Other functions and applications of EMCS are being developed and implemented in some facilities. Some of these applications are discussed in the following sections:

EMCS can be used with expert systems to make intelligent operating decisions based on stored and accumulated knowledge. For instance, integrated EMCS/expert systems can be used to diagnose faults and inefficiencies in HVAC and lighting systems. Typically, an EMCS/expert system would be programmed with basic information about the building equipment and their design operation characteristics. Over time, actual operating data would be recorded and used to train the expert system and thus improve its diagnostic accuracy. If a fault were detected, the EMCS/expert system would notify the building operator of the probable problems and suggest remedial actions or simply initiate automatically countermeasures in case of serious problems. As an example, if the indoor temperature for a specific space is noted to be rising significantly beyond the throttling range, the EMCS/expert system would examine all the relevant sensor readings (such as airflow rates, fan energy use, and cooling water flow and temperature). Then, the EMCS/expert system would indicate using its knowledge base that the probable causes for the problem are faulty fan operation or leakage in chilled water pipe (70% faulty fan and 30% chilled water pipe leakage).

Another function of EMCS that has been implemented in several buildings especially in the last decade is the maintenance of acceptable indoor air quality (IAQ) levels. Indeed, the cost of sensors for monitoring air pollution compounds such carbon monoxide (CO), carbon dioxide (CO_2), and volatile organic compounds (VOC) have decreased sufficiently enough to be incorporated in commercial, and institutional buildings. For instance, CO_2 demand controlled ventilation has been implemented in several spaces including classrooms, conference rooms, theaters, and auditoriums.

10.3.3   Design Considerations of an EMCS

When an EMCS is recommended for a facility, it is important to consider some practical issues to ensure a successful design and operation. In particular, it is recommended to:

- Allow redundancy into the control systems in case of system failure. In particular, it is important that the local control devices are able to manage and operate the system for a reasonable period of time when the central control unit malfunctions or needs servicing.
- Provide clear information about the system operation. For instance color graphic displays are preferred to numerical data for the use of facility operators and managers.
- Perform a thorough commissioning of all the components and functions of the EMCS under various operating conditions (peak cooling and peak heating modes as well as part-load conditions).
- Train the operating staff to use the EMCS to their best advantage. In particular, the operators should understand the EMCS capabilities and benefits.

The selection of an EMCS for a given facility depends typically on the required functions and on economic considerations. The desired functions and controls from an EMCS are based on several aspects including the building type, HVAC system zoning, occupancy profiles, accuracy requirements, and the objectives of the owners. For instance, owner-developers may be more interested in having an EMCS with the lowest first cost rather than a system that provides a high-quality environment which may be one of the main objectives of owner-occupants. Moreover, high-rise buildings tend to have centralized systems with built-up fans and associated controls. On the other hand, low-rise buildings use typically package rooftop fan systems with local controls.

The cost benefits of the EMCS can be evaluated after identification of all the desired controls. Some of the important factors that affect the cost of an EMCS include:

i.   System size and number of control points
ii.   Degree of automation in the control functions
iii.   Accuracy requirements for sensors and controls

To ensure that the control systems conserve energy and save operating cost, it is important to follow basic principles in their design including:

1.   The energy consuming equipment should be operated only when needed. For instance, heating temperature set-points should be set back during unoccupied periods. The heating equipment should be operated only to maintain these set-back temperatures (typically between $50^{\circ}C$ and $55^{\circ}C$ to prevent freezing damages for various components of the HVAC system).

2. Simultaneous heating and cooling should be avoided. Proper zoning and HVAC system selection can minimize – if not eliminate – the need for providing heating and cooling at the same time.
3. The outdoor air intake should be controlled. In the US, only minimum requirement ASHRAE Standard 62-1999 for ventilation need to be supplied to the building when no economizer cycles are used.
4. The heating and cooling should be provided efficiently. In particular, only actual heating and cooling load requirements should be met. In addition, free cooling/heating or low-cost energy sources should be considered first to maintain comfort within the building.

Finally, it should be noted that energy control systems should be designed to be simple and easy to operate and maintain.

10.3.4   Communication Protocols

A communication protocol consists of a set of rules that have to be applied to exchange data between two parts of an EMCS system. Most manufacturers of building automation systems have their proprietary communication protocols. Therefore, building owners are forced to purchase equipment only from the original manufacturer if he wants to expand the existing automation system. Due to the lack of interoperability between communication protocols from various manufacturers, it is almost impossible to take advantage of a facility-wide approach to control optimization and energy savings. A single manufacturer does not provide all the best possible control strategies for electrical demand limiting, heating and cooling optimization, and similar energy savings options. Moreover, several control sub-systems in a building including HVAC controls, lighting/daylighting interface, fire alarm and life safety, security, and communication systems are generally manufactured by different companies. Integrating all these subsystems is a difficult task without a common communication protocol.

As a solution to the limitations and the difficulties inherent to proprietary protocols, a number of open communication protocols have been developed in recent years. These open protocols have some interoperability capabilities and thus allow building owners and managers to keep the open door to competition on any future expansion projects. One of the most widespread and widely accepted open protocols is the BACnet, a data communication protocol for Building Automation and Control networks (Bushby and Newman, 1991). BACnet is a non-proprietary open protocol standard that supports various communication networks ranging from high speed Ethernet local area networks (LANs) to low cost networks. Since ASHRAE developed BACnet, no one specific company or consortium has an advantage or an influence on future development of the standard. Any changes for BACnet are published for public review and comment after discussions on an ASHRAE open committee that includes representatives from industry, academia, and government.

To design a BACnet device, a manufacturer needs to identify the BACnet objects and services required to achieve the intended functionality for the device. A BACnet object is a standard data structure defined with a set of properties and data types. In its current version, BACnet standard defines 20 objects such as loops, tables, schedules, commands, and programs. BACnet services are the programmed actions that use the data objects to achieve the function of the device. Services, defined by the current version of the BACnet standard, include alarm and event services (to notify of any alarm and event), file access services (to read and write files), object access services (to read or write the properties of objects), remote device management services (to troubleshoot and maintain devices), and virtual terminal services (to allow interaction between a terminal and the device). Moreover, the BACnet device has to conform to a set of specifications using a series of conformance classifications. Each conformance classification adds functional services to the device. Thus, each BACnet device design would have a protocol implementation conformance statement (PICS) prepared by the manufacturer to identify the BACnet options available in the device.

In the last few years, some manufacturers have already developed BACnet control devices. However, the integration and the application of these devices in real installations have not yet been documented.

10.4   Control Applications

Energy management and controls systems can be used to perform several functions and tasks. Early building automation systems were limited to simple functions such as simple on/off programming including duty cycling and load shedding. Currently more complex functions and controls can be achieved by EMCS. Some of these controls are now available in standard program packages such as:

- Duty cycling for motor loads to provide sequential shut-down for short periods for equipment such as supply fans of small air handling units. However for large motors, frequent duty cycling is not generally recommended due to adverse effects on belts, bearings, and motor drives.
- Demand shedding to limit electrical loads. However, it is generally difficult to identify loads that can be shed without affecting the building performance especially for HVAC systems that are not generally needed when electrical demand is high. Therefore, equipment not related to HVAC systems such plug loads or lighting fixtures are typically considered for demand shedding. In particular, programmable lighting controls can be combined with other energy-savings lighting measures including dimming and occupancy sensors.
- Partial space conditioning to allow systems to cool only a small portion of the building by controlling supply-air dampers serving various zones. With

the use of variable frequency drives, it is now possible to adjust fan speed to match small loads for most central fan systems.

When demand charges are a significant part of the electric utility bills, EMCS can be applied to reduce the electric demands especially cooling applications. Indeed, accurate and reliable controls are required to ensure cost savings since even one mistake made in one month can increase utility bills not only for the month but also for future billing periods. Therefore, software capable of storing past data and anticipating future effects related to probable weather and occupancy conditions is required to effectively operate building energy systems. Some of the demand-limiting strategies that require effective control software are described below:

- Precooling of building thermal mass: The storage capabilities of a building structure can be used to shift a portion of on-peak cooling loads to off-peak periods and thus reduce electrical demand and energy charges. This measure can be achieved by precooling the building thermal mass. Studies (Braun, 1990; Morris et al., 1994) have shown that when an effective control strategy is used, up to 35% in energy cost savings can be achieved when an effective control strategy is used to determine when and how much to precool the building. Some additional savings in operating costs can be achieved if free cooling is used for precooling when cool outdoor air temperature is introduced to the building during the night using the air handling fans. In some cases, the cost of operating air handling fans may be less than the reduction in operating costs for mechanical cooling during occupied periods.

- Thermal energy storage (TES) systems: Cooling energy can be stored in the form of either chilled water or ice using storage tanks during off-peak periods. The stored chilled water or ice can then be used to meet cooling loads during on-peak periods when electricity cost is generally high. The optimal operation of cooling plants with TES systems require anticipation of future cooling loads and effective control logic. Some of the control strategies of TES systems are discussed in Chapter 12.

- Cogeneration systems: These systems allow the simultaneous production of electricity and heat using engine generators. Combining absorption chillers with on-site engine generators can make cogeneration cost-effective. Sophisticated control strategies may be needed to operate engine generators to avoid excessive costs. Even small engine generators which are required for emergency power can be used -with proper controls- to reduce peak electrical loads and associated demand charges.

In the following sections, more detailed description of other selected applications of EMCS systems including duty cycling of motor loads, controls for outdoor air intake, optimum start of heating systems, and central heating and cooling plant operation.

10.4.1   Duty Cycling Controls

Frequent turning on and off HVAC systems (and in particular fan motors) may be actually detrimental and may not be cost-effective over the life cycle of the equipment due to added maintenance and repair costs. However, on/off cycling of motor loads can be performed safely without long term damages when minimum on and off times are respected. The National Electrical Manufacturers Association (NEMA, 1994) provides a set of recommendations for minimum on/off times of duty cycling of motor loads. Some of these recommendations are summarized in Table 10.1. It is highly recommended however, that the motor manufacturers be directly consulted to determine their suggested minimum off-time and allowable number of starts per hour.

To fully benefit from duty cycling energy cost savings (due mostly to reduction in demand charges), two methods can be used. In the first method known as parallel duty cycling, all the motors are cycled on and off at the same time. This method can provide energy cost savings when the duty period is less than the demand period (typically 15 minutes in most utility rates). However, when the duty period exceeds the demand period, there is no reduction in demand changes if all the motors are cycled on and off at the same time. In this case, it is recommended to use the second method of duty cycling called staggered duty cycling which alternates the on and off times of the motors.

Most mechanical equipment manufacturers recommend extended duty cycling periods (higher than typical demand periods of 15 minutes). Therefore, staggered duty cycling approach should be considered in most applications to ensure the safety of HVAC equipment while reducing operating costs. Example 10.2 illustrates the calculation procedure for energy cost savings incurred from staggered duty cycling measure.

Table 10.1: Allowable number of starts per hour and minimum off-time for motor loads (NEMA, 1994)

Motor Size HP (kW)	2-Pole Motors Max. Starts/hr	2-Pole Motors Min. Off-time (seconds)	4-Pole Motors Max. Starts/hr	4-Pole Motors Min. Off-time (seconds)	6-Pole Motors Max. Starts/hr	6-Pole Motors Min. Off-time (seconds)
2.0 (1.5)	11.5	77	23.0	39	26.1	35
5.0 (3.75)	8.1	83	16.3	42	18.4	37
7.5	7.0	88	13.9	44	15.8	39
15.0	5.4	100	10.7	50	12.1	44
20.0 (15.0)	4.8	110	9.6	55	10.9	48
25.0 (18.75)	4.4	115	8.8	58	10.0	51
30.0 (22.5)	4.1	120	8.2	60	9.3	53
40.0 (30.0)	3.7	130	7.4	65	8.4	57
50.0	3.4	145	6.8	72	7.7	64

Example 10.2: *Determine the reduction in the annual energy costs due to staggered duty cycling of three identical fan motors (each rated at 30 kW [40 hp]). The motor manufacturer specifies a cycle of 20 minutes on and 10 minutes off as the minimum duty cycle. The utility monthly demand charge is $10/kW. First, determine the recommended NEMA duty period assuming 2-pole motors.*

Solution:

(a) Using Table 10.2, the allowable number of starts per hour for 40 hp, 2-pole motors is 3.7 starts/hr. Thus each start should last:

Start period = (60 min/hr)/(3.7 starts/hr) = 14.4 min/start

The duty period is thus about 15 minutes. Since the duty period is the sum of off-time and on-time, and since the minimum off-time is about 2 minutes (130 seconds based on Table 10.2), the maximum on-time allowable by NEMA standard is 13 minutes. Thus, the manufacturer on-time is longer than that recommended by NEMA.

(b) The reduction in the electrical demand peak due to staggered duty cycling approach is 1/3 of the total demand of all three motors (or 1/3 * 90 kW = 30 kW). Indeed, at any given time, only two out of the three motors are operating. Thus, the annual savings in electrical demands charges is given as follows:

$$\Delta kW = 12 * 30kW = 360\,kW\,/\,yr$$

Therefore, the staggered duty cycling control provides an annual energy cost savings of $3,600.

10.4.2   Outdoor Air Intake Controls

Due to the increase in the number of occupant complaints regarding poor indoor air quality (IAQ) and the increase in buildings diagnosed with sick building syndrome, the control and measurement of outside air intake rates has come to the forefront of attention of many HVAC engineers and designers. The majority of HVAC system designers today rely on the ASHRAE "Ventilation Rate Procedure" described in ASHRAE Standard 62-1989 (ASHRAE, 1989), *Ventilation for Acceptable Indoor Air Quality*. ASHRAE Standard 62 specifies minimum ventilation rates as a function of building use and occupancy to provide adequate IAQ for conditioned spaces.

Unfortunately, the necessary monitoring equipment and control logic to maintain minimum outdoor intake rates are often nonexistent or are used improperly if they are installed. Consequently, several commercial buildings, and in particular those with Variable Air Volume (VAV) systems, have been found to have inadequate ventilation (Sterling et al. 1992). The use of appropriate airflow measurement or VAV control techniques is critical to maintain minimum outside air intake rates. In a recent work (Krarti et al., 1999b), theoretical and experimental analyses have been performed to determine the accuracy of various techniques for outside airflow measurement and control applicable to VAV systems.

Descriptions of control strategies commonly used in field are presented in the following sections. For a more complete discussion, refer to Krarti et al. (1999b). It is important to note that the following descriptions apply for VAV system operation during minimum outside air intake mode. Control at other times may differ, especially during economizer cycles.

VAV Control Techniques for Economizer Systems

In economizer systems, the size of the outside air duct must be large enough to safely provide 100% of the design flow. This large size, however, results in very low airflow velocities during minimum outside air intake rate mode which can make measurement difficult with pressure-based airflow measurement devices. Labeling of system diagrams presented here will follow the approach adopted by Kettler (1998) for consistency.

(a) Fixed Minimum Outdoor Air Damper Position

A fixed outside air damper position is a common method used to meet minimum outside airflow intake rates in VAV systems. Under design flow conditions, the outside air damper is positioned to meet the minimum outside air requirements. This predetermined damper position is then used when only minimum outside airflow is required, even as the supply fan speed is reduced.

In VAV systems, this control method does not deliver the minimum outside air intake due to variation in the static pressure of the mixing plenum (Drees et al. 1992, Mumma and Wong 1990). Outside air intake rates are much closer to a constant percentage of supply air than a constant volume flow rate, a fundamental flaw of this control method (Janu et al. 1995). Another problem with this method is stack and wind effects on the outside air intake rate (Solberg et al. 1990). In addition to the limitations inherent to the technique of using a fixed minimum outside damper position, Ke et al. (1997b) found through simulation that this method was the least effective control strategy to maintain a minimum outside airflow rate among those commonly used.

(b) Volumetric Fan Tracking

Figure 10.6 shows a schematic of the volumetric fan tracking system. The flow measurement stations, AFS-1 and AFS-2, measure the supply and return airflow rates, respectively. The return fan speed is controlled (AFC-1) to maintain a fixed differential in the return airflow rate compared to the supply airflow rate. The preset fixed differential in the return and supply airflow rates must then be made up by outside air. Damper positions for the return, exhaust, and outside air are generally set to fixed positions during minimum outside air intake mode. The fixed flow differential is typically based upon the initial system air balancing. The outside air provided to the space maintains a slight positive static pressure within the building to reduce unwanted infiltration.

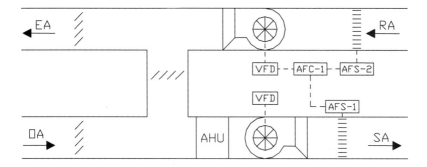

Figure 10.6: Outside airflow rate control schematic for system using volumetric fan tracking control strategy.

Volumetric tracking is one of the more common control methods used in VAV systems today (Kettler 1995, Avery 1992). The benefit of this method is that airflow rates in the supply and return ducts are generally large enough that standard flow measuring techniques can be sufficiently accurate. However, several authors have alluded to weaknesses in this control method. Elovitz (1995) states that even small measurement errors in large flow rates can translate to large errors in the calculated outside air intake rates and that a fixed differential flow is not versatile enough to account for exhaust and leakage flow rate changes.

Using a fixed position for the outside air damper also limits the flow rates of outside air for space pressurization. If the damper is not sufficiently open, it is possible that there will not be enough outside air available (Janu et al. 1995). Janu et al. (1995) also recommend that online measurement of outside air intake rates be provided and that the differential flow vary to compensate for operation of variable exhaust flows and the opening and closing of windows and doors. Finally, Kettler (1995) makes the argument that when typical measurement errors are accounted for, the outside air intake rate can vary by as much as 35%.

(c) Measurement and Control of Outside Airflow Rate with Economizer

A typical arrangement for this type of system is shown in Figure 10.7. The outside air duct is sized to allow for economizer control of the system. During minimum outside airflow intake mode, a flow measurement station (AFS-1) records the flow of outside air and controls the return and outside dampers (M-1 and M-2, respectively) to maintain the required minimum outside airflow intake rate.

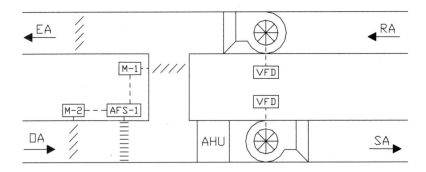

Figure 10.7: Outside airflow rate control schematic for system with economizer damper.

Due to the relatively large size of the outside air duct in this system, the measurement of the outside air intake rate at the flow measurement station (AFS-1) can be difficult with pressure-based airflow measurement devices. The accuracy of this control technique depends directly upon the accuracy with which the outside air intake rate can be measured.

(d) Plenum-Pressure Control
This method relies upon additional instrumentation such as a manometer or differential pressure transmitter to measure the pressure drop across a fixed orifice. By maintaining a constant pressure drop, the minimum outside airflow requirements can be met (Janu et al. 1995, Haines 1994, Elovitz 1995). It can be implemented either in a dedicated ductwork or in an existing economizer duct. The fixed orifice in this case is the combination of the outside air louver (L-1) and the damper installed in the outside air duct as suggested by Mumma and Wong (1990). This system is shown schematically in Figure 10.8. The pressure drop must be large enough so it can be accurately measured but not so large to create an excessive energy penalty (Ower and Pankhurst 1977, Kettler 1998). The differential pressure transmitter (DP-1) measures the pressure drop and the return air damper is controlled to maintain a constant value. Obviously, if an actuator is not located on the return air damper, one must be added.

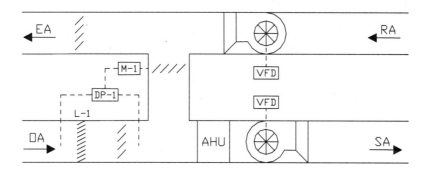

Figure 10.8: Plenum-pressure control schematic.

For a fixed damper position, the value of the loss coefficient, C, for the damper is constant. The outside airflow intake rate is related to the pressure drop across the damper by Equation 10.4 (ASHRAE 1997b):

$$V = D\sqrt{\frac{\Delta p_j}{\rho \cdot C}} \tag{10.4}$$

where:

V	=	*Velocity [fpm] (m/s)*
D	=	*Constant [1096.7] (1.4123)*

Δp_j =   *Total pressure loss [inW.G.] (Pa)*
ρ =   *Density $[lb_m/ft^3]$ (kg/m^3)*
C =   *Local loss coefficient [-]*

VAV Control Techniques for Systems with a Dedicated Outside Air Duct

The next two control strategies attempt to remedy the main disadvantage of an HVAC system equipped with only one outside air duct. By adding another duct through which only the minimum outside air must flow, the size can be made much smaller, thereby increasing the airflow velocities and thus making them easier to measure. Typically, the larger duct is used only during economizer control mode and is closed when minimum outside air intake rates are required.

(a) Measurement and Control of Dedicated Minimum Outside Duct Airflow Rate
This system is shown schematically in Figure 10.9. In economizer mode, the damper on the larger outside air duct is controlled to regulate the outside air intake rate. During minimum outside air intake mode, the dedicated outside airflow intake duct is opened while the damper in the larger outside air duct is closed. A flow measurement station (AFS-1) records the outside airflow rate and controls the return (M-1) and the dedicated outside air dampers (M-2) to maintain the minimum outside airflow intake rate. The exhaust air damper can be left in a fixed position during minimum outside airflow intake mode, or alternatively, can also be controlled from the flow measurement station (exhaust damper control not shown in Figure 10.9).

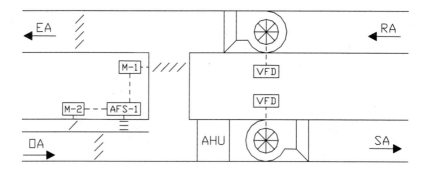

Figure 10.9: Outside airflow rate control schematic for system with dedicated minimum outside airflow duct.

(b) Outside Air Injection Fan
In this control technique, a dedicated minimum outside airflow intake duct contains a fan used to control outside airflow during minimum intake rate mode.

This system is the same as that illustrated in Figure 10.9 except a fan is installed in the dedicated minimum outside air duct. This injection fan is controlled by supply fan; when the supply fan is on, the injection fan motor is turned on. This control of the fan motor can be modified to allow the injection fan to remain off during unoccupied periods, times of building warm-up, or when the system is running in economizer mode. The fan is chosen such that it has a very flat fan curve and operates almost as a constant volume fan over the expected range of pressures (Elovitz 1995, Avery 1989). While this method can be adequate to provide the minimum required outside air intake, it is usually expensive and difficult to implement in existing buildings.

Other VAV Control Techniques

In addition to the techniques described above, other control strategies are used to control outdoor air intake. Some of these control strategies are briefly described in this section. For more information on these techniques, the reader is directed to the cited references.

- Minimum outside air damper position reset: This control strategy attempts to compensate for the main limitation of the fixed outside air damper control strategy by allowing the damper position to be reset based upon the supply fan speed. The position of the outside air damper can be found from either a linear relationship with the supply fan speed, or a higher order polynomial equation. As with the fixed minimum outside damper position control method, online measurement of the outside air intake rates are not required. However, the minimum ventilation rate may not be met if the supply airflow rate falls too low (Ke and Mumma 1997a). Additionally, since the damper and duct are often the same size, small changes in damper position translate to large changes in flow rates (a highly nonlinear relationship) and normal hysteresis can significantly affect the outside air intake rates (Drees et al. 1992). Finally, this control strategy cannot account for wind and stack effects on the system. See Solberg et al. (1990) for additional details regarding these errors.

- Supply/Return fan speed or vane position matching: The supply and return fan speeds are controlled, often off the same control signal, to match each other with a fixed differential to maintain a slight positive pressurization. The outside airflow rate is then equal to the difference between the supply and return airflow rates. However, similar to the volumetric tracking control strategy, this is only true when there is no exhaust airflow. Whenever the exhaust flow is greater than zero, the outside air intake rate will be increased and an energy penalty may result. While this control method is inexpensive and easy to implement on existing systems, it generally has been unacceptable due to mismatched fan flow characteristics over the typical range of operation (Janu et al. 1995). Elovitz (1995) has also stated that this method is not versatile enough to account for all possible circumstances encountered in building operation, such as fume hoods and the opening and closing of windows and doors.

- Direct building static pressure control: By measuring the building pressure relative to the outdoors, a closed-loop control method can be implemented to vary outside air intake and relief airflow rates. This is achieved by varying the return airflow rate or the positions of the return and relief dampers (Janu et al. 1995, Kettler 1988). The outside air intake rate is controlled by one of two methods based upon the differential pressure between the building space and the outside static pressure sensor. The first method varies the return fan speed to maintain positive space pressurization. The second method controls the return and exhaust dampers to maintain positive space pressurization. This control method is easy to implement in existing systems and has the advantage that the outside air intake rate is not a function of the supply airflow rate (Janu et al. 1995). However, Elovitz (1995) points out several drawbacks to this method. First, the differential pressure between the space and outdoors is very difficult to measure accurately due in part to wind loads and intermittent pressure changes due to the opening of windows and doors (Levenhagen 1992, Avery 1992). Secondly, the normal range of pressure differences throughout a large building can be larger than the system is trying to control. Finally, Elovitz (1995) states that the outside air intake rates are a function of the pressurization and leakage area of a building and may not remain constant over time.

- Fan capacity matching through balancing: In this method, return fans are controlled by static pressure in the ductwork rather than by tracking the supply fan. The supply and return fans are adjusted during building commissioning so they always lead/lag each other to maintain a difference in airflow. Outside air is the difference in the supply and return airflow rates. A positive building static pressure is usually maintained.

Again, when there is an exhaust airflow an energy penalty may be incurred by bringing in too much outside air. Unlike volumetric tracking, there are no flow measuring stations for fan matching through balancing. This method is cheaper and easier to implement than volumetric tracking and is more accurate that direct building pressure control. However, Levenhagen (1992) points out that the balancing contractor must ensure that the two fans are properly matched which is very difficult to achieve. Janu et al. (1995) have concluded that generally this is not possible.

- Characterization of flow through a modulated outside air damper: By characterizing the outside air intake rate as a function of both the position and pressure drop across the damper, accurate control of ventilation air can be obtained over a wide range of operating conditions. However, this process requires significant amounts of time to properly characterize the airflow rates. In addition, this method is subject to calibration drifts in transmitters and positioners, as well as looseness and hysteresis, any of which can cause substantial errors (Janu et al. 1995).

Comparative Analysis

Table 10.2 summarizes the results of an experimental comparative analysis performed by Krarti et al. (2000) to evaluate some of the control techniques for outdoor air intake under repeatable laboratory conditions. Specifically, Table 10.2 provides the average value, the standard deviation, the root mean square of the outdoor air intake flow rate, and the validity of each measurement and control method tested in a laboratory set-up. In particular, three measurement techniques are used to determine the air flow rates: averaging pitot-tube array station (P), electronic thermal anemometer (E), and CO_2 concentration balance technique (C). For more details on these measurement techniques, the testing set-up, and the experimental results, the reader is referred to Krarti et al. (2000).

Table 10.2: Summary of comparative results for the control and measurement techniques tested by Krarti et al. (2000).

System Description	Measurement Control[1]	Case	Set-point (cfm)	Validity	Outside Air Intake Rate Measurements					
					Averaging Pitot-tube Array			Electronic Thermal Anemometry		
					mean (cfm)	stdev (cfm)	RMS (cfm)	mean (cfm)	stdev (cfm)	RMS (cfm)
Fixed Damper Position	-NA-	1-A	1,600	14%	656	658	1,150	682	564	1,048
Fixed Damper Position	-NA-	1-B	2,400	23%	1,410	678	1,199	1,407	680	1,199
Fixed Damper Position	-NA-	1-C	3,200	26%	2,178	819	1,309	2,124	870	1,513
Plenum Pressure Control	-NA-	2-A	1,600	100%	1,630	69	75	1,544	70	89
Plenum Pressure Control	-NA-	2-C	3,200	100%	3,288	94	129	3,279	88	118
Direct Control with Economizer Duct	P	3-A	1,600	100%	1,635	38	52	1,546	46	71
Direct Control with Economizer Duct	P	3-C	3,200	100%	3,192	50	51	3,225	47	53
Direct Control with Economizer Duct	E	4-A	1,600	94%	1,695	55	110	1,637	55	67
Direct Control with Economizer Duct	E	4-C	3,200	100%	3,228	49	57	3,263	50	80
Volume Tracking	E	5-A	1,600	0%	2,427	439	936	2,436	458	953
Direct Control with Dedicated Duct	P	6-A	1,600	100%	1,639	39	55	1,634	47	58
Direct Control with Dedicated Duct	P	6-B	2,400	100%	2,430	41	51	2,457	51	77
Direct Control with Dedicated Duct	E	7-A	1,600	100%	1,643	36	56	1,640	43	58
Direct Control with Dedicated Duct	E	7-B	2,400	100%	2,404	39	40	2,428	42	50
Injection Fan	P	8-A	1,600	100%	1,621	31	37	1,622	36	42
Injection Fan	P	8-B	2,400	100%	2,429	28	40	2,440	36	54
Injection Fan	E	9-A	1,600	100%	1,622	38	44	1,617	36	40
Injection Fan	E	9-B	2,400	100%	2,418	33	37	2,427	32	42
Direct Control	C	-NA-[2]	1,600	75%	1,632	137	141	1,605[3]	119[3]	119[3]

[1] P = Averaging Pitot-tube Array, E = Electronic Thermal Anemometer, C = CO_2 Concentration Balance

[2] A different system setup was used for testing the concentration balance measurement technique

[3] Value is for CO_2 concentration balance measurement technique, not electronic thermal anemometry

The percentages listed in Table 10.2. in the column labeled "validity" were calculated from Eq. (10.5):

$$validity = \frac{n_v}{n} \qquad (10.5)$$

where: n_v =   the number of valid data points

Each test presented in Table 10.2 is subject to errors from the airflow measurement and the control technique used. Each 10 second data point, x_i, recorded during testing was considered valid if it met the following two conditions:

$$1) \ \left| x_i - set \ point \right| \leq \left(set \ point \cdot 10\% \right)$$

and

$$2) \ \frac{e_i}{x_i} < 15\%$$

where:
e_i = *the predicted error for the airflow measurement in the laboratory*

The first condition attempts to account for the accuracy of the control technique by requiring the data point to be within 10% of the set point. The second condition attempts to account for the accuracy of the airflow measurement technique by requiring the predicted error of the data point to be less than 15%.

In summary, accurate measurement and control of outside air intake rates in VAV systems is possible when careful attention is paid to proper installation and operation of system equipment. In systems where uniform airflow profiles exist, the use of an averaging Pitot-tube array or an electronic thermal anemometry, depending upon the expected velocities, for the direct measurement of outside airflow rates allows for direct control of minimum outside air intake rates. When these conditions are not met, the installation of a separate, dedicated minimum outside air duct, or the use of the concentration balance airflow measurement technique provide adequate alternatives. However, calculating the outside airflow rate using a temperature balance will not provide accurate results for all building operating conditions. Plenum pressure control in systems where measurement of the outside airflow rate is not possible should provide adequate control of minimum outside air intake rates. The traditional CAV control strategy of a fixed minimum outside air damper position, and the more robust volumetric fan tracking technique are not capable of accurately controlling outside airflow rates in VAV systems.

10.4.3   Optimum Start Controls

After the energy crisis of the 1970's, engineers found that building utility bills can be reduced by 12% to 34% by merely implementing an occupied thermostat setback (Bloomfiled and Fisk, 1989). During the cooling season, the unoccupied zone temperature set-point is raised while during the heating season, the set-point is lowered. In some mild climates, the indoor temperatures are allowed to float during night periods rather than defined by a night set-point. However, the winter night setback is typically set between 13 °C (55°F) and 15.5 °C (60 °F). Thus, building indoor temperature is colder than its occupied set-point in the early mornings. Due to the building thermal mass, the heating system has to be turned on earlier than the scheduled occupied time to achieve thermal comfort when people first enter the building.

The amount of time a building takes to recover from its night setback to its occupied set-point is usually referred to as the building recovery time. The length of the recovery period depends on several factors including outdoor ambient temperatures, indoor temperatures, and building thermal characteristics. Therefore, the recovery period can vary daily throughout the heating season especially in climates with sudden changes in the outdoor temperatures. However, the recovery time is typically set to be the same throughout the entire heating season by building operators to simplify the start controls of the heating system. This recovery time is defined as the earliest time the heating system needs to be started for the coldest day of the year. While this approach may achieve thermal comfort at the start of the occupancy periods throughout the heating season, it does not ensure optimal start times for the heating system especially during mild winter mornings.

With an energy management and control systems (ECMS), algorithms can be developed to determine the optimum start times and thus the best recovery periods. Several algorithms have been suggested in the literature. Typically, the recovery times are adjusted daily based on outdoor ambient temperatures and initial building zone temperatures (which may not be necessarily close to the set-point temperatures during unoccupied periods). In the following sections, some of the simplified algorithms for estimating building recovery times are presented.

Method 1: A linear relationship between recovery times, τ, and outdoor ambient temperatures, T_{amb}:

$$\tau = a_0 + a_1 T_{amb} \tag{10.6}$$

Where,

$$a_0 = \tau_{max} + \frac{\tau_{max} T_{amb,max}}{T_{amb,zero} - T_{amb,max}}$$

and,

$$a_1 = -\frac{\tau_{max}}{T_{amb,zero} - T_{amb,max}}$$

with:

- τ_{max} is the maximum recovery period
- $T_{amb,max}$ is the outdoor ambient temperature during the time when the maximum recovery period is obtained.
- $T_{amb,zero}$ is the outdoor ambient temperature during the time when the recovery period is zero.

It should be noted that the approach presented by Eq. 10.6 is relatively easy to implement since it does not require any regression analysis to determine the relation coefficients a_0 and a_1.

Method 2: A linear relationship between the recovery time, τ, and both outdoor ambient temperature, T_{amb}, and initial zone temperature, $T_{zone,initial}$

$$\tau = a_0 + a_1 T_{amb} + a_2 T_{zone,initial} \qquad (10.7)$$

Where the coefficients a_0, a_1, and a_2 are determined based on a regression analysis. This approach is first developed and implemented by Jobe and Krarti (1997). To determine the regression coefficients, it is recommended that data for at least five days be used.

Method 3: A quadratic relationship between the recovery time, τ, and both outdoor ambient temperature, T_{amb}, and initial zone temperature, $T_{zone,initial}$, using a weighting function in the form of:

$$\tau = a_0 + wa_1 T_{amb} + (1-w).[a_2 T_{zone,initial} + a_3 T_{zone,initial}^2] \quad (10.8)$$

Where the weighting parameter w is defined as follows:

$$w = 1000^{\left. -(T_{zone,initial} - T_{setnight}) \middle/ (T_{zone,final} - T_{setnight}) \right.}$$

with

- $T_{setnight}$ is the night (or unoccupied) setback temperature)
- $T_{zone,final}$ is the occupied set-point temperature

The approach presented by Eq. (10.8) is proposed by Seem et al. (1989) based on results from computer simulations.

A comparative analysis performed by Jobe and Krarti (1997) between the three approaches indicated that all three methods can reduce the start-up time for the heating system and thus save energy compared to the common approach that relies on setting the maximum recovery time throughout the entire heating season. Table 10.3 summarized the daily average reduction time in recovery period for two educational buildings located in Colorado using the three approaches discussed above.

It clear from the results presented in Table 10.3 that the approach described as method 2 provides the highest recovery time reduction while providing adequate thermal comfort within the two buildings. For the two buildings used in the comparative analysis, the recovery time was found to vary rather linearly with outdoor ambient temperatures. Therefore and since method 3 places more emphasis on indoor air temperatures [refer to Eq. (10.8)], it may not be adequate for the considered buildings.

Table 10.3: Daily average reduction in recovery periods for two buildings located in Colorado (Jobe and Krarti, 1997).

Building # Method #	Maximum Recovery Period (minutes)	Predicted Recovery Period (minutes)	Reduced start-up time (minutes)	Percent Reduction from Maximum Recovery Period
Building 1				
Method 1	90	64	26	29%
Method 2	90	53	37	41%
Method 3	90	77	13	14%
Building 2				
Method 1	75	19	56	74%
Method 2	75	19	56	74%
Method 3	75	28	47	62%

10.4.4   Cooling/Heating Central Plant Optimization

Cooling and heating central plants offer several opportunities to reduce energy operating costs through optimal or near-optimal controls for individual equipment (local optimization) and for the entire HVAC system (global optimization). While optimal controls have been developed and implemented for various components of cooling and heating central plants, global optimization remains considerably a complex endeavor and only a few strategies have been suggested and tested.

In this section, some of the local optimal control strategies are discussed. Moreover, operating strategies for entire cooling/heating plants are briefly discussed.

(a) Single Chiller Control Improvement

Before replacing an existing chiller, it may be more cost-effective to consider other cooling alternatives or simple operating strategies to improve cooling plant energy performance. In particular, a significant improvement in the overall efficiency of a chiller can be obtained through the use of automatic controls to:

- Supply chilled water at the highest temperature that meets the cooling load.

- Decrease the condenser water supply temperature (for water-cooled condensers) when the outside air wet bulb temperature is reduced.

Example 10.3 illustrates typical energy cost savings due to improved controls for a single chiller cooling plant.

Example 10.3: *A centrifugal chiller (having a capacity of 500 kW and an average seasonal COP of 4.0) operates with a leaving chilled water temperature of 4.5 °C; Determine the cost savings incurred by installing an automatic controller that allows the leaving chilled water temperature to be set on average 2.5 °C higher. Assume that the number of equivalent full-load hours for the chiller is 1,500 per year and that the electricity cost is $0.07/kWh.*

Solution: Using Figure 9.2 (refer to Chapter 9), the increase in the COP for a centrifugal chiller due to increasing the leaving chilled water temperature from 4.5 °C to 7.0 °C is about 8%. The energy use savings can be calculated using Eq. 9.6 with $SEER_e=4.0$; $SEER_r=4.0*1.08=4.32$; $N_{h,C}=1500$; and $Q_C=500$ kW; $LF_C=1.0$ (assume that the chiller is sized correctly):

$$\Delta E_C = 500kW * 1500hrs / yr * 1.0 * \left(\frac{1}{4.0} - \frac{1}{4.32} \right) = 13,890kWh / yr$$

Therefore, the annual energy cost savings are $970/yr.

(b) Controls for Multiple Chillers

When a central cooling plants consists of several chillers, a number of control alternatives exist to meet a building cooling load. Effective controls would select the best alternative for operating and sequencing the chillers to minimize the cooling plant operating costs.

Simple guidelines can be followed to operate multiple chillers at near-optimal performance. Typically, chiller operating variables such as chilled water temperature and condenser water flow rate are adjusted to ensure optimal controls. Some of the near-optimal control guidelines to operate electrically driven central chilled water systems are summarized below (ASHRAE, 1999):

- Multiple chillers should be controlled to supply identical chilled water temperatures.
- For identical chillers, the condenser water flow rates should be controlled to provide identical leaving condenser water temperatures.
- For chillers with different capacities but similar part-load performance, each chiller should be loaded at the same load fractions. The load fraction for a given chiller can be set as the ratio of its capacity to the sum total capacity of all operating chillers.

To determine the optimal chiller sequencing, a detailed analysis is generally needed to account for a number factors including the capacity and the part-load performance of each chiller and the energy use associated with all power-consuming devices such as distribution pumps.

(c) Controls for Multiple Boilers

As discussed in Chapter 8, the use of an array of small modular boilers provides a more energy efficient heating system than a single large boiler especially under part-load operation conditions. Indeed, each of the modular boilers can be operated close to its peak capacity and thus its highest energy efficiency. To optimally operate multiple boilers, it is important to know when to change the number of boilers on-line and/or off-line. The mere addition of a second boiler on-line when one boiler cannot handle the load may not provide the minimum operating cost. Indeed, the increase of firing rate (due to additional heating load) on any given boiler can cause a decrease in thermal efficiency due to higher flue-gas temperatures and thus higher thermal losses. However, the addition of a second boiler on-line increases the standing losses due to auxiliaries and the thermal losses through the added casing and piping of the second boiler. Therefore, a detailed analysis is needed to determine the changeover points for the multiple boilers. These changeover points depend on the characteristics of each boiler (ASHRAE, 1999).

10.5 Summary

In this chapter, an overview of basic components and applications of HVAC control systems has been presented. In particular, the energy cost savings incurred by various functions of energy management and control systems (EMCS) have been illustrated through selected examples and applications. In addition to being knowledgeable of the currently available control systems and applications, the energy auditor should be aware of the development in the intelligent control systems especially those applicable to HVAC systems.

Problems

10.1 In auditing a building, you find six large exhaust Fans each powered by 30-HP motor. You find that these fans can be turned off periodically with no adverse effect. You place them on a central timer such that each one is turned off for 10 minutes each hour. At any time, one of the fans is off, and the other five are running. The fans operate 10 hrs/day, 250 days/year. Assuming the company is on the following rate structure (applied monthly):

- Customer Charge: $ 345.00

- Demand Charge:
  - On-peak season: $5.50/kW
  - Off-peak season: $2.00/kW
- Energy Charge: $0.0275/kWh
- Sales Tax: 5.5%

*Note: The on-peak season is defined from May through September

10.2 The outdoor lighting can be saved on average of 2 hours per day using photo sensors. The outdoor lighting has consists of about 60 lamps rated at 1,000 W each. Determine the simple payback period of installing such sensors if they cost about $75 for each lamp. The cost of electricity is $0.05/kWh.

11
COMPRESSED AIR SYSTEMS

Abstract

This chapter outlines basic operating mechanisms of compressed air systems. Simple yet cost-effective energy efficiency measures are described to reduce waste from the production to the utilization of the compressed air. In addition, the chapter provides simplified calculation methods to estimate the energy savings of several energy efficiency measures specific compressed air systems.

11.1 Introduction

Compressed air is a commonly used utility in industrial processes and represents an important fraction of the operation cost of manufacturing facilities. It is estimated that the energy used by compressed air systems represents about 30% of the total energy consumed by electrical motors in France. Typical compressors use electricity to produce compressed air that may be needed for various industrial applications. Unfortunately, most existing compressed air systems have low efficiencies due to several factors including air leaks, inadequate selection of compressors, inappropriate uses of compressed air, and poor controls.

In this chapter, cost-effective energy conservation measures are described to reduce the operating costs of compressed air systems. First, a review of the basic principles of gas compression is provided. Then, the basic components required for the production, distribution, and utilization of compressed air are discussed. Finally, the calculation procedures for the energy savings of selected energy conservation measures are presented with illustrative examples.

11.2   Review of Basic Concepts

Figure 11.1 illustrates a simplified air compressed system operated by an electric motor. The compressed air is generally produced in a centralized location and then distributed to various locations within the facility to be used by equipment involved in either the production process or in the pneumatic control.

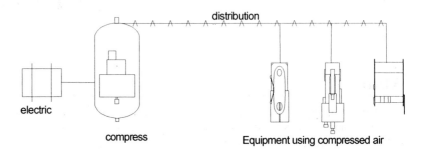

Figure 11.1: A schematic diagram for a compressed air system driven by an electric motor.

Generally, a compressed air system consists of several components including:
- One or several compressor(s) connected to a driver. The driver is typically an electric motor.
- A distribution system with piping, valves, fittings, and controls. The distribution system feeds the compressed air to operate several pieces of equipment dispersed throughout the facility.
- Other equipment such receivers, dryers, and filters.

The overall efficiency of a compressed air system depends on three stages: production, distribution, and utilization. During an audit, it is important to evaluate each of these stages in order to assess the performance and thus the potential for improving the energy efficiency of an air compressed system. A brief review of the basic principles and factors that affect the energy use for each stage for the production-distribution-utilization chain of compressed air are described below.

11.2.1   Production of Compressed Air

The basic concept of producing compressed air is relatively simple. Generally, mechanical power is provided to a compressor that increases the pressure of intake air. This intake air is typically drawn at ambient atmospheric conditions (i.e. pressure of 100 kPa or 1 bar). The compressor can be selected from several types such as centrifugal, reciprocating, or rotary screw with one or multiple stages. For small and medium sized units, screw compressors are currently the most commonly used in the industrial applications. Table 11.1 provides typical pressure, airflow rate, and mechanical power requirement ranges for different types of compressors (Herron, 1999).

Table 11.1: Typical ranges of application for various types of air compressors (Herron, 1999)

Compressor Type	Airflow Rate (m^3/s)	Absolute Pressure (MPa)	Mechanical Power Requirement (kW/L/s)
Reciprocating	0.0 - 5.0	0.340 – 275.9	0.35 – 0.39
Centrifugal	0.5 - 70.5	3.5 – 1034.3	0.46
Rotary Screw	0.5 - 16.5	0.1 – 1.8	0.33 – 0.41

A simplified energy analysis of compressed air systems can be carried out using the first law of thermodynamics applied to ideal gases. Some basic review of energy analysis applied to the compression of an ideal gas is provided to help identify the important parameters that need to be modified to reduce the energy used for the production of compressed air.

Figure 11.2 represents a piston cylinder during a compression of an ideal gas. In addition, Figure 11.2 shows the variation of the pressure as a function of the volume occupied by the gas, shown in a P-V diagram, for isothermal, adiabatic, and polytropic compressions. The relationship between temperatures and pressures at the inlet (or suction) and outlet (or discharge) of a compressor following an isothermal, an adiabatic, or a polytropic compression process can be summarized by Eq. (11.1) :

$$\frac{T_o}{T_i} = \left(\frac{P_o}{P_i} \right)^{\frac{\gamma-1}{\gamma}} \qquad (11.1)$$

Where

- $\gamma = 1$ when the compression is isothermal
- $\gamma = k$ when the compression is adiabatic (k=1.4 for dry air)
- $\gamma < k$ when the compression is polytropic (typically k=1.3 for dry air)

It should be recalled that for dry air, a simple relationship exists between the pressure, temperature, and density. This relationship is typically referred to as the equation of state and is provided by Eq. (11.2) for the compressor inlet air:

$$P_i = \rho_i . Z_{a,i} . R_a . T_i \qquad (11.2)$$

For dry air, the constant R_a can be calculated by dividing the ideal gas constant R [R=8314.4 J(kg.mole.K)] by the molar mass of air M_a (M_a=28.9). Thus, the value for the constant R_a is 287 J/(kg.K). Moreover, the compressibility factor $Z_{a,i}$ provides an indication how different the behavior of air relative to that of an ideal gas. The value of $Z_{a,i}$ ranges from zero and one and depends on the pressure and temperature of the air. For pressures above 20,000 kPa (200 atm), it can be assumed that dry air behaves like an ideal gas (i.e., $Z_{a,i}$=1).

It should be noted that an expression similar to Eq. (11.2) could be established for the compressor outlet air.

The work needed to compress a mass flow rate $\dot{m}_a$ of dry air can be estimated by applying the first law of thermodynamics to a compression process of an ideal gas. In particular, the mechanical power required for an isothermal compression can be calculated by Eq. (11.3):

$$\dot{W}_m = \dot{m}_a . R_a . T_i . Ln\left(\frac{P_o}{P_i}\right) \qquad (11.3)$$

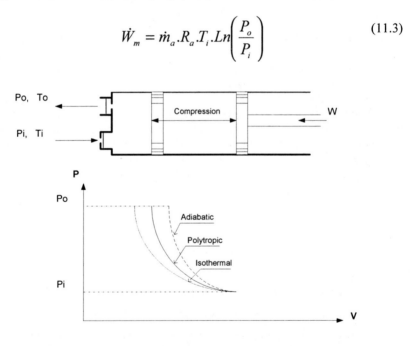

Figure 11.2: Ideal compression using isothermal, adiabatic, or polytropic processes.

The mechanical power required for an adiabatic or a polytropic compression cycle can be estimated from Eq. (11.4):

$$\dot{W}_m = \frac{\dot{m}_a.R_a.T_i.\gamma}{\gamma-1}.\left[\left(\frac{P_o}{P_i}\right)^{\frac{\gamma-1}{\gamma}}-1\right] \qquad (11.4)$$

Two important observations can be made from both Eqs (11.3) and (11.4) related to the energy efficiency of compressors:

(a) The mechanical power for a compressor increases linearly with inlet air temperature. To increase the energy efficiency of a compressed system, inlet air should be therefore as cool as possible.

(b) The mechanical power for a compressor increases with the pressure ratio. Therefore, it is important to produce compressed air with a discharge pressure limited to the maximum pressure needed by the facility. In other terms, over-pressurization should be avoided as much as possible. Moreover, the variation of the inlet air pressure can have significant impact on the mechanical power requirement by a compressor. In particular, the influence of the altitude on the ambient pressure should be account for.

Example 11.1 provides a comparative analysis of the mechanical power requirements for three compression types (i.e., isothermal, adiabatic, and polytropic). It is clear from the results of Example 11.1 that the adiabatic compression requires more energy input than all the other compression types. The case of isothermal compression provides the optimal operating conditions of the compressor. In particular, the air temperature should be maintained constant throughout the compression. Therefore, to reduce the compressor mechanical power requirements, cooling of air should be performed throughout the compression process. Unfortunately, ideal conditions of constant air temperature cannot be achieved in real compressions. However, if adequate cooling of the air is provided, a real compression can be represented by a polytropic process. As will be discussed later, inter-coolers are used to dissipate heat during compression.

Example 11.1: *Compare the mechanical power requirement to compress 1 kg/s of dry air from 100 kPa and 20 °C (ambient conditions) to 800 kPa (absolute pressure) using isothermal, adiabatic, or polytropic ($\gamma = 1.3$) compression.*

Solution: Using Eq. (11.3) with P_i=100 kPa, P_o=800 kPa, T_i= 293 K, R_a= 287 J/(kg.K); the mechanical power requirement for the isothermal compression of 1.0 kg/s can be estimated:

$$\dot{W}_m = (1.0 kg / s).(287 J / kg.K).(293K).Ln(\frac{800 kPa}{100 kPa}) = 174.86 kW$$

The mechanical power for the adiabatic compression is calculated using Eq. (11.4) with $\gamma = k = 1.4$:

$$\dot{W}_m = \frac{(1.0 kg / s).(287 J / kg.K).(293K).(1.4)}{1.4 - 1} . \left[\left(\frac{800 kPa}{100 kPa} \right)^{1.4 - 1/1.4} - 1 \right] = 238.82 kW$$

For the polytropic compression, the mechanical power requirement can be determined using again Eq. (11.4) but with $\gamma = k = 1.3$:

$$\dot{W}_m = \frac{(1.0 kg/s).(287 J / kg.K).(293K).(1.3)}{1.3 - 1} . \left[\left(\frac{800 kPa}{100 kPa} \right)^{1.3 - 1/1.3} - 1 \right] = 224.42 kW$$

As expected the isothermal compression requires less mechanical power while the adiabatic compression requires more power input than any of the three compression types. Note that the polytropic compression which is more representative of actual compression process has mechanical power requirements between those of the isothermal and adiabatic compressions.

It is interesting to note that the mechanical power input per unit of airflow rate (i.e. kW/L/s) can be calculated for the three types of compression by estimating the density of air at the inlet conditions. The ideal gas equation Eq. (11.2) can be used to determine the density of the air (with $Z_{a,i}=1$). For $T_i= 293$ K and $P_i= 100$ kPa, the density of air 1.189 kg/m3. Thus, the mechanical input for the three compression is 0.21 kW/L/s for an isothermal compression, 0.27 kW/L/s for a polytropic compression, and 0.29 kW/L/s for an adiabatic compression.

For real air compressors, it is recommended that the mechanical power or energy requirements be estimated using the assumption of adiabatic compression for several reasons including:

i.   Even though the polytropic compression is a better representation of actual compression processes, it is difficult to readily obtain the value of the exponent γ without expensive measurements. In some applications such as turbo-compressors (which increase the pressure of the air by increasing its velocity), the exponent γ can be in fact greater than k=1.4.

ii. For typical polytropic compressions, the power requirements are close to those estimated from the adiabatic compression (see results of Example 11.1).

iii. The power requirement for an adiabatic compression represents an upper limit of the power input used by real compressions and thus provides the worst case scenario.

When compressors are driven by electric motors, the electrical energy kWh_{comp} used by the compressed air systems can be determined from the following expression:

$$kWh_{comp} = \frac{\dot{W}_m.LF_{comp}.N_{h,comp}}{\eta_M} \tag{11.5}$$

Where,

- $\dot{W}_m$ is the mechanical power required by the compressor estimated using Eq. (11.3) or Eq. (11.4)
- $N_{h,comp}$ is the number of hours per year when the compressor is operated
- LF_{comp} is the average load factor of operating the compressor
- η_M is the efficiency of the motor that drives the compressor

In addition to the compressor, other accessory units are needed for the production of compressed air. These units may increase the mechanical and thus the electrical power requirement for operating compressed air systems. Some of the accessory equipment and their roles are briefly discussed below.

Filters:

The air at the inlet of the compressor should be as clean as possible to protect the compressor and its associated equipment from any damages that suspended particles in an unfiltered air can cause. Indeed, suspended particles can penetrate various parts of the air compressor system where they can either (i) obstruct small orifices (within the compressors, distribution lines, or machines using compressed air), or (ii) wear by friction some interior surfaces of the compressor (including cylinders, pistons, or rotors). The wear is even more damaging when the particles are mixed with lubricated oil.

Various air filters and air cleaning techniques can be used at the inlet of compressors depending on the size of the airborne particles. The ASHRAE handbook (1996) provides a description of a number of air cleaning techniques and their applications. A selection of an air filter for a compressor depends typically on two factors: the efficiency of the filter and the pressure drop caused by the filter. The efficiency of air filter is typically defined as the percentage of particles retained. The pressure drop through an air filter – due to either a fine screen mesh or an accumulation of dirt – decreases the capacity of the compressor. As a rule of thumb, it is estimated that every 100 mm (or 4 in) of water pressure drop (i.e. 1 kPa), the mass flow rate of compressed air is reduced approximately by 1 percent (assuming the compressor uses the same power).

It is therefore important to select proper air filters and to periodically clean or replace these filters in order to reduce the energy used by compressed air systems. Example 11.2 illustrates the increase in the energy cost for operating an air compressed system with a contaminated (i.e., dirty) air filter.

Example 11.2: *Determine the increase in energy cost of a 100-kW air compressor operating 8,000 hours per year due to a contaminated air filter. Assume that:*

a)   *the electrical motor efficiency is 90%,*
b)   *the pressure drop is 1.0 kPa for a clean air filter but becomes 3 kPa for a contaminated filter,*
c)   *the cost of electricity is $0.05/kWh*

Solution: Using the rule of thumb, it can be stated that for every 1 kPa increase in the pressure drop across the air filter, the mechanical power is increased by 1 percent (since the mechanical power is proportional to the mass flow rate). Thus, the energy cost penalty, EC_{filter}, of using a contaminated air filter can be estimated as follows:

$$EC_{filter} = \frac{100\,kW * 8000\,hrs\,/\,yr * \$0.05\,/\,kWh}{0.90} * \left(\frac{3kPa}{1kPa}\right) * 0.01 = \$1,330\,/\,yr$$

Receiving Tanks:

In some compressed air installations, receiving tanks are used and are placed either close to the compressors or near the end-use machines that have high and variable demands. Three reasons can justify the utilization of central or local air receiving tanks:

- Storage of energy in the form of compressed air to reduce unwanted cycling of the compressors.
- Regulation of the flow rate of the compressed air based on the load requirements of the end users.
- Control of the discharge pressure for the compressed air especially when piston compressors are used.

The design specifications of the receiving tanks should follow the requirements of the local authorities for pressure vessel regulations.

Dryers:

Ambient air can be moist with high water vapor content. This water vapor can condense either in the compressor, the distribution lines, or the equipment using the compressed air. To reduce or eliminate this water condensation, the compressed air should be dried using dryers. The selection of the drier type depends on two important factors:

- the quality of dry air required expressed in terms of the dew point pressure; and
- the total cost of the drier including initial costs and operating and maintenance costs.

There are a number of drying methods used with compressed air systems. In most cases, refrigerated dryers are used to bring the pressure dew point of the air to levels below 5°C. In other applications such as food processes, dessicant dryers are used to achieve very dry air with dew point pressures below 0°C.

In general, heat exchangers commonly known as after-coolers are used to reduce the water content and possibly oil vapor in the compressed air before it gets into the distribution lines and/or the receiving tanks. In particular, the compressed air is cooled using ambient air or water so condensation occurs in a separator located just after the compressor. As rule of thumb, the discharge temperature of the compressed air is typically 10°C above the water temperature (for after-coolers using water) and 15°C above the air temperature (for after-coolers using air).

Inter-Coolers:

To dissipate the compression heat, it is customary to cool the compressor cylinders and cylinder covers using heat exchange devices commonly referred to as inter-coolers. These inter-coolers can use air or water to control the temperature of the compressor. Typically, cooling of the compressed air is preferred because of convenience since it eliminates the need for water supply and thus avoids the problems associated with water distribution such as danger of freezing. However, reducing adequately the temperature of the compressor may be difficult with air cooling in some applications.

11.2.2   Distribution of Compressed Air

The distribution of compressed air to the end-use equipment is achieved by a set of piping networks connected by various fittings. There are several types of pipes that can be used to channel compressed air ranging from rigid metallic pipes to flexible plastic tubing. The selection of the pipe type depends on a number of specifications including the diameter and the pressure required for the distribution piping. Through the piping network, the flow of compressed air is typically regulated by a set of valves and pressure regulators. Moisture traps are often fitted along the distribution lines to separate any condensate from compressed air.

The pressure of compressed air at the end-use equipment is always lower that the discharge pressure from either the compressor or the receiving tanks for mostly two reasons: flow pressure drop in the distribution piping and leaks. A brief description of each source of compressed air pressure loss is provided below:

Flow Pressure Drop

Due to the flow resistance of the compressed air through the pipes, the valves, and other parts of the distribution system such as fitting and connectors, there is a pressure drop that can be estimated using the Darcy-Weisbach equation:

$$\Delta P = \rho \left(f \cdot \frac{L}{D} + \sum K \right) \frac{V^2}{2} \qquad (11.6)$$

Where,

- ΔP is the total flow pressure drop of the compressed air through the distribution system [Pa]
- ρ is the average density of compressed air through the distribution system [m^3/kg]
- f is the friction factor
- $\sum K$ is the loss coefficient due to valves and fittings in the distribution system
- L is the total length of pipe in the distribution system [m]
- D is the internal diameter of the pipe [m]
- V is the average velocity of the compresses air flow [m/s]

For more details on the calculation procedure using Eq. (11.6) to determine the flow pressure drop through pipes, the reader can refer to the ASHRAE Handbook (1997). In general, the calculations of the pressure drop for compressed air flow using equations such Eq. (11.6) are not recommended since they are complex and are not accurate due to the inherent assumptions. For instance, several assumptions are generally made to determine the friction factor and the loss coefficients for valves and fittings. Rules of thumb and charts can be used instead to determine at least the order of magnitude of compressed air pressure drop through the distribution system. However, Eq. (11.6) can be used to pinpoint some measures that can be considered to reduce the flow pressure drop through the distribution piping for a compressed air system:

- The pressure drop increases with the length of the distribution system. Therefore, an optimized piping layout can reduce the pressure drop and thus energy use of the air compressed system. In particular, it is important to reduce the number of valves and bends in the distribution network.

- The velocity of the compressed air flow should be as low as possible to reduce the pressure drop through the distribution system. This velocity reduction is recommended especially for installations where the distribution system is long. A reduction in the flow velocity can be achieved either by decreasing the mass flow rate required (i.e. end use load), lowering the temperature of the compressed air, or increasing the pipe diameter.

It should be noted that the pressure drop through the distribution system constitutes a waste of energy. Indeed, the discharge pressure has to be increased to compensate for the flow pressure drop. Eqs. (11.3) and (11.4) indicate that the increase of the compression ratio (P_o/P_i) leads to an increase in the mechanical power provided to the compressor by the driver (i.e. electrical motor). For instance, a pressure drop of 50 kPa in a compressor designed to increase the pressure from 100 kPa (1 atm) to 700 kPa (7 atm) corresponds to 3% waste in the mechanical and thus electrical power of air compressed system.

The pressure drop problems in the distribution due to flow resistance are generally inherent to the design of the piping layout and therefore should not be the focus for the energy audit of compressed air system unless the system have be replaced. Instead, the auditor should focus on the identification of measures to reduce energy waste for the existing air distribution system.

Air Leaks:

The energy losses due to leaks are generally significant in industrial compressed air installations. It is estimated that leaks can represent as much as 25 percent of the output of an industrial compressed air system (Terrell, 1999). Typically leaks can occur due to poor connectors and fittings, bad valves, or small holes in a rubber or plastic tubing or hoses. Non-producing machines with compressed air left on can also be common sources of leaks.

To identify air leaks several methods can be used during an audit walk-through. These methods can be grouped into two categories:

- Simple Procedures: Two inexpensive methods can be considered to identify location of air leaks: (i) Air leaks can be detected by walking while listening along the distribution lines. Any change in the pitch of noise due to the compressed flow inside the piping system may be an indication of leaks. This method can be applied when the background noise is not loud. (ii) Another simple and inexpensive procedure to detect air leaks is the use of water and soap in selected parts of the compressed air distribution system (especially fittings and joints between pipe sections). The formation of bubbles is a good indication of the presence of air leaks.

- Measurement Methods: The detection of air leaks can be performed using ultrasound equipment. Ultrasound leak detectors are available and can be used to scan an entire compressed air distribution system to locate easily any air leaks.

Maintenance personnel in an industrial facility can be trained to detect leaks in the compressed air systems during periods when there are limited or no production activities (i.e., week-ends, or scheduled plant shutdown periods). It is estimated that with little investment in a regular maintenance program for detecting leaks, significant savings in the operating cost of compressed air systems can be achieved.

When leaks are identified, the waste in the amount of compressed air can be estimated by simple calculations. Indeed, the waste in mass flow rate, $\Delta \dot{m}_a$, of compressed air through a hole can be estimated using the Fliegner's expression:

$$\Delta \dot{m}_a = \sqrt{\frac{2}{R_a}} . C_L . A_L . P_o . T_o^{-\frac{1}{2}} \qquad (11.7)$$

Where,

- $\Delta \dot{m}_a$ is the amount of mass flow rate in compressed air wasted through the leak [kg/s].
- P_o is the pressure of the compressed air leaving the leak [Pa]
- T_o is the temperature of the compressed air leaving the leak [K]
- C_L is the flow coefficient through the hole. It depends on the shape and size of the hole. For round holes, C_L can be set to be 0.65.
- A_L is the are of the hole [m²]

11.2.3   Utilization of Compressed Air

There are several applications for compressed air. In general, the end use equipment can be grouped into classes:

1) <u>Static End Users</u> including pneumatic control equipment such as actuators or pressure regulators. The action of these end users depends only on the presence or absence of compressed air. These end users can be found in both commercial and industrial facilities.

2) <u>Dynamic End Users</u> including assembly tools (such as screw-drivers and nut-runners) and air motors (such as tools for drilling and grinding). The action of these end users depends on the pressure and flow of compressed air. Most industrial applications of compressed air involve dynamic end use equipment.

To reduce the energy waste, the pressure delivered by the compressed air system should correspond to the highest pressure required by the end-users (with some safety factor to account for instance for the flow pressure drop as discussed in section 11.2.2). If the pressure available is higher than the pressure used by the end use equipment, energy is wasted since the unused high pressure increases the mechanical power required by the compressor [see Eqs. (11.3) and (11.4)].

In general, the use of compressed air should be avoided for applications that can be performed by other resources. For instance electrical driven tools can be used instead of air driven tools. Also, in applications involving drying, evaporation from a surface, or biological production within a large tank, blowers can be used instead of compressed air. It is estimated that if the pressure requirement is less than 200 kPa (2 atm), a blower can save 50 percent or more of the energy that would be used by compressed air (Terrell, 1999).

11.3 Common Energy Conservation Measures for Compressed Air Systems

One of the tasks involved in energy auditing is to collect data and information relevant to the design, operation, and maintenance of various systems. For compressed air systems, some of the information may need to be obtained by interviewing plant operators, processing engineers, or maintenance personnel. In particular, the following data can be gathered through one or several interviews with operating personnel:

a) Current operation controls of the compressed air system for the facility.
b) Existing problems with the quality, quantity, and pressure of the compressed air.
c) Design data and drawings and any maintenance records including compressor operating log and leak detection maintenance schedule.
d) Nameplate data for all equipment associated with the compressed air system.
e) Future plans for upgrading the compressed air system including compressor replacement, changes in the compressed air usage, and installation of instrumentation.

The data gathering can be time consuming depending on the complexity of the facility and the compressed air system. However, it is important that the energy auditor obtains enough data to understand the existing design details, current operation and maintenance procedures, and future plans to be able to propose energy conservation measures that not only save energy use and cost but also provide a more reliable compressed air system.

Description of selected energy conservation measures commonly suitable for compressed air systems are provided below with some specific calculation procedures and illustrative examples.

11.3.1   Reduction of inlet air temperature

In some compressed air installations, the compressor draws inlet air from indoors and in particular from the mechanical room or the space where they are located. This inlet air temperature can be high (30 °C or higher) and is most likely warmer than outdoor ambient air. As indicated in both Eqs. (11.3) and (11.4), higher inlet air temperatures imply higher mechanical and thus electrical energy requirements. A simple measure to reduce the inlet air temperature is to install pipes to connect the compressor to outdoors so the compressor draws ambient air. To ensure that the ambient outdoor air remains cool when it gets to the compressor, the pipes should be insulated.

The electrical energy savings, ΔkWh_{comp}, associated with a reduction of inlet air compressor can be calculated using the following expression based on Eq.(11.5):

$$\Delta kWh_{comp} = \frac{\dot{W}_m \cdot LF_{comp} \cdot N_{h,comp} \cdot (T_{i,e} - T_{i,r})}{\eta_M \cdot T_{i,e}} \qquad (11.8)$$

Where,
- $T_{i,e}$ is the annual average inlet air temperature before the retrofit [K]
- $T_{i,r}$ is the annual average inlet air temperature after the retrofit [K].

In general, $T_{i,r}$ can be assumed to be the same as the annual average outdoor temperature. An illustration of the calculation procedure using Eq. (11.8) is presented by Example 11.3.

Example 11.3: *A compressed air system has a mechanical power requirement of 100 kW with a motor efficiency of 87%. The intake air for the compressor is from the interior of the mechanical room (where the compressor is located and where the average annual temperature is 35 °C). Determine the payback period of installing an insulated piping section to connect the intake of the compressor to the ambient outdoor air. Use the following information:*
a)   *The total cost of the piping system is $850.*
b)   *The annual average outdoor air is 12 °C.*
c)   *The compressor is operating 5,000 hours per year with an average load factor of 80%.*
d)   *the cost of electricity is $0.05/kWh*

Solution: Using Eq. (11.8), the electrical energy savings due to the reduction in the intake air temperature can be estimated:

$$\Delta kWh_{comp} = \frac{100\,kW * 0.8 * 5000\,hrs/yr * [(35 - 12)K]}{0.87 * [(273 + 35)K]} = 34330\,kWh/yr$$

Therefore, the simple payback period for the installation of the piping system to connect the intake of the compressor to the outside air is:

$$SBP = \frac{\$850}{34330 kWh / yr * \$0.05 / kWh} = 0.5\, yr = 6\, months$$

11.3.2   Reduction of Discharge Pressure

When the maximum pressure required by all the end-use equipment in a facility is noticeably less than the air pressure delivered by the compressed air system, it is recommended to reduce the discharge pressure to reduce the compressor energy use. It is estimated that for every 15 kPa higher discharge pressure, 1% more energy input is required by the compressor.

Higher than needed discharge pressures have other detrimental effects on the compressed air systems. In particular, over-pressurization increases the waste of compressed air through leaks.

The electrical energy savings, $_{\Delta kWh_{comp}}$, due to the reduction in the discharge air pressure of the compressor can be calculated as follows:

$$\Delta kWh_{comp} = \frac{\%\dot{W}_m . W_m . LF_{comp} . N_{h,comp}}{\eta_M} \qquad (11.9)$$

Where, $\%W_m$ is the percent reduction in mechanical power required by the compressor. Using either Eq. (11.3) or Eq. (11.4), the percent reduction in compressor power input, $\%W_m$, can be estimated.

Example 11.4: *A compressed air system has a mechanical power requirement of 50 kW with a motor efficiency of 90%. Determine the cost savings of reducing the discharge absolute pressure from 800 kPa to 700 kPa. Assume that:*

- *The compressor is operating 4,000 hours per year with an average load factor of 70%.*
- *the cost of electricity is $0.05/kWh*

Solution: Assuming that the intake air pressure of the compressor is equal to 100 kPa (i.e., 1 atm), the reduction in the discharge pressure corresponds to a reduction in the pressure ratio P_o/P_i from 8 to 7. The percent reduction in the mechanical power requirement, $\%W_m$, can be calculated using either Eq. (11.3) or Eq. (11.4):

(a) For an isothermal compression:

$$\%W_m = \frac{Ln(8) - Ln(7)}{Ln(8)} = 6.4\%$$

Using Eq. (11.8), the electrical energy savings can be calculated:

$$\Delta kWh_{comp} = \frac{0.064 * 50kW * 4000hrs / yr * 0.70}{0.90} = 9950 kWh / yr$$

Thus, the cost savings due to a reduction in the discharge air pressure are about $500/yr.

(c) For an adiabatic compression:

$$\%W_m = \frac{(8)^{1.4-1/1.4} - (7)^{1.4-1/1.4}}{(8)^{1.4-1/1.4}} = 5.7\%$$

Using Eq. (11.8), the electrical energy savings can be calculated:

$$\Delta kWh_{comp} = \frac{0.057 * 50kW * 4000hrs / yr * 0.70}{0.90} = 8870 kWh / yr$$

Thus, the cost savings for reducing the discharge air pressure are about $450/yr.

11.3.3   Repair of Air leaks

As discussed in section 11.2.2, leaks in the distribution system result in unnecessary waste of compressed air and thus loss in the energy required to operate the compressed air system. Once the air leaks are identified and their size estimated, it is recommended to repair them as soon as possible. The energy savings associated with repair of compressed air leaks can be estimated using the following expressions:

For isothermal compressions:

$$\Delta kWh_{comp} = \frac{\Delta \dot{m}_a . N_{h,comp} . LF_{comp} . R_a . T_i . Ln\left(\dfrac{P_o}{P_i}\right)}{\eta_M} \tag{11.10}$$

For adiabatic or polytropic compressions:

$$\Delta kWh_{compm} = \frac{\Delta \dot{m}_a . N_{h,,comp} . LF_{comp} . R_a . T_i . \gamma}{(\gamma - 1).\eta_M} . \left[\left(\frac{P_o}{P_i} \right)^{\frac{\gamma - 1}{\gamma}} - 1 \right]$$ (11.11)

Where, $\Delta \dot{m}_a$ represents the mass flow rate of compressed air wasted through the leaks identified in the distribution system and can be estimated from Eq.(11.7). Example 11.5 provides an indication of the magnitude of the energy waste through a typical leak in a compressed air system. As shown in Example 11.5, the payback periods for the leak repair are generally very short.

It should be mentioned that compressed air can be wasted through the leaks even during periods where the end use tools are not in operation since air is still in the distribution system. Moreover, over-pressurization increase the amount of compressed air wasted through the leaks in addition to loss in the mechanical and electrical energy as discussed in section 11.2.1.

Example 11.5: *A leak of 5 mm has been detected in the distribution lines of a compressed air. Determine the payback period for repairing the leak. Assume an isothermal compression. Use the following information:*

a) *The total cost of the leak repair is $150.*
b) *The annual average compressed air temperature and absolute pressure are 20 °C and 900 kPa, respectively.*
c) *The compressor is operating 3000 hours per year with an average load factor of 70%.*
d) *The annual average ambient air temperature and pressure are 15 °C and 100 kPa respectively.*
e) *The electrical motor efficiency is 90%.*
f) *The cost of electricity is $0.05/kWh.*

Solution: Using Eq. (11.7), the waste in the compressed mass flow rate through the leak can be estimated:

$$\Delta \dot{m}_a = \sqrt{\frac{2}{287}} * 0.65 \ * 0.0000196 \ * 800000 \ .(273 + 20)^{-\frac{1}{2}} = 0.050 \text{kg} / \text{s}$$

For an isothermal compression, the annual electrical energy waste by the leak is calculated using Eq. (11.10):

$$\Delta kWh_{comp} = \frac{(0.050 \text{kg}/\text{s}) * 3000 \text{hrs}/\text{yr} * 0.70 * 287 \text{J}/\text{kg.K.}(273 + 15) * \text{Ln}\left(\frac{800}{100}\right)}{0.90} = 20050 \text{kWh}/\text{yr}$$

Therefore, the simple payback period for the piping system that connects the intake of the compressor to the outside air is:

$$SBP = \frac{\$150}{20050 \text{kWh} / \text{yr} * \$0.05 / \text{kWh}} = 0.15 \text{yr} = 2 \text{months}$$

11.3.4   Other Energy Conservation Measures

Other energy conservation measures that can be considered for compressed air systems are listed below:

- Replacement of inefficient compressors with new and high-efficiency compressors.

- Reduction of the compressed air usage and air pressure requirements by making some modifications to the processes.

- Installation of heat recovery systems to use the compression heat within the facility for either water heating or building space heating.

- Installation of automatic controls to optimize the operation of several compressors by reducing part load operations.

- Use of booster compressors to provide higher discharge pressures. Booster compressors can be more economical if the air with the highest pressure represents a small fraction of the total compressed air used in the facility. Without booster compressors, the primary compressor will have to compress the entire amount of air to the maximum desired pressure.

It should be mentioned that detailed technical and economical analyses should be carried out to ensure that the above listed measures are both cost-effective and non-disruptive to the proper operation of the entire air compressed system.

11.4 Summary

This chapter provides basic information on the operation procedures for compressed air systems typically used in industrial facilities but also in some commercial buildings. Several energy conservation measures have been presented. In addition, simplified analysis methods based on fundamental thermodynamic principles have been provided to estimate energy and cost savings from easy to implement operation and maintenance measures specific to compressed air systems. For industrial applications, significant energy savings can be obtained by improving the energy performance of compressed air systems in all the phases of production, distribution, and utilization.

Problems

11.1 In an industrial facility located in Denver CO, a compressed air system has a mechanical power requirement of 250 kW with a motor efficiency of 85%. The average compressed air temperature and absolute pressure are 25°C and 900 kPa, respectively. The compressor is operated 5,000 hours/yr with an average load factor of 80%. If the cost of electricity is $0.08/kWh, determine the energy and cost savings by reducing the discharge absolute pressure to (i) 800 kPa, and (ii) 750 KPa. [Use both isothermal and adiabatic compression models].

11.2 For the same conditions defined in Problem 11.1, estimate the energy and cost savings from repairing a leak. Assume the size varies from 1/8 in to 2 in [in an increment of 1/4 of an inch]. If the cost of repairing leak is $1,250, determine the smallest leak worth repairing if the longest payback period that is acceptable is one year.

11.3 A facility requires three levels of compressed air streams: 2 kg/s of 500 kPa (5 atm) air, 4 kg/s of 800 kPa (8 atm) air, and 5 kg/s of 900 kPa (9 atm) air. The existing installation generates only compressed air stream at 900 kPa. Estimate the energy and cost savings if three separate compressors are used (each to compress air at one of the three levels). The facility is operated 60 hrs/week and 300 days/year. The cost of electricity is $ 0.08/kwh.

11.4 A compressor is operated 5,000 hrs per year to generate 800 kPa air. A heat recovery can be used to increase the inlet compressor air temperature to 35°C – instead of compressing air from 100 kPa and 15°C.

(a) Estimate the energy and cost savings for using higher inlet temperature. Assume that the electricity cost is $0.07/kWh.

(b) Determine the required cost for the heat recovery system if the payback period does not exceed 5 years.

(c) In an economy of 6% and with the installed cost found in (b) for the heat recovery system, determine the threshold value of electricity price for which it is no longer cost-effective to invest in the heat recovery system. Use the LCC analysis.

12

THERMAL ENERGY STORAGE SYSTEMS

Abstract

This chapter provides a description of basic principles of designing and operating thermal energy storage (TES) systems for space cooling applications. First, an overview of the various types of cooling thermal energy storage systems is presented. Then various control strategies for TES systems are discussed to minimize the operating costs of cooling plants. A simplified calculation procedure is provided to determine the energy cost savings incurred from installing TES systems. Finally, typical operating cost savings incurred by implementing optimal controls to operate TES systems are presented.

12.1 Introduction

Thermal energy storage (TES) is generally defined as the temporary storage of energy for later use when heating or cooling is needed. For heating applications, heat storage systems are used with energy stored at high temperatures (above 20 °C). For cooling applications, energy is stored at low temperatures (below 20 °C). The concept of TES is not new and was used a few centuries ago to cool churches using blocks of ice that were stored in the cellar.

Recently, TES technology has been shown to be especially effective in reducing the operating cost of cooling plant equipment. By operating the refrigeration equipment during off-peak hours to recharge the storage system and discharge the storage during on-peak hours, a significant fraction of the on-peak electrical demand and energy consumption is shifted to off-peak periods. Cost savings are realized because utility rates favor leveled energy consumption patterns. The variable energy rates reflect the high cost of providing energy during relatively short on-peak periods. Hence, these rates constitute an incentive to reduce or avoid operation of the cooling plant during on-peak periods by the cool storage system. Large differential between on- and off-peak energy and peak consumption rates generally make cool storage systems economically feasible.

Some electric utility companies actually encourage the use of TES systems to reduce the cost required to generate on-peak electric power. Indeed, the need of building new generation plants needed to meet the demand during on-peak hours can be eliminated by promoting the use of off-peak power. Thus, utilities have initiated different rate structures to penalize the use of electric power during on-peak periods. In addition to the differential charges for on-peak vs. off-peak energy rates, the utilities have imposed demand charges, based on the monthly peak demand. These demand charges are intended to recover fixed costs such as investing in new generation plants, and transmission or distribution lines. In the US, it is estimated that 35% of the electric peak demand is due to cooling. Therefore, the use of TES systems can be a good alternative to delay the use of chillers to meet space cooling especially for commercial buildings.

TES systems can also be used to reduce the size and thus the initial cost of the cooling equipment especially in applications such as churches, where peak loads occur only for limited period during a year. For instance, a church can have a peak cooling load of 300 kW over a three-hour period which occurs once a week. Instead of using a chiller of 300 kW to operate for three hours in order to provide the required 900 kWh of cooling load, a 90-kW cooling system can be installed. The same cooling load (900 kWh) can then be produced by operating the 90-kW cooling system for a longer period (at least a 10-hour period to account for any storage losses).

Due to compressor efficiency and storage losses, cooling plants with TES systems may actually consume more energy that cooling plants without TES systems. However, TES systems, if designed and controlled properly can reduce the overall operating costs of cooling plants. To determine the potential of operating cost reduction, the auditor should carefully consider the various factors that affect the design and operation of TES systems. In the following sections, an overview of the types of TES systems as well as factors that affect both the design and the operation of cooling plants with TES systems is provided. Finally, some simplified calculation examples as well as results of parametric analyses, reported in the literature, are presented to illustrate some typical cost savings incurred due to adequately installing and properly controlling TES systems for space cooling applications.

12.2 TYPES OF TES SYSTEMS

Thermal energy storage can be achieved by two mechanisms:

1. Sensible energy storage by increasing (for heating applications) or decreasing (for cooling applications) the temperature of the storage medium (water, for instance).

2. Latent energy storage by changing the phase of the storage medium (Phase Change Materials – PCM –, eutectic salt solutions, or ice-water mixtures).

For cooling applications, there are several types of TES systems that have been installed in various commercial buildings and industrial applications. Among these TES systems are:

i. Chilled Water Storage systems: These systems consist typically of tanks where chilled water (temperature above freezing point) is stored before it is used during off-peak periods. There is no change of phase for the water in these systems, thus they can store a limited density of energy. Water is selected since it has the highest specific heat of all common materials (4.18 kJ/kg·C). Typically, a tank volume varying from 0.09 to 0.17 m^3 is required to store 1 kWh of energy using chilled water.

ii. Eutectic Salts: In these systems, a solution of salts is used to store energy at low temperatures. The advantage of these systems is that temperature below 0°C can be achieved before the solution is frozen. In addition, some salts have heat of fusion comparable to that of ice. It should be noted that the solution of salts needs to be mixed in a controlled ratio to ensure that the mixture melts completely and has the same composition in both liquid and solid phases. For eutectic salts, the volume requirement for the storage tank is estimated to be 0.05 m^3/kWh.

iii. Ice Storage Systems: In these systems, the water is transformed into ice which is stored in tanks. Therefore, the water can be present in two forms (liquid and solid) inside the tank. Typically, the ice is made during the off-peak periods (charging) and is melted during on-peak periods (discharging). Ice storage systems have higher energy density compared to chilled water systems. Thus, the volume of the storage tank required for ice systems is significantly less than that for chilled water systems (almost one-fourth). In addition, ice storage systems allow for innovative HVAC system design such as cold air distribution systems which have lower initial costs compared to conventional distribution systems. Common ice storage systems include:

- Ice harvesters [Figure 1(a)]. In these systems, thin ice layers are formed around vertical plates (evaporators) that are sprayed with water which is pumped from the tank. The ice layers are harvested to the storage tank by circulating hot gases through the evaporator. The ice mixed with water is stored in the tank to obtain what is often referred to as ice slurry. The volume requirement for the storage tank used in ice harvesters is about 0.025 m^3/kWh.

- Internal melt ice-on-coil storage systems [see Figure 1(b)]. In these systems, direct expansion coils are fitted inside the storage tank which is filled with water. Brine solution (mixture of water and ethylene glycol) is typically circulated through the coils with a temperature in the range of –6 °C to –3°C. In the charging mode, ice layers are formed around the coils. In the discharging mode, the ice is melted by circulating warm brine solution in the coils to be cooled in order to provide space cooling. The volume of the storage tank required for internal melt ice-on-coil systems varies from 0.019 to 0.023 m^3/kWh.

- External melt ice-on-coil storage systems. They are similar to the internal melt ice-on-coil system in that the ice is made around coils filled with brine solution. However, the water that results from melting ice in the storage tank is directly used to provide space cooling. Typically, a volume of 0.023 m^3/kWh is used to size storage tanks for external melt ice-on-coil systems.
- Containerized ice storage systems [see Figure 1(c)]. In these systems, small containers of various shape (typically spherical) filled with water are used inside a tank to store energy. The water inside the containers is frozen by directly cooling the solution inside the tank (which acts as the evaporator). The typical volume requirement for containerized ice storage systems is 0.048 m^3/kWh.

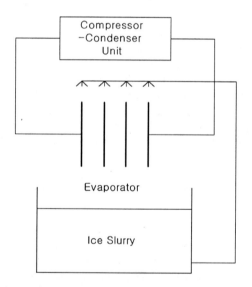

(a) Ice harvester storage system

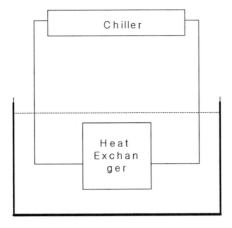

(b) Ice-on-coil with internal melt storage system

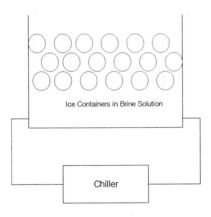

(c) Containerized ice storage system

Figure 12.1: Commonly installed ice storage systems

12.3 Principles of TES Systems

Figure 12.2 illustrates a typical configuration for a cooling plant with a TES system. Instead of one chiller, some cooling plants may have a base-load chiller that provides cooling up to a threshold load (determined by the capacity of the base chiller). Any additional cooling loads are either met directly by a second chiller (TES chiller) or the storage system. This second chiller is used to charge the TES system during unoccupied periods (or off-peak hours).

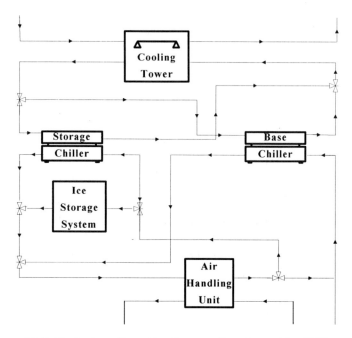

Figure 12.2: Typical configuration for a cooling plant with a TES system.

As discussed in Chapter 9, the energy efficiency of a cooling system is characterized by its coefficient of Performance (COP) which is defined as the ratio of the heat extracted divided by the energy input required. The maximum theoretical value for COP can be estimated using the ideal Carnot cycle COP which can be expressed in terms of the absolute temperature of the evaporator, T_C (the lowest temperature in the cycle), and the condenser, T_H (the highest temperature in the cycle) as follows:

$$COP_{Carnot} = \frac{T_C}{T_H - T_C} \qquad (12.1)$$

It can be seen from Eq. (12.1), the lower the evaporator temperature, the lower the COP. Therefore, and the chiller energy efficiency is reduced when it is operated at low temperatures. In particular, the chiller operates at low efficiencies when it is used to charge an ice storage tank rather than to meet directly the space cooling load. In addition, there are energy losses from the storage tanks when ice is kept for long periods of time without being utilized. Therefore, the energy used to meet a cooling load through a TES system may actually be higher than that consumed by a chiller used to provide cooling directly. It is therefore important to pinpoint that the main advantage in the use of cooling plants with TES systems is not primarily to save energy but rather to reduce the electrical demand during on-peak periods and/or to provide additional cooling beyond the chiller capacity. In most cases, TES systems are used primarily to reduce the cost of operating a cooling plant while maintaining indoor thermal comfort. The factors that affect the operation of cooling plants with TES systems include the following:

- TES and cooling plant performance during charging/discharging.
- Control strategies used to operate the TES system.
- Utility rate structures (real-time-pricing and time-of-use rates including ratchet clauses).
- Cooling load profile and non-cooling electrical load profile.

Some of the effects of these factors on operating cost savings of cooling plants are discussed in the following sections.

12.4 Charging/Discharging of TES systems

The operation of the TES systems may lead to partial – rather than full – charging and discharging of the storage tank. The performance of TES systems under partial charging/discharging can affect the overall performance of the cooling plant especially for ice storage systems characterized by ice breakage effects for low charging levels. Typically, charging is performed during nighttime (off-peak period) and discharging occurs during the daytime (on-peak period). In some instances, the ice storage tank is not fully charged and/or not fully discharged. Partial charging and discharging cycles may occur when the peak load is not high or when the charging/discharging time is limited as is the case for some real time pricing rates (Krarti et al. 1999c, and Henze and Krarti 1999). In other cases, the chiller may be too small to fully charge the entire tank during one charging period (i.e., one day). In these cases, the storage tank operates almost exclusively under partial charging and discharging cycles.

During partial charging and discharging sequences, thin ice layers can be trapped between water layers within an internal melt ice-on-coil storage tank. In particular, the coils are surrounded by ice which can be trapped by a water layer during a charging cycle. Figure 12.1 shows a cross-section view of one coil during charging cycle when the ice layers for adjacent coils do not overlap. When the tank is discharged again, the heat transfer between the brine circulating within the coils and the surrounding ice is relatively high. Similarly, Figure 12.2 illustrates a cross-section view of the coil during discharging cycle.

Partial charging/discharging cycles affect the TES performance which is generally difficult to model. Some of the effects that characterize ice storage systems especially under partial charging/discharging operation include:

- Effects of water flow within the tank during charging or discharging cycles.
- Gravitational effects that deform the ice formations around the coils.
- Effects of ice breakage at the end of discharging cycles.

Only few detailed models have been developed to predict the behavior and determine the thermal performance of ice storage systems. Most of these models typically can simulate the ice tank thermal performance during full charging and discharging cycles, but not during partial charging and discharging cycles. In particular, the model developed by Jekel et al. (1993) is based on fundamental principles of mass and energy conservation applied to the coil simulated as one node segment. The accuracy of the model is relatively marginal with an average error of 11% when compared to the manufacturer's data (Drees, 1994). Strand et al. (1994) have proposed a simplified model for an internal melt ice-on-coil storage system using a non-linear correlation to predict the heat transfer rate between the brine solution and the ice/water interface. The correlation coefficients were obtained from data specific to one manufacturer's model. Therefore, the model of Strand et al. cannot handle different ice tank designs. The model developed by Drees and Braun (1995) is based on the model proposed by Jekel et al. (1993) and uses thermal network technique to find the radius of the ice. The model developed by Carey et al. (1995) for the indirect ice storage system was integrated in the TRNSYS computer simulation program. However, the model has some restrictions. For instance, the ice/water interface is not allowed to be larger than one-half of the distance between the coils. The numerical model developed by Neto and Krarti (1997a) is more complete and has been thoroughly validated against measured data (Neto and Krarti 1997b). Neto and Krarti's model uses thermal network and iterative approach technique to find the radius of the ice during charging cycle and the radius of the water during discharging cycle. In addition, Neto and Krarti's model takes into account the overlapping phenomenon of the ice and water due to layer superposition.

However, all the models listed above simulate only full charging or discharging cycles, and cannot be applied to model accurately the thermal performance of an ice tank under partial charging and discharging cycles. Only the models developed by Vick et al. (1996), West and Braun (1999), and Kiatreungwattan (1998) can simulate partial charging and discharging cycles. All the three models are based on simplifying assumptions to simulate the performance TES systems. For instance some models neglect the overlapping of the ice and water layers which reduces the heat transfer between the brine solution and the ice/water. However, all three models can be used to evaluate the effects of partial charging/discharging operation on the overall performance of the cooling plant.

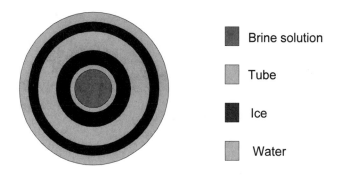

Figure 12.3: Cross-section of the coil during charging period.

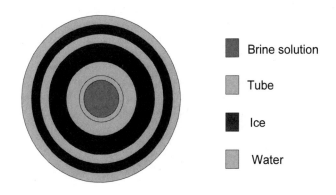

Figure 12.4: Cross section of the coil during discharging period.

Based on the results of experiments performed under controlled laboratory set-up, Kiatreungwattan (1998) compared the chiller energy use (expressed in kWh) operating with a single cycle of charging with that obtained for a sequence of partial charging and discharging cycles. The volume of the ice at the final state is set to be the same for both sequences (i.e., full charge and partial charge). Figures 12.3 and 12.4 show the chiller thermal output and electrical energy use for respectively, full charging cycle, and a sequence of partial charging and discharging cycles (corresponding to a sequence of charging-discharging-charging cycles). Table 12.1 provides a summary of the measured performance results for both charging/discharging sequences. The results indicate that one full charging cycle consumes significantly less electrical energy than a sequence of partial charging and discharging cycles with a total energy use of 238.5 kWh for full charging cycle and 531.1 kWh for the sequence of partial charging and discharging cycles. The increase in the chiller electric energy use for the partial charging/discharging operation is attributed to two main factors:

(i)   longer chiller operation time (the charging period is 8.7 hours for one single full charging cycle but is 12.7 hours for the partial charging/discharging sequence)

(ii)   more ice made (measured by the total charging energy which is 130.0 ton-hrs for full charging vs. 247.5 for the sequence of partial charging/discharging).

In addition, the results of Table 12.1 indicate that the average chiller efficiency – estimated by dividing the total chiller electrical power use by the total chiller load – is better for full charging cycle (average 1.26 kW/ton) than for a sequence of partial charging and discharging cycles (1.53 kW/ton). This result is attributed to part-load performance of the chiller. On average, the load on the chiller operating under partial charging/discharging sequence is lower than that experienced by the chiller under full charging cycle.

An average charging rate (expressed in tons) is estimated as the ratio of total charging energy (ton-hrs) over the cumulative charging time (hrs). The average charging rate is an indicator of the average heat transfer effectiveness of the ice storage tank. Table 12.1 indicates that the average charging rate is higher for a partial charging/discharging cycle (19.5 tons) compared to that for a full charging cycle (15 tons). This result is due to the fact that after discharging, the ice layer is reduced and better heat transfer is achieved between the brine and the water.

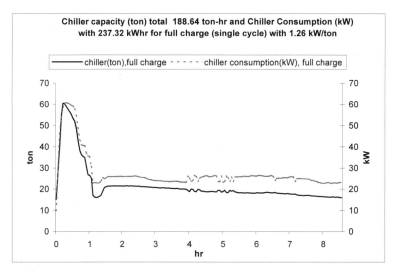

Figure 12.5: Chiller capacity and chiller power consumption during a single cycle of full charging obtained from an experimental study by Kiatreungwattan (1998).

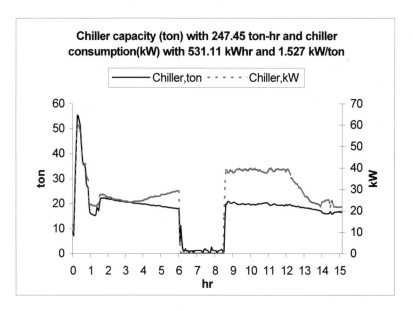

Figure 12.6: Chiller capacity and chiller power consumption during partial charging and discharging cycles obtained from an experimental study by Kiatreungwattan (1998).

Table 12.1: Average charging rate and KW/ton for the experiments performed by Kiatreungwattan (1998).

	One single full charging cycle	Sequence of partial charging/discharging cycles
Total charging time (hr)	8.65	12.70
Total charging load (ton-hr)	130.04	247.45
Average Charging rate (ton)	15.03	19.48
Total chiller consumption (kWh)	238.47	531.11
Total chiller capacity (Ton-hr)	189.67	347.62
Average chiller efficiency (kW/ton)	1.257	1.527

Thus, the experimental results indicate that the average heat transfer effectiveness of the ice tank is higher when operating with partial charging/discharging cycles than when operating with a full single charging cycle. However, the same results indicate that the chiller operates less efficiently (higher average kW/ton) under partial charging/discharging operating conditions.

In addition to the experimental analysis, simulation results obtained using numerical models (Vick et al., 1996; West and Braun, 1999; and Kiatreungwattan, 1998) indicate that several variables affect the heat transfer effectiveness of the ice storage tank and the efficiency of the chiller under partial charging/discharging cycles. In particular, Kiatreungwattan, (1998) found that:

- The average heat transfer effectiveness of the ice storage tank improves (by up to 10%) when a sequence of partial charging and discharging cycles is considered instead of one single full charging or discharging cycle.
- The average heat transfer effectiveness of the ice storage tank decreases (by up to 15%) when shorter time length for the discharging cycle is used -in a sequence with only one partial discharging cycle-.
- The average heat transfer effectiveness of the ice storage tank increases with higher brine flowrate. For instance, when the brine flowrate is increased from 60 gpm to 100 gpm, the charging effectiveness is

increased by 28%. However, the increased flowrate may increase the pumping energy which is not considered in this analysis.

In summary, the reported results indicate that a sequence of partial charging and discharging cycles provides higher average heat transfer effectiveness than a single charging cycle. Meanwhile, the chiller efficiency for the same sequence of partial charging and discharging cycles are slightly lower than that for a single full charging cycle.

12.5   TES Control Strategies

Once installed, the TES system needs to be properly operated and controlled in order to achieve the desired cost savings. Several control strategies have been proposed and actually used to operate existing TES installations. The control of a TES system depends on the operating mode considered full vs. partial storage as discussed below.

12.5.1   Full storage

This operating strategy is also called load shifting and consists of generating the entire on-peak cooling load during off-peak periods when no significant cooling load exists. Therefore, the TES system operates at full capacity and the chiller does not operate at all during the on-peak hours. Thus and in order to implement a full storage operating mode, the TES system has to be sized properly so it can hold enough energy to meet the cooling load for the entire design day on-peak hours. Full storage strategy is best suited for applications where the length of the on-peak cooling period is short compared to the off-peak period when the TES system can be charged. It can be an effective operating strategy when the on-peak demand charges are high. Moreover, the control under a full storage operating strategy is simple since all is needed is a timer clock to define charging and discharging periods. However, the full storage strategy requires large chiller and storage capacities and thus high initial costs.

12.5.2   Partial Storage

It can be defined as an operating strategy when the TES system meets only part of the on-peak cooling load. The remainder of the load is provided directly by the chiller. Partial storage requires lower initial costs than full storage since both the chiller and the storage for partial storage are of smaller size. However, partial storage operating strategies require more complex controls that full storage systems. Some of these controls are discussed below.

Chiller-Priority Control

The simplest of TES control strategies used is the *chiller-priority control*. For this strategy, the chiller runs continuously under conventional chiller control (direct cooling) possibly subject to a demand-limit while the storage provides the remaining cooling capacity if required.

The simplicity lies in the fact that the conventional chiller control is not altered yielding high average part loads and a smooth demand curve, however, with the disadvantage that the meltdown of the ice is not controlled to allow for maximal demand reduction. In addition, the time-of-day dependent energy rate structure is not accounted for.

Constant-Proportion Control

This control strategy considers that the storage meets a constant fraction of the cooling load under all conditions. Thus, neither the chiller nor the storage have priority in providing cooling. This simple control strategy provides a greater demand reduction than chiller-priority control since the chiller capacity fraction used for a particular month will track the fraction of the annual cooling design load the building experiences during that month. For example, in a month in which 50% of the annual design load occurs, only 50% of the chiller capacity will be requested.

Constant-proportion control is rather easy to implement in practice by assigning a fixed fraction of the total temperature difference between brine supply and return flow to be realized by the storage and the remainder by the chiller. Finding the best load fraction for each application is a matter of trial and error. Caution should be exercised to be sure that the chiller can always meet the remaining load fraction.

Storage-Priority Control

As the name reflects, *storage-priority control* requires melting as much ice as possible during the on-peak period. It is generally defined as that control strategy that aims at fully discharging the available storage capacity over the next available on-peak period. Thus, both the simultaneous operation of the chiller plus the storage and the terminal state-of-charge are specified; yet how this is accomplished in detail is not known (Tamblyn 1985).

The following implementation of storage priority has been suggested by Henze et al. (1997b): The chiller is base loaded during off-peak hours to recharge the ice storage for the next on-peak period. The chiller operates in one of two modes during the on-peak period. In the first mode, the chiller operates at a reduced capacity in parallel with the storage during on-peak hours so that at the end of the on-peak period the storage inventory is just depleted. If this is not possible without prematurely depleting storage, the control switches to the second mode in which the storage provides a constant proportion of the load in each on-peak hour similar to constant-proportion control.

Optimal Controls

Recently, several optimal control strategies have been proposed in order to minimize energy use and operating cost while maintaining occupant comfort. The main requirement for an optimal control strategy is typically the forecasting of cooling load and weather. The key references for optimal operating controls are Braun (1992), Drees and Braun (1996), Henze, et al. (1997a), and Henze, et al. (1997b).

For instance, the optimal control in Henze, et al. (1997a) is defined as that sequence of control actions that minimizes the total operating cost (i.e., including demand and energy charges) of the cooling plant over the simulation period. The optimization technique used is the predictive optimal controller that is proposed by Henze et al. (1997b) and is based on a "closed-loop" optimization approach. In particular, the prediction of the weather, price, and loads is updated at the beginning of each time step (typically one hour) over the optimization period. The planning horizon for the proposed controller consists of a fixed-length moving window over an entire simulation period. At each time step, only the action of the first hour is executed. In this approach, a model for the cooling plant and the TES system (the "planning" model) is needed to perform the closed-loop optimization calculation. The operating cost savings incurred by the optimal control is determined using the actual plant behavior. A more detailed description and discussion of the predictive optimal controller is provided in Henze et al. (1997b).

12.5.3   Utility Rates

There are typically two categories of utility rates which can provide some incentives for considering the installation of TES systems in cooling plants. The first rate structure is common and is known as the time-of-use (TOU) rate while the second is relatively new and is currently available from only a limited number of utilities: the real-time-pricing (RTP) rate. The reader is referred to Chapter 2 for detailed information on electric utility rate structures.

TOU Rates

These rates are currently common for most electric utility companies. They typically penalize energy use during predefined on-peak periods. Indeed, the day is divided into two or more periods during which the charges for power demand and/or energy use are set. Typically, the hours when cooling is needed are part of the on-peak period since the demand for electrical power is the highest. The charges for both energy and demand are greater during the on-peak period in an attempt by the utilities to level off the electrical power demand curve to avoid the need to operate power generation plants during short periods of time.

RTP Rates

It is believed that RTP rates will be common when utilities are deregulated. Generally the utilities determine their RTP rates based on the actual marginal costs of generating, transmitting, and distributing electricity. Although the details of RTP rate implementations may vary widely, they have several common features. Typically, the energy prices are set for a single day and are provided to the customer the preceding day. In addition, the RTP rates do not generally change in real time, but are rather constant for periods ranging from one-half hour to five hours (Norford et al., 1996).

Figure 5 shows the RTP rates used in the simulation analysis. The selected rates represent two possible scenarios of RTP rates. The sinusoidal cyclic rate is a fictitious rate and could represent an RTP rate for a week with low cost differential between on- and off-peak hours. The actual rate is extracted from real RTP rates that a US utility applied during 1994. The rate is for a week with high cost differential between on- and off-peak hours.

Chapter 2 provides a more detailed discussion of electric utility rates typically available in the US.

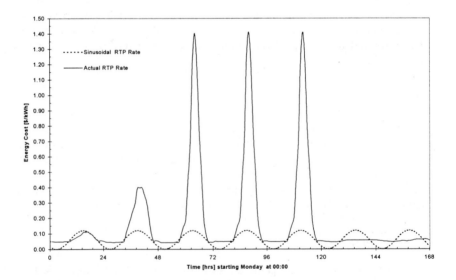

Figure 12.7: RTP rate profiles for the sinusoidal cyclic rate and the non-cyclic actual rate.

12.6   Measures for Reducing Operating Costs

As discussed in the introduction, TES systems can be utilized to reduce the operating costs of cooling systems. Typically, three measures which involve TES systems are available. These measures are:

♦ Install a TES system in the existing cooling plant.
♦ Install a TES system and replace the existing chiller (with a smaller capacity).
♦ Improve the existing operating controls of the TES systems.

To assess the cost effectiveness of any of the above listed measures, a detailed analysis is generally needed. Indeed, the utility rates which provide the major incentives for TES systems are set on an hourly and/or seasonal basis. Therefore, the use of a dynamic energy analysis tool is typically warranted to determine the overall operating cost of cooling plants with TES systems. However, simplified calculation methods can be sometimes used to especially assess whether or not a TES system is economically viable.

In this section, a simplified analysis method is presented to evaluate the cost-effectiveness of installing a TES system. In addition a summary of typical results reported in the literature for selected operating cost savings strategies suitable for TES systems is provided.

12.6.1   Simplified Feasibility Analysis of TES System

The electrical power demand reduction due to the use of TES systems depends on the control strategy selected and can be estimated using the following simplified expression:

$$\Delta kW_{TES} = \left(\frac{\dot{Q}_C}{SEER_{CHW}} \right)_e - \left(\frac{(1-X).\dot{Q}_C}{SEER_{CHW}} \right)_r \qquad (12.2)$$

With the electric demand reduction calculated above, savings in the demand charges can be estimated.

Similarly, the energy cost savings calculation incurred from the use of TES system can be calculated as follows:

$$\Delta EC_{TES} = \dot{Q}_C.N_{h,c}^{TES} \left(\frac{C_{on-pk}}{SEER_{CHW}} - \frac{C_{off-pk}}{SEER_{ICE}} \right) \qquad (12.3)$$

Where, the indices *e* and *r* indicate the values of the parameters respectively before and after retrofitting the cooling unit (i.e. adding the TES system).

➢ SEER is the seasonal efficiency ratio of the cooling unit. When available, the average seasonal COP can be used instead of the SEER. Typically the $SEER_{CHW}$ for producing chilled water (to directly cool the space) is higher than $SEER_{ICE}$ for making ice (to charge the TES system).

➢ $\dot{Q}_C$ is the rated capacity of the cooling system

➢ X is the fraction of the on-peak cooling load (occurring during the hour when maximum electrical power demand is obtained) shifted to off-peak period.

➢ $N_{h,C}^{TES}$ is the number of equivalent on-peak cooling full-load cooling hours which have been shifted during off-peak periods by using the TES system.

Example 12.1 illustrates a sample of calculation to determine the cost-effectiveness of installing a TES system for a commercial building.

Example 12.1: *Consider the cooling load profile for an office building shown in Fig. 12.8. The cooling is provided by a chiller having a capacity of 1000 kW and with an average seasonal COP of 3.5. The non-cooling profile experienced by the same office building is illustrated in Fig. 12.9. It is proposed to install an ice storage system. When making ice the chiller has an average COP of 3.0. Determine the simple payback period of installing an ice storage system if the cost of electricity is as follows:*

▪ *Energy Charges: $0.07/kWh for on-peak hours (between 10:00 and 15:00 during week-days) and only $ 0.02/Kwh during other hours.*
▪ *Demand charges: $15/kW during on-peak hours and $ 0/kW during off-peak hours. The demand charges are assessed on a monthly basis.*

The installed cost of the TES system is $100/kW. Assume that the number of typical cooling days during the entire year both before and after the installation of the TES system is 250 days. The TES system is operated with demand-leveling control so that the power demand during on-peak hours never exceeds 500 kW (which is the maximum non-cooling load).

Solution: In this example, the entire cooling load during on-peak have to be shifted and thus in Eq. (12.2) the fraction X=1. Therefore, the savings in the electric power demand can be calculated using $SEER_{CHW}$=3.5; and Q_C=1000 kW:

$$\Delta kW_{TES} = 1000kW * \left(\frac{1}{3.5}\right) = 286kW$$

Thus, the cost savings due to demand charges are: 286 kW*$15/kW*12 months/yr = $51 480/yr.

The energy cost savings of using the TES can be calculated using Eq. (12.3) with the assumption that there are 250 typical cooling days per year:

$$\Delta EC_{TES} = 1000\,kW * 5hrs\,/\,day * 250\,days\,/\,yr\left(\frac{\$0.07\,/\,kWh}{3.5} - \frac{\$0.02\,/\,kWh}{3.0}\right) = \$16,667\,/\,yr$$

The size of the ice storage tank is such that it can hold the cooling energy required during on-peak period, that is 1,000 kW*5 hrs=5,000 kWh. Therefore, the simple payback period for installing TES is estimated as follows:

$$SPB = \frac{\$100\,/\,kWh * 5000\,kWh}{(\$51480 + \$16667)} = 7.3\,years$$

A Life Cycle Cost analysis may be required to determine if the investment in a TES system is really warranted.

It should be noted that more savings can be obtained if the chiller has to be replaced. Indeed, when the TES is installed the chiller capacity can be reduced. For the case of this example, a chiller with 500 kW cooling capacity is sufficient (instead of 1,000-kW chiller) to charge the TES system and cool the office building during off-peak periods.

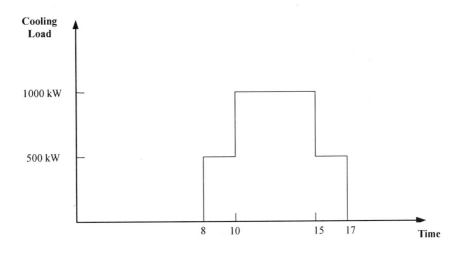

Figure 12.8: Cooling Load Profile for an office building (see Example 12.1)

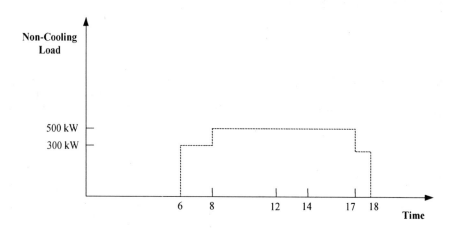

Figure 12.9: Electrical non-cooling load -profile for the office building used in Example 12.1

12.6.2   TES Control Improvement

In this section, typical results are reported on operating cost savings due to improvement of TES controls based on a detailed analyses (Krarti et al. 1999c). As discussed earlier, the performance of cooling plants with TES systems is affected by several factors. The effects of some of these factors on the cost savings of various TES control strategies are discussed below.

Effect of Plant size

The size of the cooling plant equipment including the capacity of both the chiller and the storage tank can significantly affect the operating cost savings incurred by using one of the control strategies discussed in section 12.3. Some of these effects were investigated systematically using different sizes of chillers and storage tank for both an office building and a hotel (Krarti et al. 1999c). Similar findings were reported for both buildings during both peak and swing periods. In this section, only the results for the swing season cooling load profile for the hotel are presented.

The size of the chillers and the storage tank for various design cases (for the hotel) considered in the analysis are provided in Table 12.2. The performance of all the control strategies is also summarized in Table 12.2, using chiller-priority control as a base-case (i.e., all the cost savings are relative to the savings incurred for the same cooling plant design operated with chiller priority control.

Table 12.2: Effect of cooling plant design on the performance of TES control strategies for a hotel. The savings are compared to a plant with chiller controlled TES.

System		Base Case	Reduced Storage Size	Increased Chiller Size	Reduced Base-load Chiller
Base-load chiller size	[tons]	550	550	620	350
TES chiller (chw) size	[tons]	160	160	270	160
TES chiller (ice) size	[tons]	120	120	190	120
Storage tank size	[ton-hours]	2500	1600	2500	2500
Savings					
Constant proportion	[%]	+0.4	+0.4	0.0	-1.5
Storage priority	[%]	-0.2	-0.3	-0.1	-6.7
Optimal control	[%]	-3.5	-3.4	-1.5	-11.8

In particular, the results from Table 12.2 indicate that:

- The optimal controller out-performs all the conventional controls independently of the cooling plant design.

- All the control strategies provide a significant cost saving relative to the chiller priority, for the case of a reduced base-load chiller. However, the optimal controller produces a smaller saving when the sizes of both chillers are increased (i.e., in the case of Increased Chiller Size).

- The optimal controller provides a significant cost saving (11.8 %) when two favorable conditions are present:

    (i) the base-load chiller is small (or the cooling load to be shifted is significant relative to the storage size) and
    (ii) the TES chiller size is small compared to the storage capacity. Table 12.1 shows that meeting condition (ii) by itself does not translate into increased cost savings (i.e., Base case vs. Reduced Storage).

Effect of the Cooling Load Profile

Table 12.3 summarizes the operating cost savings of the optimal controller and the conventional control strategies relative to the performance of the chiller-priority control during the peak and swing seasons for both an office and a hotel (Krarti et al. 1999c). The results presented in Table 12.3 are based on the actual RTP rate presented in Fig. 12.3.

Table 12.3 indicates that the optimal controller and the two conventional controls (i.e., constant-proportion and storage-priority) provide more operating cost savings for the office rather than for the hotel. This result is due to the fact that the cooling load profiles for the office building present distinguishable periods of high and low cooling demands unlike those for the hotel.

Other interesting features can be noted from the results shown in Table 12.2. First, the savings percentages for all the control strategies are significantly reduced when the non-cooling electrical loads are included in the total operating cost of the cooling plant. This reduction is due to the small contribution of the cooling load to the total building electrical load especially for the swing seasons. Second, the optimal controller outperforms the chiller-priority more significantly during the swing season rather than during the peak season when the non-cooling electrical loads are not considered. This result is attributed to two factors: (i) the storage-priority control generally performs poorly during swing seasons as pointed out by Braun (1992) and Henze et al. (1997b) since the storage is used only slightly when the cooling loads are low, and (ii) during the swing season, while the cooling load to be shifted is smaller, the economic benefit for using the storage is greater since only a fraction of the cooling load during the peak season can be shifted from high to low rates. However, Table 12.3 shows that when the non-cooling electrical loads are considered, the optimal controller seems to perform better during the peak season for the office building. The fact that the cooling load contribution for the office building is much lower during the swing season than that during the peak season explains the lower percentage for the cost savings during the swing season. Meanwhile for the hotel, the cooling contribution to the total electrical load remains almost the same for the two seasons.

Table 12.3: Effect of the cooling load profile on performance of the TES control strategies (Krarti et al., 1999c).

COOLING LOAD	OFFICE PEAK		OFFICE SWING		HOTEL PEAK		HOTEL SWING	
Non-Cooling Loads	Without	With	Without	With	Without	With	Without	With
Constant Priority	-5.5%	-1.4%	-4.4%	-0.6%	+0.2%	0.0%	+0.1%	0.0%
Storage Priority	-6.7%	-1.7%	-10.6%	-1.5%	-1.0%	-0.2%	-1.0%	-0.2%
Optimal Control	-17.3%	-4.4%	-20.0%	-2.8%	-2.8%	-0.5%	-3.7%	-0.7%

12.7 Summary

Some of the benefits of TES systems are discussed in this Chapter. In particular, it has been indicated that TES systems, if properly designed and controlled, can save significant operating costs for the cooling plants of commercial and institutional buildings. Operation of TES systems can be affected by several factors including primarily the electrical utility rates. In the future, with the full implementation of a deregulated electricity market, TES systems are expected to offer even greater opportunity to save energy costs for cooling plants.

Problems

12.1 Consider a building with 250 tons peak cooling load. In the peak-cooling day, the cooling load occurs during 8 hours with a load factor of 75 %. During the entire year, there are 1, 000 equivalent full-load hours. Cooling is required only during the on-peak period of the utility rate structure.

The utility rate structure includes (i) demand charge: $10/Kw with 12 months ratchet period, (ii) energy charge: the energy charge differential between on-peak and off-peak is $0.05/kWh. The installed cost of a chiller is $60/ton. The cost of a storage system is $60/ton-hr. Consider that the efficiency of the chiller is on average 1 kW/ton.

(1) Determine the cost-effectiveness (using a simple payback analysis) of installing (a) a full storage system and (b) a partial storage system with 50 % undersized chiller. For the partial storage system, assume that only 400 of equivalent full-load hours (relative to the new chiller size) have to be supplied by the chiller during on-peak period.

(2) Determine the effect of the number of hours that the chiller has to be used during on-peak period on the cost-effectiveness of the partial storage system.

12.2 Comment on the advantages and disadvantages of both full and partial storage systems.

13
COGENERATION SYSTEMS

Abstract

This chapter outlines the basic concepts of cogeneration systems and their applications for both commercial and industrial sectors. First, an overview of the various types of cogeneration systems is presented. Then, some guidelines are summarized to determine the cost-effectiveness of cogeneration systems. In particular, a simplified calculation procedure is provided to determine the energy cost savings incurred from installing cogeneration systems. Finally, practical considerations are provided for financing cogeneration systems.

13.1 Introduction

The process of generating both electricity and thermal energy is generally referred to as cogeneration. While the concept of cogeneration is not new, it has been applied to a wide range of commercial buildings only recently. Indeed, until the 1980's, cogeneration systems were used only in large industrial or institutional facilities with high electric demand (typically over 1000 kW). After the energy crisis of 1973 during which the fuel and electricity prices increased significantly (by a factor of five), the US government passed in 1978 the National Energy Act (NEA) which includes the Public Regulatory Policies Act (PURPA). The impact of the PURPA on cogeneration has been significant. Indeed, PURPA regulations obliged the local utilities to purchase cogenerated electricity and to provide supplementary or back-up power to any qualified cogeneration facilities. The Energy Policy Act of 1992 increased the appeal of cogeneration systems even more by opening up transmission line access and retail wheeling.

In addition to the favorable regulations, the development of energy-efficient cogeneration systems and small pre-engineered packaged cogeneration units has provided the needed incentives to encourage the implementation of systems capable of generating electricity and heat for commercial, institutional, and even residential applications. Currently, cogeneration systems are available over a wide range of sizes from less than 50 kW (micro-systems) to over 100 MW.

Moreover, advances in controls have helped in developing better procedures to operate and integrate the various components of a cogeneration system (including prime mover, electrical generator, and heat recovery systems).

The main advantage of cogeneration systems is their overall energy efficiency. If both electricity and thermal energy are fully utilized, a cogeneration plant may have an overall efficiency of 70% which is significantly higher than the efficiency of 35% for a typical electric power plant. This higher overall efficiency may justify the economical feasibility of cogeneration systems especially for some applications such as hospitals or university campuses where both thermal energy and electricity are needed throughout the year.

To evaluate the feasibility of a cogeneration system for a particular facility, an auditor should consider several technical and economical aspects as well as have a good understanding of the governmental regulations and legislation related to electrical power generation and its environmental impacts. While this chapter provides some discussion of the main regulatory considerations and financial options for cogeneration in the US, the main focus of the chapter is on the engineering principles that can be used to evaluate the energy and cost savings due to a cogeneration system.

13.2   Types of Cogeneration Systems

Currently, there are several types of cogeneration systems depending on the technology used and the size. In general, three categories of cogeneration systems can be considered:

(i)   Conventional Cogeneration Systems: These systems consist of large cogeneration units (more than 1,000 kW) and require a thorough design process to select the size of all equipment and components (i.e., prime movers, electrical generators, and heat recovery systems).

(ii)   Packaged Cogeneration Systems: These systems are small (below 1,000 kW) and are easy to design and install since they are pre-engineered and pre-assembled units.

(iii)   Distributed Generation Technologies: Some cogeneration systems can use a number of technologies that have been recently developed to produce both electricity and heat including fuel cells.

13.2.1   Conventional Cogeneration Systems

A conventional cogeneration plant consists of several pieces of equipment to produce electricity and heat (in the form of either steam or hot water). The number and the type of equipment in a cogeneration plant depends on the size of the system and the procedure used to generate electricity and heat. However, the main components of a typical cogeneration system include:

1. A prime mover: this is the most important equipment in a cogeneration system. It is typically a turbine that generates mechanical power using a primary source of fuel. There are three turbine types that are commonly used in cogeneration plants: turbines operated by steam generated from boilers, gas turbines fueled by natural gas or light petroleum products, and internal combustion engines fueled by natural gas or distillate fuel oils.

2. A generator: a device that converts the mechanical power to electrical energy.

3. A heat recovery system: which consists of heat exchangers that can recover heat from exhaust or engine cooling and convert it into a useful form – typically hot water.

To operate a cogeneration plant, a robust control system is needed to ensure that all the individual pieces of equipment provide the expected performance. Two basic operation cycles are used to generate electricity and heat: either a bottoming cycle or a topping cycle.

Bottoming Cycle: In this cycle, the generation of heat is given the priority to supply process heating to the facility. Thermal energy is produced directly from fuel combustion (in the prime mover). Heat is then recovered and fed to the generator to produce electricity as illustrated in Figure 13.1(a). Industrial plants characterized by high-temperature heat requirements (such as steel, aluminum, glass, and paper industries) typically use bottoming cycle cogeneration systems.

Topping Cycle: Unlike the bottoming cycle, the generation of electricity takes precedence over the production of heat as indicated in Figure 13.1(b). The waste heat is then recovered and converted to either steam or hot water. Most existing cogeneration systems are based on topping cycles. A hybrid of a topping cycle commonly used by several industrial facilities and even by electrical utilities is the combined cycle as depicted in Figure 13.1(c). In this cycle, a gas turbine is typically used to produce electricity. The exhaust gas is then fed to a heat recovery steam generator to generate more electricity – using a steam turbine. For a cogeneration plant, a small portion of steam can be converted into a useful form of thermal energy.

A detailed analysis is needed to determine the size of various equipment and to determine if the cogeneration plant is cost-effective. However, pre-engineered and pre-assembled cogeneration systems are now available with reduced costs and equipped with advanced controls. These packaged systems range in size from 600 kW to 30 kW and even smaller sizes down to 5 kW (micro-cogeneration systems). The packaged cogeneration systems are suitable for small buildings such as office buildings, schools, and small industrial facilities.

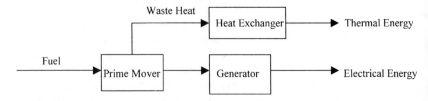

Figure 13.1 (a): Topping Cycle Cogeneration System

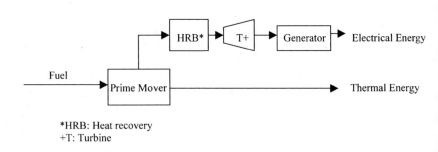

*HRB: Heat recovery
+T: Turbine

Figure 13.1 (b): Bottoming Cycle Cogeneration System

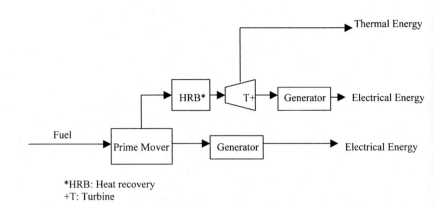

*HRB: Heat recovery
+T: Turbine

Figure 13.1 (c): Combined Cycle Cogeneration System

13.2.2   Packaged Cogeneration Systems

For cogeneration facilities requiring small systems ranging from less than 50 kW to about 1 MW, pre-engineered and factory-assembled cogeneration units are currently available with reduced construction, installation, and operation costs. In addition, small packaged systems with capacities ranging from 4 to 25 kW have been developed and can be installed in short period of time with little interruption of service. Almost all packaged cogeneration systems are equipped with advanced controls to improve the reliability and the energy efficiency of the units.

Packaged cogeneration units typically use reciprocating engines and are often sold as turn-key installations. In particular, manufacturers are generally responsible for the testing and installation of the complete cogeneration systems. The development of packaged systems has enlarged the appeal of cogeneration to a wide range of facility types including office buildings, restaurants, homes, and multi-family complexes. In addition, packaged cogeneration systems are now typically more cost-effective than conventional systems for small and medium hospitals, schools, and hotels. However, each facility has to be thoroughly evaluated to determine the economical feasibility of any packaged cogeneration system.

13.2.3   Distributed Generation Technologies

Distributed generation is a relatively recent approach proposed to produce electricity using small and modular generators. The small generators with capacities in the range of 1 kW to 10 MW can be assembled and relocated in strategic locations (typically near customer sites) to improve power quality, reliability, and flexibility in order to to meet a wide range of customer and distribution system needs. Some technologies have emerged in the last decade that allow to generate electricity with reduced waste, cost, and environmental impact which may make the future of distribution generation promising especially in a competitive deregulated market. Among these technologies are fuel cells, micro-turbines, combustion turbines, gas engines, and diesel engines. Table 13.1 summarizes some distribution generation technologies with their current efficiencies, sizes, and applications (American Gas, 1999).

Table 13.1: Characteristics of current distributed generation technologies

Type	Efficiency (%)	Size	Applications
Combustion Turbine	24-40	500 kW- 30 MW	Cogeneration (commercial/industrial), transmission and distribution support
Diesel Engine	36-42	50 kW – 6 MW	Standby, remote and peak shaving power
Gas Engine	28-38	5 kW – 2 MW	Cogeneration (commercial/industrial), peak shaving and primary power
Micro-turbine	21-40	25 kW – 300 kW	Cogeneration (commercial/light industrial), primary power
Fuel Cell	40-65	1 kW – 3 MW	Cogeneration (residential/commercial), primary power

Perhaps the recent developments in the fuel cells represent the best opportunity for distributed generation to come of age and to play a significant part in the 21st century electricity market.

The principle of the fuel cell was first demonstrated over 150 years ago. In its simplest form, the fuel cell is like a battery with two electrodes in an electrolyte medium, which serves to carry electrons released at one electrode (anode) to the other electrode (cathode). Typical fuel cells use hydrogen (derived from hydrocarbons) and oxygen (from the air) to produce electrical power with other by-products (such as water, carbon dioxide, and heat). High efficiencies (up 73%) can be achieved using fuel cells. While space applications have profited from the fuel cell technology, the real revolution of this technology will come from commercial, residential, and industrial applications.

Table 13.2 summarizes various types of fuel cells that are under development. Each fuel cell type is characterized by its electrolyte, fuel (source of hydrogen), oxidant (source of oxygen), and operating temperature range. The PAFC using phosphoric acid is the fuel cell type closest to commercialization.

Table 13.2: Types of Fuel Cells

Fuel Cell Name	Electrolyte	Fuel	Oxidant	Operating Temperatures ($^{\circ}$C)
PAFC	Phosphoric acid	Pure hydrogen	Clear air (without CO2)	200
AFC	Alkaline	Pure hydrogen	Pure oxygen and water	60-120
SPFC	Solid polymer	Pure hydrogen	Pure oxygen	60-100
MCFC	Molten Carbonate	Hydrocarbons	Air and oxygen	650
SOFC	Solid oxide	Any fuel	Air	900-1000

13.3 Evaluation of Cogeneration Systems

To evaluate the technical and the economical feasibility of a cogeneration system, it is important to collect accurate data about the facility and its energy consumption. In particular, current and projected future energy consumption and costs need to be available. For a detailed evaluation analysis, hourly electrical and thermal energy data are required. However, monthly and even yearly energy data can be sufficient for a preliminary feasibility analysis of cogeneration systems.

In this section, basic considerations are discussed to evaluate the feasibility of cogeneration systems.

13.3.1 Efficiency of Cogeneration Systems

To account for the fact that a typical cogeneration system produces both electrical power, E_e, and thermal energy, E_t, from fuel energy, FU, as illustrated in Figure 13.3, the overall thermodynamic efficiency, $\eta_{overall}$, of the cogeneration system is defined as follows:

$$\eta_{overall} = \frac{E_e + E_t}{FU} \tag{13.1}$$

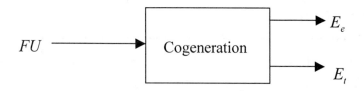

Figure 13.2: Energy input/output for a typical cogeneration system.

It should be noted that for a cogeneration facility to meet the criteria specified by the PURPA, it has to comply with certain efficiency standards. These standards use a "PURPA efficiency", η_{PURPA}, which is defined by Eq. (13.2):

$$\eta_{PURPA} = \frac{E_e + E_t / 2}{FU} \qquad (13.2)$$

PURPA states that to be a Qualified Facility (QF) for cogeneration, the efficiency, η_{PURPA}, has to be at least:

(a) 45% for the cogeneration facilities with a useful thermal energy fraction that is larger than 5%.

(b) 42.5% for the cogeneration facilities with a useful thermal fraction that is larger than 15%.

The main purpose of the PURPA efficiency is to insure that sufficient amount of thermal energy is produced so that the cogeneration facility is more efficient than the electric utility. A number of regulations within PURPA provide incentives to develop cogeneration facilities. Among these incentives are the legal obligations of the electric utilities toward the cogeneration facilities set by section 210 of PURPA (FERC, 1978). For instance, the electric utility has to:

- Purchase cogenerated electrical energy from the Qualified Facilities (QFs).
- Sell electrical energy to the QFs.
- Provide access for the QFs to transmission grid.

Example 13.1 illustrates how to estimate overall thermodynamic and PURPA efficiencies for conventional cogeneration systems.

Example 13.1: *Consider a 20-MW cogeneration power plant in a campus complex. An energy balance analysis indicates the following energy fluxes for the power plant:*

- *Electricity generation 33%*
- *Condenser losses 30%*
- *Stack losses 30%*
- *Radiation losses 7%*

It is estimated that all the condenser losses but only 12% of the stack losses can be recovered. Determine both the overall thermodynamic efficiency as well as the PURPA efficiency of the power plant.

Solution: First, the recovered thermal energy is determined:

$$E_t = \text{Condenser Losses} + \text{Part of the Stack Losses}$$

This thermal energy output can be expressed in terms of the fuel use (FU) of the power plant:

$$E_t = 30\% * FU + 12\% * [30\% * FU] = 0.34 * FU$$

The energy flow for the campus power plant is summarized in the diagram below:

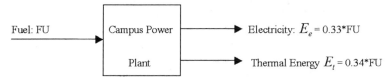

Thus, the overall thermodynamic efficiency of the power plant can be easily determined using Eq (13.1):

$$\eta_{overall} = \frac{E_e + E_t}{FU} = \frac{0.33 * FU + 0.34 * FU}{FU} = 0.67$$

The PURPA efficiency can be calculated using Eq. (13.2),

$$\eta_{PURPA} = \frac{E_e + E_t / 2}{FU} = \frac{0.33 * FU + 0.34 * FU / 2}{FU} = 0.50$$

> Therefore, the campus power plant meets the PURPA criteria ($\eta_{PURPA} > 45\%$) and is thus a qualified cogeneration facility.

13.3.2   Simplified Feasibility Analysis of Cogeneration Systems

To determine if a cogeneration system is cost-effective, simplified analysis procedures can be used first. A further evaluation with more detailed energy analysis tools may be warranted to determine the optimal design specifications of the cogeneration system.

Example 13.2 illustrates one simplified calculation procedure that can be used to determine the cost-effectiveness of installing a cogeneration system for a hospital building.

Example 13.2: *Consider a 60-kW cogeneration system that produces electricity and hot water with the following efficiencies: (a) 26% for the electricity generation, and (b) 83% for the combined heat and electricity generation. Determine the annual savings of operating the cogeneration system compared to a conventional system that consists of purchasing electricity at a rate of $0.08/kWh and producing heat from a boiler with 70% efficiency. The cost of fuel is $5/MMBtu. The maintenance cost of the cogeneration system is estimated at $1.00 per hour of operation. Assume that all the generated thermal energy and electricity are utilized during 6,500 hrs/year.*

Determine the payback period of the cogeneration system if the installation cost is $2,500/kW.

Solution: First, the cost of operating the cogeneration system is compared to that of the conventional system on an hourly basis:

(a) Cogeneration System: For each hour, 60-kW of electricity is generated (at an efficiency of 26%) with fuel requirements of 0.787 MMBtu [=60 kW*0.003413 MMBtu/kW/0.26]. In the same time, a thermal energy of 0.449 MMBtu [=0.787 MMBtu *(0.83-.26)] is obtained. The hourly flow of energy for the cogeneration system is summarized in the diagram below:

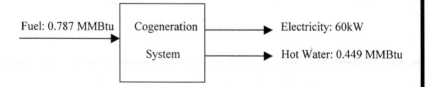

Thus, the cost of operating the cogeneration on an hourly basis can be estimated as follows:

Fuel Cost:	0.787 MMBtu/hr * $ 5/MMBtu =	$3.93/hr
Maintenance Cost:		$1.00/hr
Total Cost:		$4.93/hr

(b) Conventional System: For this system, the 60 kW electricity is directly purchased from the utility, while the 0.449 MMBtu of hot water is generated using a boiler with an efficiency of 0.65. Thus the costs associated with utilizing a conventional system are as follows:

Electricity Cost:	60 kWh/hr * 0.08 =	$4.80/hr
Fuel Cost (Boiler):	(0.449 MMBtu/hr)/0.65 * $ 5/MMBt	$3.45/hr
Total Cost:		$8.25/hr

Therefore, the annual savings associated with using the cogeneration system are:

$$\Delta Cost = (\$8.25/hr - \$4.93/hr) * 6500\ hr/yr = \$21,580/yr$$

Thus, the simple payback period for the cogeneration system:

$$SPB = \frac{\$2500/kW * 60kW}{\$21,580} = 7.0\ years$$

A Life Cycle Cost analysis may be required to determine if the investment on the cogeneration system is really warranted.

For more detailed evaluation, it is important to determine the main goal of the cogeneration system. Ideally, the cogeneration system can be sized to match exactly both the electrical and thermal loads. Unfortunately, there is almost never an exact match. Therefore, the cogeneration system has to be designed to meet specific load requirements such as the base-load thermal demand, base-load electrical demand, peak thermal demand, or peak electrical demand. The main features of each design scenario are briefly described below:

Base-load Cogeneration Systems: produce only a portion of the facility's electrical and thermal requirements. Thus, production of supplemental thermal energy (using a boiler for instance) and purchase of additional electrical energy are generally required. Base-load cogeneration systems are suitable for facilities characterized by variable thermal and electrical loads but not willing or able to sell electrical power.

Thermal-Tracking Cogeneration Systems: are those systems that produce all the thermal energy required by a facility. In case the generated electrical energy exceeds the electrical demand, the facility has to sell power to the utility. In case the generated electrical energy is lower than the electrical demand, additional power has to be purchased from the utility. Thermal-tracking cogeneration systems are increasingly becoming attractive to small buildings that have to pay higher utility rates than large industrial and commercial facilities.

Electricity-Tracking Cogeneration Systems: are designed to match electrical loads. Any supplemental energy requirements are produced through boilers. These systems are typically suitable for large industrial facilities with fairly high and constant electrical loads and lower but variable thermal loads.

Peak-Shaving Cogeneration Systems: In the case where the cost associated with peak electrical demand is high, it may be cost-effective to design cogeneration systems specifically for peak shaving even though these systems may operates only few hours (less than 1,000 hours per year).

Example 13.3 illustrates a monthly analysis to determine the optimum size for a base-load cogeneration system that is designed without selling any generated electrical power sales. The following equations have been used to carry out the analysis summarized in Table 13.4.

The monthly electrical energy, kWh_{cogen}, and the monthly thermal energy, TE_{cogen}, produced by a cogeneration system of capacity kW_{cogen}, are estimated using Eq.(13.3) and Eq.(13.4), respectively:

$$kWh_{cogen} = Min\{24.N_d.kW_{cogen}; kWh_{actual}\} \qquad (13.3)$$

$$TE_{cogen} = Min\{24.N_d.kW_{cogen}; TE_{actual}\} \qquad (13.4)$$

Where,

- N_d is the number of days in the month,
- kW_{cogen} is the electrical power capacity of the cogeneration system,
- kWh_{actual} is the actual electrical energy used by the facility during the month
- TE_{actual} is the actual thermal energy used by the facility during the month.

Example 13.3: *Provide a simple payback period analysis for implementing a cogeneration system to be installed in a hospital. Use the following characteristics for the cogeneration system:*

Fuel input rate: 10,000 Btu/kWh
Heat recovery rate: 5,500 Btu/kWh
Maintenance cost: $0.02/ kWh
Maximum electrical output: 200 kW or 300 kW
Installed equipment cost: $1000/kW

Table 13.3 summarizes the energy usage and cost of the hospital. Assume that the boiler(s) efficiency is 70%. For this analysis, assume also that the cogeneration system requires diesel fuel only (the other option is dual fuel). Assume the heating value of diesel fuel is 140, 000 Btu/ gal.

Solution: For each month, the energy cost incurred with a cogeneration system is calculated using a step-by-step procedure based on Eq. (13.3) and Eq. (13.4). Table 13.4 summarizes the results of the step-by-step analysis performed for the month of January. Table 13.5 provides the results for all the months with the payback period for each cogeneration size. In this example, the smaller cogeneration system (200 kW) is more cost-effective since there is no option to sell excess generated power to the utility. However, a detailed economic analysis should be carried out to optimize the size of the cogeneration system.

Table 13.3: Monthly utility data for the hospital used in Example 13.3

Month	\multicolumn{3}{Electricity}			Fuel Oil	
	(kWh)	(kW)	($)	(Gallon)	($)
January	226,400	546	2,7020	20,659	14,911
February	273,600	572	28,949	20,555	12,639
March	280,800	564	31,048	16,713	9,670
April	228,000	526	25,251	10,235	4,742
May	246,000	692	28,755	12,193	5,347
June	301,200	884	36,604	12,352	9,001
July	346,800	1,040	45,031	20,604	3,122
August	403,200	944	46,374	17,276	5,711
September	303,600	860	36,541	10,457	3,762
October	276,000	872	33,559	10,890	3,726
November	272,400	662	28,042	13,478	5,255
December	276,000	524	25,041	17,661	7,808
Total	3,434,000		393,215	183,073	85,694

Table 13.4: Details of step-by-step analysis performed for the month of January in Example 13.3

	Cogeneration	System
Energy/Cost Requirements	**200-kW**	**300-kW**
Electrical Energy Requirements (kWh)	226,400	226,400
Thermal Energy requirements (MMBtu)	2,024	2,024
Cogenerated Electrical Energy, kWh_{cogen} (kWh)	148,808	223,400
Cogenerated Thermal Energy, TE_{cogen} (MMBtu)	808	1,228
Electrical Energy to be purchased from Utility (kWh)	77,600	3,000
Thermal Energy to be directly generated (MMBtu)	1,206	796
Fuel Use for Cogeneration (Gal)	10,628	15,957
Fuel Energy for Direct Generation of Thermal Energy (Gal)	8,614	5,686
Total Fuel use requirements (Gal)	19,242	21,643
Cost of Utility Electrical Energy ($)	9,258	557
Cost of Fuel ($)	13,893	15,627
Cost of Maintenance ($)	2,976	4,468
Total Cost ($)	26,127	20,450

Table 13.5: Cogeneration cost savings and payback periods

	Putput(kW)	200	300
	Conventional E cost	Cogen E Cost	Cogen Cost, Cogen
Month	($/month)	($/month)	($/month)
January	41,931	26,127	20,450
February	41,588	28,920	27,687
March	40,718	27,108	21,753
April	29,993	14,527	14,813
May	34,102	20,178	14,121
June	45,605	31,658	26,034
July	49,153	32,167	24,142
August	52,085	37,815	31,536
September	40,303	26,388	20,414
October	37,285	22,687	16,339
November	33,297	21,581	16,512
December	32,849	22,097	17,892
Total ($)	478,909	311,251	251,692
Equipment cost ($)	0	200,000	300,000
Payback Period (yr)	--	1.19	1.32

13.3.3   Financial Options

To finance a cogeneration system, several financial options are generally available. Selecting the most favorable financial arrangement is critical to the success of a cogeneration project. A number of factors affect the selection of the best financial arrangement for a given cogeneration project. These factors include ownership arrangements, risk tolerance, tax laws, credit markets, and cogeneration regulations. In the US, the most common financial approaches for cogeneration facilities are the following:

(i)   Conventional Ownership and Operation: In this financing structure, the owner of the cogeneration facility either totally or partially funds the project from internal sources. In the case of partial funding, the owner can borrow the remainder funds from a conventional lending institution. Operation and maintenance of the cogeneration system can be performed by an external contractor.

(ii)   Joint Venture Partnership: This structure is an alternative to the conventional ownership and operation with a shared financing and ownership with a second partner such as an electric utility. Indeed, PURPA regulations provide the option for an electric utility to own up to 50% of a cogeneration facility. The joint venture financing structure reduces the risks for both partners but may increase the complexity of the various contracts between all the involved parties: including the owner and its partner, gas provider, electric utility, lending institution, and possible operation and maintenance contractor.

(iii)   Leasing: In this financing option, a company builds the cogeneration facility with a leasing agreement from the owner to use part or all the thermal and electrical energy output of the cogeneration plant. The construction of the cogeneration system by the lessor (i.e., the builder of the facility) can be financed through funds from lenders and/or investors. The owner is generally heavily involved in the construction phase of the cogeneration facility.

(iv)   Third-Party Ownership: This financing structure is similar to that described for the Leasing case. However, in the third-party ownership, the owner is not involved in both the financing and construction of the cogeneration facility. Instead, a third party or a lessee develops the project, and arranges for gas/fuel supply, electrical power and heat sales, and operation and maintenance agreements. The finances can be arranged by a lessor through funds from investors and/or lenders.

(v)   Guaranteed Savings Contracts: Using this financing option, a developer first builds and maintains the cogeneration facility. Then, the developer enters into a guaranteed savings contract with the energy consumer (the owner). This contract is typically made for a period ranging from 5 to 10 years with a guaranteed fixed savings per year. This type of financial structure is common for small cogeneration systems (i.e., packaged units) since it shifts all the financing and operation risks from the owner to the facility developer (i.e., the guaranteed savings contractor).

13.4 Summary

In this chapter, existing types and designs for cogeneration systems are briefly discussed. Moreover, simplified feasibility analyses of cogeneration systems are described with some illustrative examples. In the future, cogeneration is expected to become more attractive for small buildings especially with new developments in fuel cell technologies and microprocessor-based control systems. These developments will make small cogeneration systems cost-effective, reliable, and efficient for even non-traditional cogeneration applications such as residential complexes and office buildings.

Problems

13.1 Provide a simple payback analysis for implementing a cogeneration system in a hospital. Assume the following characteristics for the cogeneration system:

- Fuel input rate: 8,000 Btu/kWh
- Heat recovery rate: 4,800 Btu/kWh
- Maintenance cost: $0.02/ kWh
- Maximum electrical output: 200 kW, 300kW, 400 kW, 500 kW, 600 kW, 700 kW or 800 kW
- Installed equipment cost: $1000/kW, or $2000/kW

Table 13.3 summarizes the energy usage and cost of the hospital. Assume that the boiler(s) efficiency is 75%. The heating value of diesel fuel is 140, 000 *Btu/gal.*

Present of the results in one graph: the payback period vs. the equipment cost for various equipment sizes.

13.2 Same as Problem 13.1. Assuming that the electric cost varies from $0.05/kWh to $0.15/kWh (including demand charge), determine the variation of the cogeneration system payback period vs. electricity cost. For this question, assume that the capacity and the cost of the cogeneration system are 300 kW and $2000/kW, respectively.

13.3 Same as Problem 13.1. If the life of the cogeneration system is 40 years, provide the optimal size of the cogeneration system for the hospital when the average interest rate is 8% and the inflation rate is 4%.

13.4 A 400-kW cogeneration has an overall efficiency of 86%. The cogeneration system generation produces hot water at a thermal efficiency of 24%. The cogeneration system is operated 6,500 hrs/yr. Determine the cost-effectiveness of the cogeneration system if its installation costs is $60,000. The cost of electricity – purchased directly from the utility – is $0.08/kWh. The cost of natural gas is $1.00/therm. Without the cogeneration system, a gas-fired boiler with an efficiency is 80% is utilized to generate hot water. The maintenance cost of the cogeneration system is $0.015/kWh. Perform a LCC analysis with a life cycle of 10 years and discount rate of 4%.

14

HEAT RECOVERY SYSTEMS

Abstract

Common heat recovery systems are described in this chapter. In particular, energy performance and applications are outlined for air-to-air heat exchanges. Simplified analysis methods, illustrated with examples, are presented to determine the cost-effectiveness of installing air-to-air heat recovery systems.

14.1 Introduction

Several processes inherent to the operation the heating ventilating and air conditioning (HVAC) systems result in heat rejection to the outside. All or part of this heat can be recovered and used to perform other useful functions. Improvements in air-to-air heat exchangers have made the recovery of waste heat cost-effective for some building systems. In industrial applications, the concept of heat recovery is relatively old and common since heat is a by-product of several manufacturing processes. For instance, waste incinerators are commonly used in industrial facilities to generate steam or hot water.

Both sensible and latent heat can be recovered from various HVAC systems such as exhaust air ducts, chillers, heat pumps, and cogeneration systems. The recovery of sensible heat results generally in temperature increase of a fluid (such as outdoor intake air). Meanwhile, the latent heat affects mainly the humidity level of air streams. In some cases, the addition of latent heat can also change the air temperature when a phase change occurs. Specifically, when the humid air condenses due to a contact with a cold surface, the air temperature increases. However, when the humid air is evaporated, the air temperature is decreased. Several heat recovery devices allow sensible heat recovery including air-to-air plate heat exchangers, heat pipes, and glycol heat reclaim systems. Latent heat is recovered using desiccant systems.

In this chapter, a brief description of common heat recovery systems used in buildings is provided. In particular, the thermal efficiencies and the applications of various heat reclaim devices are presented. Moreover, simplified calculation procedures are outlined to estimate the potential energy use and cost savings due to addition or improvements of heat recovery devices.

14.2   Types of Heat Recovery Systems

Waste heat recovery can be achieved by heat exchangers that can take several forms and shapes depending on the systems involved in the exchange of thermal energy. In particular, heat exchangers can be grouped into three categories depending on the temperature involved:

- Low-temperature heat exchangers with fluid temperatures less than 230 °C (450 °F). Applications of low-temperature heat exchangers are common in buildings such as preheating of ventilation air with exhaust air.
- Medium-temperature heat exchangers with fluid temperatures ranging from 230 °C (450 °F) and 650 °C (1,200 °F). Examples of medium-temperature heat recovery systems include incinerators.
- High-temperature heat exchangers with fluid temperatures above 650 °C (1,200 °F). Generally, the use of high-temperature heat exchangers is specific to industrial processes such as in steel/aluminum furnaces.

Moreover, the fluids involved in the exchange of heat can determine the type of waste heat recovery system that is the most suitable for a given application. Three common types of heat exchangers are considered:
1. Gas-to-Gas waste heat recovery systems that include heat pipes, rotary thermal wheels, liquid coupled heat exchangers, and plate-fin heat exchangers.
2. Gas-to-Liquid waste heat recovery systems such as fire-tube or water-tube boilers, heat pipes, and economizers.
3. Liquid-to-Liquid waste heat recovery systems including shell-and-tube heat exchangers and plate heat exchangers.

There are several types of heat recovery systems that can be considered to reclaim waste heat in buildings. Some of the commonly used air-to-air heat recovery systems and their applications are described in the following sections.

These systems reclaim heat typically between intake and exhaust air streams and consist of plates, fins, or coils that are placed and extended in both intake and exhaust ducts. The air-to-air heat exchangers can be used to heat intake air during winter and cool it during summer when conditions are favorable. The energy efficiency of the air-to-air heat exchangers depends on the configuration and the temperature difference. Typically, the air-to-air heat exchangers have an energy efficiency that ranges from 45% to 65%.

1) <u>Plate-air-to-air heat exchangers:</u> These heat recovery systems have the advantage that the exhaust air does not mix with intake air and thus provide an effective method to retrieve heat virtually free of cross-contamination. The plate air-to-air heat exchangers are attractive for buildings that require large amounts of outside air. Therefore, the commercial applications for

these heat recovery systems include hospitals and restaurants. Figure 14.1 illustrates one type of plate air-to-air heat exchanger.

2) Heat-pipes: Based on a concept developed for nuclear energy applications during the 1940s, heat pipes provide simple and effective devices to reclaim heat. A heat pipe consists of a copper tube lined with a wick media and filled with refrigerant. When one end of the heat pipe is heated – by placing it in the exhaust air stream, for instance – the refrigerant is vaporized and flows to other end to provide heat to the intake air by condensation of the refrigerant. Typically, the heat pipe has a recovery rate that ranges from 50% to 70%. Even though heat pipes are more expensive than the plate air-to-air heat exchangers, their maintenance requirements are small since they have no moving parts. The useful life expectancy of heat pipes can be over 25 years. Figure 14.2 presents a basic configuration of a heat exchanger equipped with heat pipes.

3) Rotary thermal wheels: include a rotating cylinder filled with an air-permeable medium with large internal surface area. The medium can be appropriately selected to either recover sensible heat only or to reclaim total heat (i.e., sensible and latent heat). Typically, the air streams flow in a counter-flow configuration as depicted in Figure 14.3 to increase the heat transfer effectiveness. Generally, rotary thermal wheels include a purge section to reduce cross-contamination between air streams. This cross-contamination occurs by carryover air is entrained within the rotating heat exchanger medium or by leakage due to differential pressure across the two air streams.

4) Glycol loop heat exchangers: consist generally of finned-tube water coils placed in the supply and exhaust air streams. These coils are parts of a closed loop system that transfer heat from one air stream to the other using glycol solution (an antifreeze solution). These systems, often referred to as run-around coils, are suitable for sensible heat recovery applications. Figure 14.4 illustrates one configuration of run-around coil systems that is commonly used to preheat or precool fresh outdoor air with exhaust air.

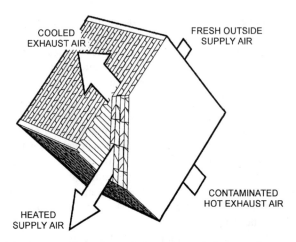

Figure 14.1: One configuration of plate air-to-air heat exchanger (*Source:* ASHRAE, <u>Systems and Equipment Handbook</u>, 2000. With permission.)

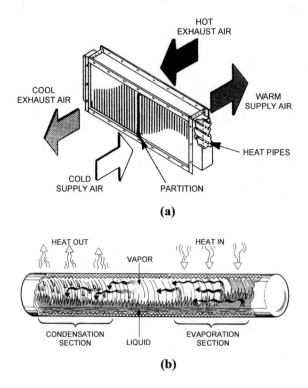

Figure 14.2 (a) Air-to-air heat exchanger equipped with heat pipes, and (b) construction details of one heat pipe (*Source:* ASHRAE, <u>Systems and Equipment Handbook</u>, 2000. With permission.)

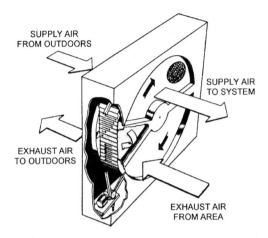

Figure 14.3: Basic components of a rotary thermal wheel (*Source:* ASHRAE, <u>Systems and Equipment Handbook</u>, 2000. With permission.)

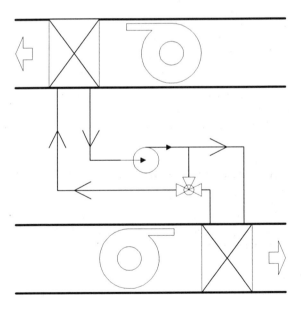

Figure 14.4: Basic set-up of a glycol loop heat exchanger

Table 14.1 summarizes some of the characteristics of the four heat recovery systems discussed above.

Table 14.1: Characteristics of Air-to-Air Waste Heat Recovery Systems Commonly used in Building Applications

	Plate HX	Heat Pipe	Rotary Wheel	Run-Around Coil
Temperature Range	-56 °C to 815 oC -70 °F to 1500 °F	-40 °C to 35 °C -40 °F to 95 °F	-56 °C to 93 °C -70 °F to 200 °F	-45 °C to 480 °C -50 °F to 900 °F
Type of Heat Transfer	Sensible/Total	Sensible	Sensible/Total	Sensible

Effectiveness Range:
  Sensible
  Total
	50% to 80%
	55% to 85%
		45% to 65%	50% to 80%
			55% to 85 %
				55% to 65%

Heat Rate Control	By-pass dampers/ducts	Change tilt angle	Change wheel speed	By-pass valve or change fluid flow
Advantages	(i) Has no moving parts (ii) Can be easily cleaned (iii) Provides low-pressure drop	(i) Has no moving parts (ii) Provides flexibility for fan location	(i) Has a compact size (ii) Allows latent heat transfer (iii) Provides Low-pressure drop	(i) Allows a separation between two air streams (ii) Provides flexibility for fan location
Main disadvantage	not suitable for latent heat transfer	low effectiveness values	potential cross-air contamination	hard to optimize its performance

14.3 Performance of Heat Recovery Systems

The performance of heat recovery systems can be determined using laboratory testing. For instance, ASHRAE Standard 84 (ASHRAE, 1991) presents a systematic procedure for testing and evaluating the performance of air-to-air heat exchangers under controllable laboratory conditions. Moreover, ARI Standard 1060 (ARI, 1997) provides an industry-established standard to rate and verify the performance air-to-air heat exchangers for use in energy recovery ventilation equipment. However, the in-situ performance of heat recovery systems can be significantly different from that obtained through laboratory testing. Indeed, balances in mass airflow as required by both ASHRAE and ARI standards are generally not achieved in field operation of heat recovery systems. Therefore, the actual – rather than the rated –

performance of the heat recovery systems should be used to determine their cost-effectiveness in retrofit applications.

The performance of air-to-air heat exchangers depends on the type of heat and mass transfer involved and is typically measured in terms of:

- Sensible energy transfer using dry-bulb temperature.
- Latent energy transfer using humidity ratio.
- Total energy transfer using enthalpy.

Figure 14.5 illustrates a basic model for a heat exchanger between two streams of fluids (typically air). In Fig. 14.5, the following parameters are indicated:

- X_i (i=1,2,3, and 4) represents one of three possible characteristics of the fluid: temperature (for sensible heat), humidity ratio (for latent heat), and enthalpy (for total heat).
- m_{in} is the mass flow rate of the fluid stream from which heat is recovered.
- m_{out} is the mass flow rate of the fluid stream that recovers heat.

To characterize the heat recovery capability of a heat exchanger, an index referred to as the effectiveness is usually provided and is defined as the ratio of the actual heat transfer and the maximum possible heat transfer that can occur between the two streams. Using the heat balance applied to the fluid streams as indicated in Fig. 14.1, it can be shown that the effectiveness can estimate using the following expression:

$$\varepsilon = \frac{m_{in}.(X_2 - X_1)}{Min[m_{in}, m_{out}].(X_4 - X_1)} \tag{14.1}$$

It should be noted, when only sensible heat is recovered, the expression of Eq.(14.1) involves temperature values as indicated in Eq.(14.2):

$$\varepsilon = \frac{m_{in}.c_{p,in}.(T_2 - T_1)}{Min[m_{in}.c_{p,in}; m_{out}.c_{p,out}].(T_4 - T_1)} \tag{14.2}$$

For air-to-air heat exchangers, when the two streams have the same mass flow (such as the case of make-up air systems), the expression for the effectiveness (sometimes referred to as the efficiency) can be further simplified to:

$$\varepsilon = \frac{(T_2 - T_1)}{(T_4 - T_1)} \tag{14.3}$$

In general, the effectiveness of a heat recovery system depends on two main factors including:

1. The contact surface between the heat exchanger and the fluid streams. The higher this contact surface, the more heat is recovered. However, the increase in the surface area – using fins, plates, and coils – required to improve the heat recovery effectiveness results in an increase in the pressure drop across the heat exchanger which has to be overcome by additional fan/pump power. Another factor that significantly affects the performance of a heat recovery system is fouling, which is basically the undesired accumulation of particulate such as dust on the contact surfaces. Fouling results in additional resistance to heat transfer and thus reduces the heat exchanger performance. Moreover, fouling increases surface roughness and increases the pressure drop, and thus may lead to higher fan or pump energy requirements. Therefore, it is important to regularly clean the heat exchanger surfaces in order to maintain a good performance of the heat exchanger.
2. Temperature difference between the two fluid streams. The higher the temperature difference, the more efficient is the heat recovery system. For instance, in cold climates where the temperature difference between air intake and air exhaust is higher than 30 °F (20 °C), the effectiveness of heat recovery systems is higher than in mild climates where the temperature difference may not exceeds 15 °F (10 °C).

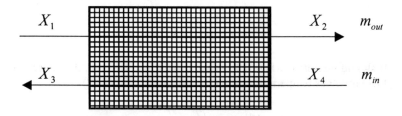

Figure 14.5: Typical Air Handling Unit for an air HVAC system

14.4 Simplified Analysis Methods

To assess the feasibility of heat recovery systems, simplified analysis methods can be used. These simplified methods are based on fundamental thermodynamics and heat transfer principles. Example 14.1 presents a simplified energy and cost analysis of reusing waste hot water for a thermal process, while Example 14.2 provides a method to assess the cost-effectiveness of air-to-air heat exchanger for a laboratory make-up air system. Finally, Example 14.3 outlines a bin-based analysis method to estimate energy and cost savings for rotary thermal wheel installed in an air handling unit serving a hospital building. For more accurate analysis of the performance of heat recovery systems, detailed simulation tools may be used.

Example 14.1: *A thermal process within a facility rejects 10,000 lbm/hr of hot water (T_{out} =190 °F). An energy audit of the facility revealed that instead of using an open water system which feeds fresh cold water (T_{in}=55 °F) to the boiler, a closed water loop system can be set-up (with a total cost of $90,000) to reuse the hot water wasted by the facility at the end of the thermal process as depicted in the sketch below. Determine the simple payback period for the closed-loop installation if the cost of fuel is $6/MMBtu and the boiler efficiency is 80%. Assume that the facility is operated 4,000 hours per year.*

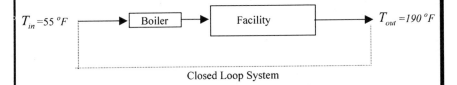

T_{in}=55 °F → Boiler → Facility → T_{out} =190 °F

Closed Loop System

Solution: First the annual fuel savings is estimated based on the energy saved by using water at T_{out} =190 °F rather than T_{in}=55 °F:

$$\Delta Q = \frac{\dot{m}c_p.(T_{out} - T_{in})}{\eta_b} = \frac{(10,000lbm/hr)*(1.0Btu/hr.°F)*[(190-55)°F]}{0.80} = 1.687*10^6 Btu/hr$$

Then, the total fuel cost savings during one year can be estimated:

$$\Delta E = \Delta Q * N_h * Cost = 1.687 MMBtu/hr * 4000hrs/yr * \$6/MMBtu = \$40,500/yr$$

Thus, the simple payback for installing the closed loop water system is:

$$SP = \frac{Initial\_Cost}{Annual\_Savings} = \frac{\$70,000}{\$40,500} = 2.22 years$$

Therefore, the project of installing a closed water loop system is cost-effective and should be considered to reduce the facility operating cost.

Example 14.2: *A heat exchanger is considered to recover heat from the exhaust air of a 5,000-cfm laboratory make-up air system located in Denver, CO. The indoor temperature within the laboratory is kept at T_{in}=70 °F. The exhaust air temperature is T_{ea}=120 °F.*

(a) *Estimate the amount of outdoor air that needs to by-pass the heat recovery system under winter design conditions (T_{oa}=-10 °F) to ensure that the supply air temperature is equal to the indoor temperature (i.e., $T_{sa} = T_{in}$).*

(b) *Determine the simple payback period if the installation cost of the heat recovery system is $1/cfm. For Denver, the average winter season outdoor temperature is T_{oa}=42.4 °F and the cost of gas is $0.10/CCF. The heat recovery system is operated 24 hours/day during 271 days/year (winter season). The heat content of gas is 840 Btu/ft^3 (Denver is located at an altitude of about 5280 ft).*

Solution: (a) The by-pass factor f can be determined using the fact that the supply temperature provided to the laboratory is the result of mixing two air streams as depicted in the diagram below:

■ outdoor air [representing a fraction f of the total air supply and kept at the temperature T_{oa}=-10 °F], and

■ heated air [coming from the heat recovery system at the temperature T_{ha} (to be determined), and representing a fraction (1-f) of the total supply air]

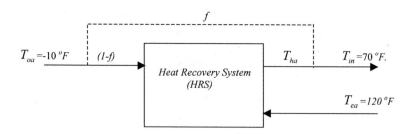

First, the heated air temperature is calculated using the definition of the heat recovery system effectiveness provided by Eq. (14.3):

$$T_{ha} = T_{oa} + \varepsilon.(T_{ea} - T_{oa}) = -10° F + 0.70*(120° F + 10° F) = 81° F$$

Then, the fraction, f, can be determined by setting $T_{sa} = T_{in}$:

$$T_{in} = T_{oa}.f + T_{ha}.(1-f)$$

Thus, the fraction, f, is:

$$f = \frac{(T_{in} - T_{ha})}{(T_{oa} - T_{ha})} = \frac{70-81}{-10-81} = 0.12$$

(c) Using the average winter conditions, the energy rate saved by the heat recovery system can be estimated as following:

$$\Delta E = \dot{m}_a.c_p.(T_{sa} - \overline{T}_{oa})$$

Where, $T_{sa} = T_{in} = 70^o F$ (assuming that the waste energy is recovered only to temper the make-up air; more thermal energy can actually be recovered if it can be used for other purposes such as space heating) and $\overline{T}_{oa} = 42.4^o F$. Thus:

$$\Delta E = (5,000cfm)*(0.91Btu/hr.^o F.cfm)*(70^o F - 42.4^o F) = 124,200 Btu/hr$$

The savings in fuel use, ΔFU, can then be calculated using a gas boiler efficiency of 80% and a number of operating hours of $N_h = 6,504$ *hrs/yr* (=24 hrs/day * 271 days/yr):

$$\Delta FU = \frac{\Delta E.N_h}{\eta_b} = \frac{(124,200 Btu/hr)*(6,504 hrs/yr)]}{0.80} = 1.007*10^6 Btu/yr = 21,021 CCF/yr$$

Thus, the simple payback period for installing the waste heat recovery system is:

$$SP = \frac{Initial\_Cost}{Annual\_Savings} = \frac{\$1/cfm*5,000cfm}{\$0.1/CCF*21,021CCF/yr} = 2.1 \ \ years$$

Therefore, the installation of heat recovery system in the laboratory is cost-effective.

Example 14.3: *A thermal wheel is considered to recover heat from the exhaust air of a 20,000 air-handling unit conditioning a hospital located in Denver, CO. The indoor temperature within the hospital is kept at $T_{in} = 68^o F$*

during the winter and T_{in}=75 °F during the summer. The effectiveness of the thermal wheel is 75% and is operated during both heating and cooling season 24 hours/day (365 days/year).

Determine the annual cost savings of using the thermal wheel if the gas-fired boiler efficiency is 80% and the chiller COP is 3.5. Use a bin analysis to estimate the annual savings in heating, cooling, and fan energy use. Estimate the payback period for installing the thermal wheel if its initial cost is $30,000. The gas cost is $1/MMBtu and electricity cost is $0.05/kWh. The static pressure added by the thermal wheel on both supply and exhaust fans is 0.55 in of water.

Solution: (a) The annual energy savings can be estimated using a bin analysis based on the following 5 °F-bin data for Denver, CO (ASHRAE, 1997):

Temperature Range (°F)	Average Temperature (°F)	Hours of Occurence (# of hrs)
-10 to -6	-8	4
-5 to 0	-3	32
0 to 4	2	31
5 to 9	7	74
10 to 14	12	110
15 to 19	17	207
20 to 24	22	500
25 to 29	27	706
30 to 34	32	700
35 to 39	37	660
40 to 44	42	665
45 to 49	47	671
50 to 54	52	741
55 to 59	57	663
60 to 64	62	655
65 to 69	67	708
70 to 74	72	514
75 to 79	77	353
80 to 84	82	360
85 to 89	87	258
90 to 94	92	131
95 to 99	97	17

The energy analysis of the thermal wheel can be divided into three parts:

- Heating Energy Savings: The annual heating energy savings can be estimated using a bin analysis. Each bin is characterized by an average outdoor temperature, $\overline{T}_{oa,b}$, and a number of hours of occurrence, N_b :

$$\Delta E_H = \dot{m}_a . c_p . \sum_b N_b . (T_{in} - \overline{T}_{oa,b})$$

Thus:

$$\Delta E_H = (5,000 \ cfm) * (0.91 \ Btu/hr.^o F.cfm) * \sum_b N_b * (68^o F - \overline{T}_{oa,b})$$

The heating will be needed as long as the outdoor temperature is below the building heating balance temperature, assumed to be 60°F in this case (due to internal gains -see Chapter 6 for more details-), that is $\overline{T}_{oa,b} \le 57^o F$.

The fuel use savings can be obtained from the heating energy savings $\Delta \dot{E}_H$, and the boiler efficiency η_b, as indicated below:

$$\Delta FU_H = \frac{\Delta E_H}{\eta_b} = \frac{\Delta E_H}{0.80}$$

Cooling Energy Savings: The sensible cooling energy savings can be estimated by following the same bin analysis considered for the heating energy savings. Thus, cooling energy use savings for each bin can be estimated as indicated below:

$$\Delta E_C = \dot{m}_a . c_p . N_b . \eta_{th} . (\overline{T}_{oa,b} - T_{in,c})$$

Therefore:

$$\Delta E_C = (20,000 \ cfm) * (0.91 \ Btu/hr.^o F.cfm) * N_b * 0.75 * (\overline{T}_{oa,b} - 75^o F)$$

The electricity savings can be obtained from the cooling energy savings ΔE_C, and the chiller COP as indicated below:

$$\Delta kWh_C = \frac{\Delta E_C}{COP} * \frac{1.0kWh}{3,412Btu}$$

Fan Energy Penalty: Since the thermal wheel adds static pressure on both the supply and exhaust fans, additional fan energy is needed to move the air through the duct system. This additional electrical power for one fan can be estimated in terms of horsepower (see Chapter 7):

$$\Delta Hp_{fan} = \frac{cfm.\Delta P_s}{6,356*\eta_s} = \frac{20,000cfm*(0.45in)}{6,356*0.65} = 2.18\,Hp$$

Thus, the total fan energy penalty for each bin is determined as follows:

$$\Delta kWh_{fan} = (1.0Hp + 2*2.18Hp)*(0.746kW/Hp)*N_b$$

The calculation of the energy savings for both heating and cooling as well as the fan energy penalty has to be performed for each bin as indicated above. The cost savings can be then easily obtained based on the fuel and electricity rates which are respectively: $3/MMBtu and $0.05/kWh.

The results of the bin analysis are summarized in the Table below:

Temperature Range (°F)	Fuel Use Savings (MMBtu)	Cooling Electricity Savings (kWh)	Fan Energy Penalty (kWh)	Total Cost Savings ($)
-5 to -1	27	0	88	76
0 to 4	90	0	320	254
5 to 9	68	0	260	190
10 to 14	101	0	424	283
15 to 19	143	0	656	395
20 to 24	313	0	1595	860
25 to 29	395	0	2259	1073
30 to 34	438	0	2851	1171
35 to 39	401	0	3031	1051
40 to 44	295	0	2659	752
45 to 49	257	0	2871	628
50 to 54	214	0	3131	485
55 to 59	143	0	3047	277
60 to 64	0	0	0	0

65 to 69	0	0	0	0
70 to 74	0	0	0	0
75 to 79	0	892	1559	-33
80 to 84	0	2784	1391	70
85 to 89	0	3223	940	114
90 to 94	0	2293	472	91
95 to 99	0	75	12	3

Total Annual Cost Savings :   7,739

It should be noted that the thermal wheel is not operated during the period when the outdoor temperature is in the range of 60 °F to 74 °F (when free cooling can be obtained). The results of the bin analysis indicate clearly that the thermal wheel should not also be operated when the temperature is in the range of 75 °F to 79 °F.

From the annual cost savings of $7,739, the simple payback period can be estimated to be:

$$SP = \frac{Initial\_Cost}{Annual\_Savings} = \frac{\$30,000}{\$7,739} = 3.9 \quad years$$

Therefore, the installation of the thermal wheel is cost-effective.

14.5 Summary

In this chapter, selected types of heat recovery systems are discussed. Simplified analysis methods are presented to evaluate the cost-effectiveness of air-to-air heat recovery systems in some HVAC applications. In the future, it is expected that heat recovery systems will be more commonly integrated with other HVAC equipment such as chillers, cogeneration engines, and heat pumps. Typically, heat recovery systems are cost-effective even though they are fairly expensive to install.

Problems

14.1 Provide a simple payback period of installing a heat wheel system in a 80,000 cfm constant volume AHU. During the winter, air is exhausted at 72 °F, and during the summer at 76 °F. The heat recovery system operates during both heating and cooling seasons with 78% efficiency.

The boiler is 75 percent efficient and uses natural gas above 50 °F ambient and diesel below 50 °F ambient. The chiller uses 0.8 kW/ton. Assume that the energy costs are $0.09/kWh, $0.90/gal (diesel), and $0.70/CCF (gas).

The heat recovery system costs $1.50 per cfm and increases the static pressure in both supply and exhaust ducts by 0.75 in WG. Assume that the fan static efficiency is 75%. To move the wheel, a 1.5 hp motor is required.

To solve this problem, use bin calculation for Denver climate.

14.2 Use the data provided in Problem 14.1. Determine the critical cost of fuel - diesel and natural gas - (assuming the cost of electricity remains constant) when using of heat wheel is not economical if the interest rate is 7% and inflation rate is 3% and the life cycle is 20 years.

15

WATER MANAGEMENT

Abstract

This chapter provides common water and energy conservation measures for indoor and outdoor water usage. Since the cost of water can be a significant fraction of the total utility bill, water management should be considered during an energy audit of a building. Water-saving fixtures are discussed in this chapter as well as measures to reduce irrigation water needs and to reuse waste water.

15.1 Introduction

In recent years, the cost of water usage has increased significantly and represents an important fraction of the total utility bills especially for residential buildings. In some western US cities where the population growth has been high, the cost of water has increased by more than 400% during the past 10 years. In the future, it is expected that the cost of water will increase at higher rates than the cost of energy and the consumer price index (CPI). Figure 15.1 provides the annual variation of the CPI and the cost rate of various utilities for the US from 1970 to 1995 (EPA, 1995). With cable television, water and sewer have the highest increases in annual cost rates since 1984. Therefore, it is worthwhile to explore potential savings in water use expenditures during a building energy audit.

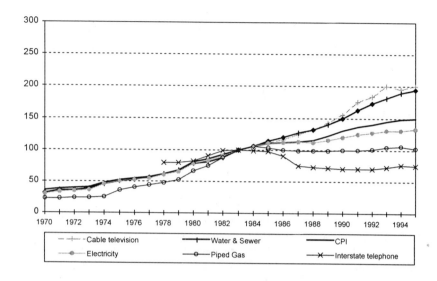

Figure 15.1: Evolution of the annual cost rates for various utilities in the
US (Conchilla, 1999)

Under the Energy Policy Act of 1992, the US government has recognized
the need for water management and requires Federal agencies to implement any
water conservation measure with a payback period of 10 years or less. The
Federal Energy Management Program (FEMP) has been established to help
Federal agencies identify and implement cost-effective conservation measures to
improve both energy and water efficiency in Federal facilities. The technical
assistance offered by FEMP includes development of water conservation plans,
training information resources, and software tools. In particular, FEMP has
developed WATERGY, computer software, that can be used to estimate
potential water and associated energy savings for buildings.

There are several water conservation strategies that can be considered for
buildings. These strategies can be grouped into three main categories:

(i)   Indoor water management with the use of water-efficient plumbing
        systems (such as low-flow showerheads and water-efficient
        dishwashers and washing machines).

(ii)   Outdoor water management associated with irrigation and landscaping
        (including the use of low-flow sprinkler heads, irrigation control
        systems, and xeriscape).

(iii)   Recycling of water usage by installing processing systems that reuse
        water.

Some of the proven water conservation technologies and techniques are described in the following sections with a special emphasis on indoor water management.

15.2   Indoor Water Management

The use of water-conserving fixtures and appliances constitutes one of the most common methods of water conservation, particularly in residential buildings. In general, the retrofit of toilets, showerheads, and faucets with water-efficient fixtures can be performed with little or no change in lifestyle for the building occupants. Similarly, water-saving appliances such as dishwashers and clothes washers can provide an effective method to reduce indoor water usage in buildings. Another common and generally cost-effective method to conserve water is repairing leaks. It is estimated that up to 10% of water is wasted due to leaks (deMonsabert, 1996). In addition to water savings, energy use reduction can be achieved when the water has to be heated as in case of domestic hot water applications (i.e., showering and hand washing). In the following sections, selected water-conserving technologies are described with illustrative examples to showcase the potential of water and associated energy savings due to implementation of these technologies.

15.2.1   Water-Efficient Plumbing Fixtures

Water, distributed through plumbing systems within buildings, is used for a variety of purposes such hand washing, showering, and toilet flushing. In recent years, water-efficient plumbing fixtures and equipment have been developed to promote water conservation. Table 15.4 summarizes typical US household end use of water with and without conservation. The average US home can reduce inside water usage by about 32% by installing water-efficient fixtures and appliances and by reducing leaks.

Table 15.1: Average US household indoor water end use with and without conservation (*Source:* AWWA 1999)

End Use	Without Conservation Gallons/capita/day	With Conservation gallons/capita/day	Savings gallons/capita/day
Toilets	20.1 (27.7%)	9.6 (19.3%)	10.5 (52%)
Clothes Washers	15.1 (20.9%)	10.6 (21.4 %)	4.5 (30%)
Showers	12.6 (17.3%)	10.0 (20.1%)	2.6 (21%)
Faucets	11.1 (15.3%)	10.8 (21.9%)	0.3 (2%)
Leaks	10.0 (13.8%)	5.0 (10.1%)	5.0 (50%)
Other Domestic	1.5 (2.1%)	1.5 (3.1%)	0 (0%)
Baths	1.2 (1.6%)	1.2 (2.4%)	0 (0%)
Dish Washers	1.0 (1.3%)	1.0 (2.0%)	0 (0%)
Total	72.5 (100%)	49.6 (100%)	22.9 (32%)

In this section, some of the proven water-efficient products are briefly presented with some calculation examples that illustrate how to estimate the cost-effectiveness of installing water-efficient fixtures. As a general rule, it is recommended to test the performance of water-efficient products to ensure user satisfaction before any retrofit or replacement projects.

Water-Saving Showerheads:
The water flow rate from showerheads depends on the actual inlet water pressure. In accordance with the Energy Policy Act of 1992, the showerhead flow rates are reported at an inlet water pressure of 80 psi. The water flow rate is about 4.0 gpm (gallons per minute) for older showerheads, and is 2.2 gpm for newer showerheads. The best available water-efficient showerheads have flow rates as low as 1.5 gpm. In addition to savings in water usage, water-efficient showerheads provide savings in heating energy cost. The calculation procedure for the energy use savings due to reduction in the water volume to be heated is presented in section 15.2.3 and is illustrated in Example 15.1.

Water-Saving Toilets
Typical existing toilets have a flush rate of 3.5 gpf (gallons per flush). After 1996, toilets manufactured in the US are required to have flush rates of at least 1.6 gpf. Therefore, significant water savings can be achieved by retrofitting existing toilets especially when they become leaky. Leaks in both flush-valve and gravity tank toilets are common and are often invisible. The use of dye tablet testing helps the detection of toilet water leaks.

Water-Saving Faucets

To reduce water usage for hand washing, low-flow and self-closing faucets can be used. Low-flow faucets have aerators that add air to the water spray to lower the flow rate. High efficiency aerators can reduce the water flow rates from 4 gpm to less than 1 gpm. Self-closing faucets are metered and are off automatically after a specified time (typically 10 seconds) or when the user moves away from the bathroom sink (as detected by a sensor placed on the faucet). The water flow rates of self-closing faucets can be as low as 0.25 gpc (gallons per cycle).

Example 15.1: *Determine the annual energy, water, and cost savings associated with replacing an existing showerhead (having a water flow rate of 2.5 gpm) with a low-flow showerhead (1.6 gpm). An electric water heater is used for domestic hot water heating with an efficiency of 95%. The temperature of showerhead water is 110 °F. The inlet water temperature for the heater is 55 °F The showerhead use is 10 minutes per shower, 2 showers per day, and 300 days per year. Assume that the electricity price is $0.07/kWh and that the combined water and waste water cost is $4/1,000 gallons.*

Solution: The annual savings in water usage due to replacing a showerhead using 2.5 gpm by another using only 1.6 gpm can be estimated as follows:

$$\Delta m = 2 * [(2.5 - 1.6)gpm] * 10\,min/\,day * 300\,days\,/\,yr = 5,400\,gal\,/\,yr$$

The energy savings incurred from the reduction in the hot water usage can be estimated as indicated below:

$$\Delta E = 5,400\,gal\,/\,yr * 8.33\,Btu\,/\,gal.° F[(110 - 55)° F]\,/\,0.95 = 2.604 * 10^6\,Btu\,/\,yr$$

Therefore, the annual cost savings in energy use and in water use are, respectively:

$$\Delta Cost = [(2.604 x 106\,Btu\,/\,yr)\,/(3.413\,Btu\,/\,Wh)] * 1000 W\,/\,kW * \$0.07\,/\,kWh = \$53\,/\,yr$$

and,
$$\Delta Cost = 5,400\,gal\,/\,yr * \$4.0\,/\,1000\,gal = \$22\,/\,yr$$

Thus, the total annual savings incurred from the water-efficient showerhead are $75. These savings make the investment in water-saving showerheads virtually certain to be cost-effective.

Repair Water Leaks

It is important to repair leaks in water fixtures even if these leaks consist of few water drips per minute. Over long periods of time, the amount of water wasted from these drips can be significant as indicated in Table 15.3. The daily, monthly, and annual water wasted due to leaks can be estimated using Table 15.3 by simply counting the number of drips in one minute from the leaky fixture. It should be noted that a leak of 300 drips per minute (i.e., 5 drips per second) corresponds to steady water flow.

Table 15.2: Volumes of water wasted from small leaks

Number of Drips per Minute	Water Wasted per Day (gal/day)	Water Wasted per Month (gal/month)	Water Wasted per Year (gal/year)
1	0.14	4.3	52.6
5	0.72	21.6	262.8
10	1.44	43.2	525.6
20	2.88	86.4	1,051.2
50	7.20	216.0	2,628.0
100	14.40	432.0	5,256.0
200	28.80	864.0	10,512.0
300	43.20	1,296.0	15,768.0

Water/Energy Efficient Appliances

In addition to water-efficient fixtures, water can be saved in residential buildings by using water-efficient appliances such clothes washers and dishwashers. The reduction of water needed to clean dishes or clothes can actually increase the energy efficiency of the household appliances. Indeed, a large fraction of the electrical energy used by both clothes washers and dishwashers is attributed to heating the water (85% for clothes washers and 80% for dishwashers). Typical water and energy performance of conventional and efficient models available for residential clothes washers and dishwashers are summarized in Table 15.5. Example 15.2 provides an estimation of the potential water and energy savings due to the use of efficient clothes washers.

Table 15.3: Water and energy efficiencies of residential dishwashers and clothes washers

Performance	Dishwashers	Clothes Washers
Conventional Models:		
Water Use (gal/load)	14.0	55.0
Energy Factor*	0.46	1.18
Efficient Models:		
Water Use (gal/load)	8.5	42.0
Energy Factor*	0.52-0.71	2.50-4.18

*Note: Energy Factor is a measure of the energy efficiency of the appliance. For dishwashers, the energy factor is the number of full wash cycles per kWh. For clothes washers it is the volume of clothes washed (in ft^3) per kWh.

Example 15.2: *Estimate the annual cost savings incurred by replacing an existing clothes washer (having 2.65 ft^3 tub volume) with a water/energy efficiency appliance that has an energy factor of 2.50 ft^3/kWh and uses 42 gallons per load. The washer is operated based on 400 cycles (loads) per year. The water is heated using an electric heater. Assume that the electricity cost is $0.07/kWh and that water/sewer cost is $5/1000 gallons.*

Solution: Using the information provided in Table 15.3, the water savings due to using an efficient clothes washer is 13.0 gallons per load. Based on 400 loads per year, the annual water savings are 5,200 gallons which amounts to cost savings of $26.

The annual electrical energy savings per load can be calculated using the energy factors as indicated in Table 15.3:

$$\Delta E = 2.65\,ft/load * [(1/1.18 - 1/2.50)kWh/ft^3] * 400\,loads/yr = 474.3\,kWh/yr$$

Thus, the annual electrical energy cost savings are $33. Therefore, the total annual cost savings achieved by using an energy/water efficient clothes washer are $59. The cost-effectiveness of the efficient clothes washer depends on the cost differential in purchasing price (between the efficient and the conventional clothes washer models). For instance, if the cost differential is $200, the payback period for using an efficient clothes washer can be estimated to be:

$$Payback = \frac{\$200}{\$59} = 3.4\,years$$

In this case, the use of a water/energy-saving clothes washer is cost-effective.

15.2.2   Domestic Hot Water Usage

In most buildings, hot water is used for hand washing and showering. To heat the water, electric or gas boilers or heaters are generally used. The energy input required to heat the water can be estimated using a basic heat balance equation as expressed by Eq. (15.1):

$$Q_w = m_w c_{w,p} (T_{w,i} - T_{w,o}) \qquad (15.1)$$

Where:
- m_w is the mass of the water to be heated. The hot water requirements depend on the building type. ASHRAE Applications Handbook (1999) provides typical hot water use for various building types.
- $C_{w,p}$ is the specific heat of water.
- T_{in} is the water temperature entering the heater. Typically, this temperature is close to the deep ground temperature (i.e., well temperature).
- T_{out} is the water temperature delivered by the heater and depends on the end-use of the hot water.

Based on Eq. (15.1) there are three approaches to reduce the energy required to heat the water. These approaches are:

(i)   Reduction of the amount of water to be heated (i.e. m_w). This approach can be achieved using water-saving fixtures and appliances as discussed in section 15.2.1. Further reduction in domestic hot water use can be realized through changes in water use habits of building occupants.

(ii)   Reduction in the delivery temperature. The desired delivery temperature depends mostly on the end-use of the hot water. It should be noted that a decrease of the water temperature occurs through the distribution systems (i.e., hot water pipes). The magnitude of this temperature decrease depends on several factors such as the length and the insulation level of the piping system. For hand washing, the delivery temperature is typically about 120 °F. For dish washing, the delivery temperature can be as high as 160 °F. However, booster heaters are recommended for use in areas where high delivery temperatures are needed such as dishwashers. A detailed analysis of water/energy efficient appliances can be found in Koomey et al. (1994).

Table 15.4: Typical hot water temperature for common residential applications

Applications	Hot Water Temperatures
Dishwashers	140-160 °F
Showers	105-120 °F
Faucet Flows	80-120 °F
Clothes Washers	78-93 °F

(iii)   Increase in the overall efficiency of the hot water heater. This efficiency depends on the fuel used to operate the heater and on factors such as the insulation level of the hot water storage tank. The National Appliance Energy Conservation Act (NAECA) of 1987 defined minimum acceptable efficiencies for various types and sizes of water heaters as indicated in Table 15.5. The energy factor (EF) is a measure of the water heater efficiency and is defined as the ratio of the energy content of the heated water to the total daily energy used by the water heater.

Table 15.5: NAECA Minimum Efficiency Standards for Water Heaters

Water Heater Fuel Type	Minimum Acceptable Energy factor (EF)
Electric	0.93-(0.00132 x rated storage volume in gallons)
Gas	0.62 – (0.0019 x rated storage volume in gallons)
Oil	0.59 – (0.0019 x rated storage volume in gallons)

Example 15.2:

(a) *Estimate the annual energy cost of domestic hot water heating for an office building occupied by 1000 people. Each person uses on average 3 gal/day of hot water. The water temperature at the bathroom sink is 120 °F. Due to the length of the distribution system, an average of 20 °F decrease in water temperature occurs before the hot water reaches the bathroom sink. The heater is an oil-fired boiler with an efficiency of 75%. The cost of fuel oil is $0.80/gal. Assume that the entering water temperature is 55 °F and that the building is occupied 300 days per year.*

(b) *Determine the cost savings if the losses in hot water temperature in the distribution system are virtually eliminated by insulating the pipes.*

Solution:

(a) Using Eq. (15.1), the annual fuel use by the water heater can be estimated. First, the water mass can be determined

$$m_w = 3.0 gal/pers/day * 1000 pers * 300 days/yr * 8.33 lmb/gal = 7.497x10^6 lbm/yr$$

Therefore, the fuel use required to heat the water from 55°F to 140°F (=120°F + 20°F) is:

$$FU = 7.497x10^6 lbm/yr * 1.0 Btu/lmb°F * [(140-55)°F)]/(0.75*138x10^3 Btu/gal)$$

Or

$$FU = 6,157 gals/yr$$

Thus, the annual fuel energy cost for water heating is:

$$Cost = [6,157 gals/yr] * \$0.80/gal = \$4,927/yr$$

When the temperature decrease across the piping system is reduced from 20°F to 10 °F (so that the actual hot water temperature from the heater is reduced from 140°F to 130°F), a reduction in the annual cost of heating the water results and is estimated to be:

$$\Delta Cost = [1-(120-55)/(140-55)] * \$4,927/yr = \$1,180/yr$$

It should be noted that the losses through the distribution systems are generally referred to as parasite losses and can represent a significant fraction of the fuel use requirements for water heating.

An economic evaluation analysis should be conducted to determine the cost-effectiveness of insulating the piping system.

15.3 Outdoor Water Management

Outdoor water conservation includes mostly innovative strategies for landscaping and irrigation of lawns and trees. Specifically, water savings can be achieved by reducing the over-watering of lawns using adequate irrigation control systems, or by replacing all or part of a landscape with less water-dependent components such as rocks and indigenous vegetation, a method known as xeriscaping. Other methods to conserve water outdoor use including swimming pools and HVAC equipment such evaporative cooling systems.

15.3.1   Irrigation and Landscaping

In addition to its esthetics, vegetation consisting of trees, shrubs, and/or turfgrass can have a positive impact on the energy use in buildings by reducing cooling loads, especially in hot and arid climates. However, with water costs rising, it is important to reduce the irrigation cost for a vegetated landscape. The amount of water necessary for irrigation depends on several factors including the type of plant and the climate conditions.

In a study conducted in a residential neighborhood consisting of 228 single-family homes near Boulder, CO, Mayer (1995) found that 78% of the total water used in the test neighborhood during the summer is attributed to lawn irrigation. Therefore, outdoor water management provides a significant potential to reduce water use for buildings. Some of the practical recommendations to reduce irrigation water use include:

- Water lawns and plants only when needed. The installation of tensiometers to sense the soil moisture content helps to determine when to water.
- Install automatic irrigation systems that provide water during early mornings or late evenings to reduce evaporation.
- Use of a drip system to water plants.
- Add mulch and water retaining organic matter to conserve soil moisture.
- Install windbreaks and fences to protect the plants against winds and reduce evaporatranpiration.
- Install rain gutters and collect water from downspouts to irrigate lawns and garden plants.
- Select trees, shrubs, and groundcovers based on their adaptability to the local soil and climate.

The amount of irrigation water use is generally difficult to determine exactly and depends on the type of vegetation and the local precipitation. Typically, the water needed by a plant is directly related to its potential evapotranspiration (ET) rate. The ET rate of a plant measures the amount of water released through evaporation and transpiration of moisture from the leaves. Figure 15.2 illustrates various factors that may affect the ET and the local climate around a typical residential building (Conchilla, 1999). The maximum possible ET rate can be estimated using one of several calculation procedures including the Penman method (Pereira, 1996) based on climate driving forces (i.e., solar radiation, wind speed, ambient temperature).

Table 15.6: Annual Normal Precipitation and ET Rates for Selected US Locations

Location	ET for Turfgras* (inches)	ET for common Trees* (inches)	Precipitation* (inches)
Phoenix, AZ	48.10	30.05	3.77
Austin, TX	38.86	24.29	21.88
San Francisco, CA	26.90	16.82	3.17
Boulder/Denver, CO	26.96	16.85	13.92
Boston, MA	24.13	15.08	22.38

Note: The values are provided for a growing season assumed to extend from April to October for all locations.

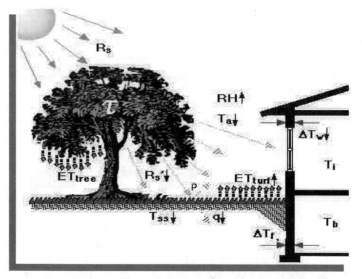

Figure 15.2: Effects of ET and other factors on the local climate around a typical house

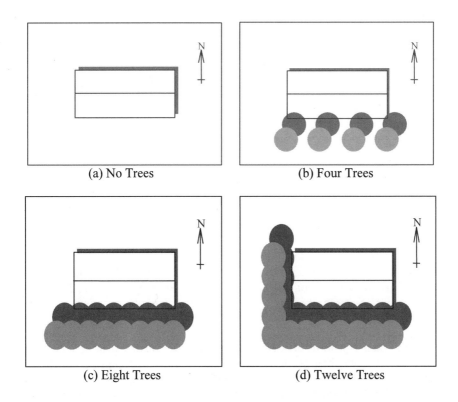

Figure 15.3: Number and location of trees around a house considered in the analysis performed by Conchilla (1999).

Soil, vegetation, and trees, while they consume water, can actually save some heating and cooling energy with well-designed landscaping. Indeed, some computer simulation studies indicated that trees well positioned around a house can save up to 50% in energy use for cooling. Table 15.5 provides the magnitude of energy savings in the cooling loads for a typical residential dwelling surrounded by a variable number of trees in selected US locations based on a simulation analysis (Conchilla, 1999). The placement configurations of the trees are illustrated in Figure 15.3. The savings in building cooling energy use attributed to vegetation are the effects of the cooler microclimate that trees and soil cover create due to shading and evapotranspiration (ET) effects. A Number of studies (Huang 1987, Taha 1991, Akbari 1993, Akbari et al. 1997, and Conchilla 1999) showed that summer time air temperatures can be 2°F to 6 °F (1°C to 3 °C) cooler in tree-shaded areas than in treeless locations. During the winter, the trees provide windbreaks to shield the buildings from wind effects such as air infiltration and thus reduce heating energy use.

Table 15.7: Percent Savings of annual cooling loads for single-family homes for various vegetation types*

Location	Turfgrass	4 Trees	8 Trees	12 Trees
Phoenix, AZ	1.2 %	4.5 %	8.0 %	13.1 %
Austin, TX	2.9 %	4.8 %	9.2 %	14.7 %
San Francisco, CA	18.7 %	27.2 %	40.3 %	57.0 %
Denver, CO	3.4 %	10.4 %	18.7 %	30.7 %
Boston, MA	8.9 %	8.6 %	19.7 %	34.8 %

*Note: These savings are estimated relative to a bare soil with no groundcover or trees.

15.3.2   Waste Water Reuse

The reuse options for building applications are generally limited to graywater use and rainwater harvesting. However, the available options depend on the type and location of the building, and the legal regulations applicable to water reuse. Sewage water treatment facilities are other options that are available but require large investments that are too costly to be considered for individual buildings.

Graywater is a form of waste water with lesser quality than potable water but higher quality than black water (which is water that contains significant concentrations of organic waste). Sources of black water include water that is used for flushing toilets, washing in the kitchen sink, and dish washing. Graywater comes from other sources such washing machines, baths, and showers and is suitable for reuse in toilet flushing. In addition, graywater can be used instead of potable water to supply some of the irrigation needs of a typical domestic dwelling landscaped with vegetation. It is believed that graywater can actually be beneficial for plants since it often includes nitrogen and phosphorus which are plant nutrients. However, graywater may also contain sodium and chloride which can be harmful to some plants. Therefore, it is important to chemically analyze the content of graywater before it is used to irrigate the vegetation around the building.

Several graywater recycling systems are available ranging from simple and low-cost systems to sophisticated and high-cost systems. For instance, a small water storage tank can be easily connected to a washing machine to recycle the rinse water from one load to be used in the wash cycle for the next load. The most effective systems include settling tanks and sand filters for treatment of the graywater.

Another method to conserve water used for irrigation is rainwater harvesting, especially in areas where rainfall is scarce. The harvesting of rainwater is suitable for both large and small landscapes and can be easily planned in the design of a landscape. There is a wide range of harvesting systems to collect and distribute water. Simple systems consist of catchment areas and distribution systems. The catchment areas are places from which water can be harvested such as sloped roofs. The distribution systems such as gutters and downspouts help direct water to landscape holding areas which can consist of planted areas with edges to retain water. More sophisticated water harvesting systems include tanks that can store water between rainfall events or periods. These systems result in larger water savings but require higher construction costs and are generally more suitable for large landscapes such as parks, schools, and commercial buildings.

15.4 Summary

This chapter outlined some measures to reduce water usage in buildings. In particular, landscaping and waste water reuse, and water-saving plumbing fixtures can provide substantial water and energy reduction opportunities. The auditor should perform water use analysis and evaluate any potential water management measures for the building.

Problems

15.1 An office building has 200 occupants, each of whom uses 3.5 gallons of hot water per day for 250 days each year. The temperature of the water as it enters the heater is 55 °F (an annual average). The water must be heated to 150°F in order to compensate for a 20°F temperature drop during storage and distribution, and still be delivered at the tap at 130°F. The hot water is generated by an oil-fired boiler (using fuel # 2) with an annual efficiency of 0.60. The cost of fuel is $0.80 per gallon. Determine:

(a) The fuel cost savings if the delivery temperature is reduced to 90°F.

(b) The fuel cost savings if hot water average usage is reduced by two-thirds.

15.2 Determine the best insulation thickness – using simple payback period analysis – to insulate a hot water storage tank with a capacity of 500 gallons. The cost of insulation is as follows: $2.60/sqft for 1-in fiberglass insulation, $2.95/sqft for 2-in fiberglass insulation, and $3.60/sqft for 3-in fiberglass insulation. The average ambient temperature surrounding the tank is 65°F. Assume appropriate dimensions for the 500-gallon tank.

15.3 An air compressor has a water cooling system that requires a flow rate of 3.5 gal/min. Water enters the compressor at 68°F and leaves at 110°F. The air compressor is operated 5,000 hours per year. Calculate the annual water and energy savings if the water could be used as boiler makeup water. The boiler is gas-fired and has an average seasonal efficiency of 80%. The water and sewage cost $0.002/gal and natural gas costs $1.10/therm.

16

METHODS FOR
ESTIMATING ENERGY
SAVINGS

Abstract

This chapter overviews some of the methods that can be used to estimate energy and cost savings incurred from implementing energy conservation measures. These methods are especially used in energy projects financed through performance contracting and are often referred to as measurement and verification (M&V) tools. The methods and their applications are briefly described with some examples to illustrate how savings are estimated. In addition, some of the protocols used in the M&V applications are discussed.

16.1 Introduction

After an energy audit of a facility, a set of energy conservation measures (ECMs) are typically recommended. Unfortunately, several of the ECMs that are cost-effective are often not implemented due to a number of factors. The most common reason for not implementing ECMs is the lack of internal funding sources (available to owners and/or managers of the buildings). Indeed, energy projects have to compete for limited funds against other projects that are perceived to have more visible impacts, such as improvements in productivity within the facility.

Over the last decade, a new mechanism for funding energy projects has been proposed to improve energy efficiency of existing buildings. This mechanism, often called performance contracting, can be structured using various approaches. The most common approach for performance contracting consists of the following steps:

- A vendor or contractor proposes an energy project to a facility owner or manager after conducting an energy audit. This energy project would save energy use and most likely energy costs and thus would reduce the facility operating costs.

■ The vendor/contractor funds the energy project using typically borrowed moneys from a lending institution.

■ The vendor/contractor and facility owner/manager agree on a procedure to repay the borrowed funds from energy cost savings that may result from the implementation of the energy project.

An important feature of performance contracting is the need for a proven protocol for measuring and verifying energy cost savings. This measurement and verification protocol has to be accepted by all the parties involved in the performance contracting project: the vendor/contractor, the facility owner/manager, and the lending institution. For different reasons, all the parties have to ensure that cost savings have indeed incurred from the implementation of the energy project and are properly estimated.

The predicted energy savings for energy projects based on an energy audit analysis are generally different from the actual savings measured after implementation of the energy conservation retrofits. For instance, Greeley et al. (1990) found in a study of over 1700 commercial building energy retrofits, that a small fraction (about 16%) of the energy projects have predicted savings within 20% of the measured results. Therefore, accepted and flexible methods to measure and verify savings are needed to encourage investments in building energy efficiency.

Direct "measurements" of energy savings from energy efficiency retrofits or operational changes are almost impossible to perform since several factors can affect energy use such as weather conditions, levels of occupancy, and HVAC operating procedures. For instance, Eto (1988) found that during abnormally cold and warm weather years, energy consumption for a commercial building can be respectively, 28% higher and 26% lower than the average weather year energy use. Thus, energy savings cannot be easily obtained by merely comparing the building energy consumption before (pre) and after (post) retrofit periods.

Over the last few years, several measurement and verification (M&V) protocols have been developed and applied with various degrees of success. Among the methods proposed for the measurement of energy savings are those proposed by the National Association of Energy Service Companies (NAESCO, 1993), the Federal Energy management Program (FEMP, 1992), the American Society of Heating Refrigeration and Air Conditioning Engineers (ASHRAE, 1997), the Texas LoanSTAR program (Reddy et al., 1994), and the North American Energy Measurement and Verification Protocol (NEMVP) sponsored by DOE and later updated and renamed the International Performance Measurement and Verification Protocol (IPMVP, 1997).

In this chapter, general procedures and methods for measuring and verifying energy savings are presented. Some of these methods are illustrated with calculation examples or with applications reported in the literature.

16.2   General Procedure

To estimate the energy savings incurred by an energy project, it is important to first identify the implementation period of the project, that is the construction phase where the facility is subject to operational or physical changes due to the retrofit. Figure 16.1 illustrates an example of the variation of the electrical energy use in a facility that has been retrofitted from constant volume to a variable air volume HVAC system. The time-series plot of the facility energy use clearly indicated the duration of the construction period, the end of the pre-retrofit period, and the start of the post-retrofit period. The duration of the construction period depends on the nature of the retrofit project and can range from a few hours to several months.

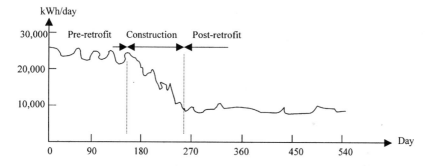

Figure 16.1: Daily variation of a building energy consumption showing pre-retrofit, construction, and post-retrofit periods after CV to VAV conversion.

The general procedure for estimating the actual energy savings, ΔE_{actual}, from a retrofit energy project is based on the calculation of the difference between the pre-retrofit energy consumption predicted from a model and the post-retrofit energy consumption obtained directly from measurement (Kissock et al, 1998):

$$\Delta E_{actual} = \sum_{j=1}^{N} \Delta E_j = \sum_{j=1}^{N} (\tilde{E}_{pre,j} - E_{post,j}) \qquad (16.1)$$

where
- N is the number of post-retrofit measurements (for instance, during one year, N=365 daily data can be used).

- $\widetilde{E}_{pre,j}$ is the energy use predicted from a pre-retrofit model of the facility using the weather and operating conditions observed during the post-retrofit period.

- $E_{post,j}$ is the energy used by the facility measured during the post-retrofit period.

Therefore, it is important to develop a pre-retrofit energy use model for the facility before estimating the retrofit energy savings. This pre-retrofit model helps determine the base-line energy use of the facility based on the weather and operating conditions during the post-retrofit period.

In some instances, the energy savings estimated using Eq. (16.1) may not be representative of average or typical energy savings from the retrofit project. For instance, the measured energy use during the post-retrofit period may coincide with abnormal weather conditions and thus may lead to retrofit energy savings that are not representative of average weather conditions. In this case, a post-retrofit energy use model for the building can be used instead of measured data to estimate normalized energy savings that would occur under typical weather and operating conditions. The normalized energy savings, ΔE_{norm}, can be calculated as follows:

$$\Delta E_{norm} = \sum_{j=1}^{N} \Delta \widetilde{E}_{j} = \sum_{j=1}^{N} (\widetilde{E}_{pre,j} - \widetilde{E}_{post,j}) \qquad (16.2)$$

- $\widetilde{E}_{pre,j}$ is the energy use predicted from a pre-retrofit model of the facility using normalized weather and operating conditions.

- $\widetilde{E}_{post,j}$ is the energy use predicted from a post-retrofit model of the facility using normalized weather and operating conditions.

It should be noted that the energy savings calculation procedures expressed by Eq. (16.1) and Eq. (16.2) can be applied to sub-systems of a building (such as lighting systems, motors, chillers, etc.) as well as to the entire facility. In recent years, several techniques have been proposed to estimate various energy end-uses of a facility. Some of these approaches are discussed in section 16.3.

There are several approaches to estimate the energy savings from energy retrofits. These methods are used as part of the M&V protocol in performance contracting projects and range from simplified engineering approaches to detailed simulation and measurement techniques. For specific projects, the method to be used in M&V of savings depends on the desired depth of the verification, the accuracy level of the estimation, and on the accepted cost of the total M&V project. In general, the cost of the M&V procedure depends on the metering equipment needed to obtain detailed data on the energy consumption of the facility and its end-uses. The installation cost of the metering equipment can represents up to 5% of the total energy project costs, especially if no energy management system is available in the facility. A number of suggestions have been proposed to reduce the cost of metering equipment including:

- Use of statistical sampling techniques to reduce the metering requirements such as the case for lighting retrofit projects where only a selected number of lighting circuits are metered. A complete metering of the entire lighting circuits within a facility, while feasible, can be very costly.

- End-use metering for specific systems directly affected by the retrofit project. For a CV to VAV conversion project, for example, metering may be needed only for the retrofitted air-handling unit.

- Limited metering requirements with stipulated calculation procedures to verify savings. For instance, the energy savings incurred from a replacement of motors can be estimated using simplified engineering methods that may require only the metering of operation hours.

Some of the most commonly used methods for verifying savings are briefly presented in the following section.

16.3   Energy Savings Estimation Models

16.3.1   Simplified Engineering Methods

These methods are commonly used in performance contracting projects that include retrofits of lighting systems and/or motors. The calculation methods have to be accepted by all the parties involved in the retrofit project to be representative of actual savings. The step-by-step calculation procedure is generally part of the written agreement between the parties involved in the energy project. Examples 16.1 and 16.2 provide illustrations of some of the simplified engineering methods that one can use to verify savings in lighting and motor retrofit projects. The reader is referred to Chapter 5 for further details about the calculation procedures to estimate the energy and cost savings for lighting retrofit projects.

Example 16.1: *A medium office building has 800 luminaires equipped with four 40-W lamps and standard magnetic ballasts. The manager of the building agreed with an energy service company to replace these luminaires with four 32-W lamps and electronic ballasts. In the agreement, a specific calculation procedure has been defined to verify the energy savings due to the lighting retrofit. In particular, it was agreed that the lighting system is operated 8 hours per day, 5 days per week, 50 weeks per year. In addition, the wattage rating is set to be 192 W for the standard luminaire and 140 W for the energy efficient luminaire.*

Estimate the annual energy savings stipulated by the agreement. Determine the cost savings due to the lighting retrofit if the electricity cost is $0.07/kWh.

Solution: The reduction in the electrical energy input for all the luminaires is first determined:

$$\Delta k W_{light} = [(0.192 - 0.140) kW] * 800 = 41.6 \ kW$$

The total number of hours for operating the lighting system during one year can be estimated as follows

$$\Delta t_{light} = 8 hrs / day * 5 days / week * 50 weeks / year = 2,000 hrs / yr$$

Thus, the annual energy savings due to the lighting retrofit are:

$$\Delta kWh_{light} = 41.6 kW * 2,000 \ hrs / yr = 83,200 \ kWh / yr$$

Therefore, the annual cost savings that resulted from the stipulated savings are:

$$\Delta C_{light} = \$0.07 / kWh * 83,200 kWh / yr = \$5,824 / yr$$

Example 16.2: *An energy service company agreed to convert all the 12 constant volume air-handling units (AHUs) within a commercial building into variable air volume (VAV) systems by installing variable frequency drives (VFDs) and VAV terminal boxes. To verify fan energy savings, it was agreed to meter only one AHU. Metering all the 12 AHUs was determined to be not economically feasible. The metering of one AHU revealed the following: (a) the supply fan is rated at $kW_{e, fan}$ = 40 kW and is operated for 6,000 hours/year, and (b) the VFD operates according to the following load profile:*

> *at 100% of its speed, 500 hours per year (i.e. 8.33% of total operating hours),*
> *at 80% of its capacity, 3,500 hours per year (i.e. 58.33% of total operating hours),*
> *at 60% of its capacity, 1,500 hours per year (i.e. 25.0% of total operating hours), and*
> *at 40% of its capacity, 500 hours per year (i.e. 8.33% of total operating hours).*

It was agreed that the energy use of one motor is proportional to its speed squared (The reader is referred to Chapter 7 for more details on fan performance). Determine the fan energy and cost savings for the CV to VAV conversion for all the AHUs as stipulated in the agreement which states that the annual fan energy savings for the all the AHUs is simply 12 times the fan energy savings estimated from the metered AHU. Assume that electricity costs $0.07/kWh.

Solution: The electrical energy input of the supply fan motor is first determined for the metered AHU by calculating a weighted average fan rating (see Example 7.3 for more details):

$$\overline{kW}_{e,fan}^{VSD}=40kW*(1.0*0.083+0.64*0.583+0.36*0.25+0.16*0.083)=22.4kW$$

The total number of hours for operating the fan during one year is $N_{f,h}=6,000$ hours. The electrical energy use for one supply fan can be estimated as indicated below:

$$kWh_{e,fan}^{CV} = 40kW * 6,000hrs = 240,000kWh / yr$$

and,

$$kWh_{e,fan}^{VSD} = 22.4kW * 6,000hrs = 134,400kWh / yr$$

Thus, the annual fan energy savings due to the CV to VAV conversion for all the 12 AHUs are:

$$\Delta kWh_{AHUs} = 12*[(240,000-134,400)kWh / yr] = 1,267,200kWh / yr$$

The annual fan energy cost savings that resulted from all the 12 retrofitted AHUs are:

$$\Delta C_{light} = \$0.07/ kWh*1,267,200kWh/ yr = \$88,704/ yr$$

16.3.2   Regression Analysis Models

Using metered data, models for building energy use for pre- and/or post-retrofit periods can be established using regression analysis. The first developed regression model for building energy use estimation has been an application of the variable base degree days (VBDD) method [the reader is referred to Chapters 4 and 6 for a more detailed description of the VBDD method]. Indeed, Fels (1986) proposed the PRIncepton Scorekeeping Method (PRISM) to correlate the monthly utility bills and the outdoor temperatures to estimate the energy use for heating and cooling and estimate any energy savings for residential retrofit measures. A similar approach has been used to develop a regression analysis method, FASER, for large groups of commercial buildings (OmniComp, 1984).

In recent years, general regression approaches have been proposed to establish baseline models for commercial buildings energy use which can be used to estimate retrofit savings. Two main regression models have been developed and applied successfully to predict energy use in commercial buildings:

- Single-variable regression analysis models that assume that the building energy use is driven by one variable (typically, the ambient temperature).
- Multi-variable regression models that account in addition to temperature for other independent variables (such as solar radiation, humidity ratio, and internal gain) to predict the building energy use.

A brief description of these two regression analysis models is provided in the following sections.

Single-Variable Regression Analysis Models

These models constitute the main procedures adopted by the International Performance Measurement and Verification Protocol (IPMVP, 1997). A simple linear correlation is assumed to exist between the building energy use and one independent variable. The ambient temperature is typically selected as the independent variable especially to predict commercial/residential building heating and cooling energy use. Degree days with properly selected balance temperature can be used as another option for the independent variable.

Ambient-temperature based regression models have been shown to predict building energy use with an acceptable level of accuracy even for daily data sets (Kissock et al., 1992; Katipamula et al., 1994; Kissock and Fels, 1995) and can be used to estimate energy savings (Claridge et al. 1991, and Fels and Keating 1993). Four basic functional forms of the single-variable regression models have been proposed for measuring energy savings in commercial and residential buildings. The selection of the function form depends on the application and the building characteristics. Figure 16.2 illustrates the four basic functional forms commonly used for ambient-temperature linear regression models. The regression models, also called change-point or segmented-linear models, combine both search methods and least-squares regression techniques to obtain the best-fit correlation coefficients. Each change-point regression model is characterized by the number of the correlation coefficients. Therefore, the two-parameter model has two correlation coefficients (β_o and β_1) and consists of a simple linear regression model between building energy use and ambient temperature. Table 16.1 summarizes the mathematical expressions of four change-point models and their applications. In general, the change-point regression models are more suitable for predicting heating rather than cooling energy use. Indeed, these regression models assume steady-state conditions and are insensitive to the building dynamic effects, solar effects, and non-linear HVAC system controls such as on-off schedules.

Other types of single-variable regression models have been applied to predict the energy use of HVAC equipment such as pumps, fans, and chillers. For instance, Phelan et al. (1996) used linear and quadratic regression models to obtain correlation between the electrical energy used by fans and pumps and the fluid mass flow rate.

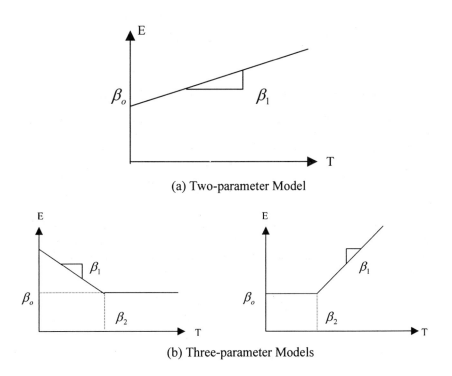

(a) Two-parameter Model

(b) Three-parameter Models

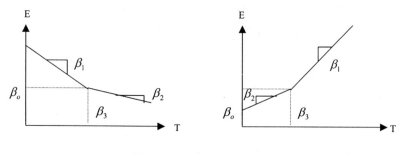

(c) Four-parameter Models

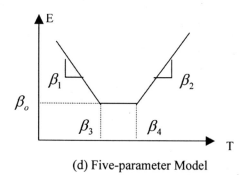

(d) Five-parameter Model

Figure 16.2: Basic forms of single-variable regression models

Multi-Variable Regression Analysis Models

These regression models use several independent variables to predict building energy use and/or energy savings due to retrofit projects. Several studies indicated that the multi-variable regression models provide better predictions of monthly, daily, and even hourly energy use of large commercial buildings than the single-variable models (Haberl and Claridge, 1987; Katipamula et al. 1994, 1995, and 1998). In addition to outdoor temperature, multi-variable regression models use internal gain, solar radiation, and humidity ratio as independent variables. For instance, the cooling energy use for commercial buildings conditioned with VAV systems can be obtained using the following functional form of a multi-variate regression model (Katipamula et al., 1998):

$$E = \beta_0 + \beta_1 T_a + \beta_2 I + \beta_3 I T_a + \beta_4 T_{dp}^+ + \beta_5 q_i + \beta_6 q_s \quad (16.3)$$

Where:

- β_0, through β_6 are regression coefficients.

- I is an indicator variable that is used to model the change in slope of the energy use variation as function of the outdoor temperature.
- T_a is the ambient outdoor dry-bulb temperature
- T_{dp}^+ is the outdoor dew-point temperature
- q_i is the internal sensible heat gain
- q_s is the total global horizontal solar radiation

Table 16.1: Mathematical expressions and applications of change-point regression models

Model Type	Mathematical Expression	Applications
Two-Parameter (2-P)	$E = \beta_0 + \beta_1.T$	Buildings with constant air volume systems and simple controls
Three-Parameter (3-P)	Heating: $E = \beta_0 + \beta_1.(\beta_2 - T)^+$ Cooling: $E = \beta_0 + \beta_1.(T - \beta_2)^+$	Buildings with envelope-driven heating or cooling loads (i.e., residential and small commercial buildings)
Four-Parameter (4-P)	Heating: $E = \beta_0 + \beta_1.(\beta_3 - T)^+ - \beta_2.(T - \beta_3)^+$ Cooling: $E = \beta_0 - \beta_1.(\beta_3 - T)^+ + \beta_2.(T - \beta_3)^+$	Buildings with variable-air-volume systems and/or with high latent loads. Also, buildings with non-linear control features (such as economizer cycles and hot deck reset schedules)
Five-Parameter (5-P)	$E = \beta_0 + \beta_1.(\beta_3 - T)^+ + \beta_2.(T - \beta_4)^+$	Buildings with systems that use the same energy source for both heating and cooling (i.e., heat pumps, electric heating and cooling systems)

For commercial buildings conditioned with a constant-volume HVAC systems, the cooling energy use can be predicted using a simplified model of Eq. (16.3) as follows:

$$E = \beta_0 + \beta_1 T_a + \beta_2 T_{dp}^+ + \beta_3 q_i + \beta_4 q_s \qquad (16.4)$$

The models represented by Eq. (16.3) and Eq. (16.4) have been applied to predict the cooling energy consumption in several commercial buildings using various time scale resolutions: monthly, daily, hourly, and hour-of-day (HOD). The HOD predictions, which require a significant modeling effort since the energy use data have to be regrouped in hourly bins corresponding to each hour of the day and 24 individual hourly models have to be obtained, are found to have the better accuracy (Katipamula et al., 1994). Table 16.3 indicates some of the advantages and disadvantages of the different multi-variable regression modeling approaches. Generally, monthly utility bills can be used to develop monthly regression models but metering is required to establish daily, hourly, and HOD models.

Table 16.2: Typical advantages and disadvantages of multi-variable regression models with different time resolution (Katipamula et al., 1998)

Advantages /disadvantages	Monthly	Daily	Hourly	HOD
Modeling Effort	Minimum	Minimum	Moderate	Difficult
Metering Needs	None	Required	Required	Required
Data Requirements	At leat 12 months	At least 3 months	At least 3 months	At least 3 months
Application to Savings Estimation	In some cases	In most cases	All cases	All cases
Prediction Accuracy	Low	High	Moderate	High

It should be noted that the multi-variable regression models discussed above are developed without retaining the time series nature of the data. Other regression models can be considered to preserve the time-variation of the building energy use. For instance, Fourier series models can be used to capture the daily and seasonal variations of commercial buildings energy use (Dhar et al., 1998).

Multi-variable regression models have been applied not only to estimate total building energy use but also to predict the behavior of individual pieces of HVAC equipment such as chillers, fans, and pumps. In particular, polynomials models have been widely used to model chiller energy use as a function of part-load ratio, evaporator leaving temperature, and condenser entering temperature (LBL, 1982). Other regression models for chillers have been obtained based on fundamental engineering principles (Gordon and Ng, 1994).

16.3.3   Dynamic Models

The regression models discussed in section 16.3.2 cannot generally account for transient effects such as thermal mass that can cause short-term temperature fluctuations during warm-up or cool-down periods of a building. To capture building energy use transient effects, dynamic models are typically recommended. Most of the dynamic models that are based on a physical representation of the building energy systems are complex in nature and require detailed calibration procedures. These models can be grouped into four major types: (a) Thermal Network Models, (b) Time-Series Models, (c) Differential equation Models, (d) Modal Models. Some of these models are briefly discussed in Chapter 4. For more detailed evaluation of the four types of dynamic models, the reader is referred to studies by Rabl (1988) and Reddy (1989).

In recent years, other types of dynamic models based on connectionist approaches have been applied to predict building energy use and estimate retrofit energy savings. An international competition (Kreider and Haberl, 1994; and Haberl and Thamilseran, 1996) has indicated that the connectionist approach provides superior accuracy for predicting both long-term and short-term energy use in buildings compared to the traditional methods. Techniques based on neural networks (NNs) have been developed and applied to forecast building energy use for both short and long term periods (Kreider and Wang, 1992; Anstett and Kreider, 1993; Gibson and Kraft, 1993; Curtiss, 1997; and Kreider et al., 1995 and 1997). In particular, NNs have been used to predict hourly building thermal and electrical energy use for a period of one day, one week, and even one month. Typically, weather data, occupancy profiles, and day types have been considered as input parameters for the NNs to predict building energy use.

A typical neural network consists of several layers of neurons that are connected to each other. A connection is a unique information transport link from one sending to one receiving neuron. Figure 16.3 shows a schematic diagram of the structure of a typical neural network. The first and last layers of neurons are called input and output layers; between them are one or more hidden layers. The connections between the layers are determined using a training data set to "learn" the weights between various neurons.

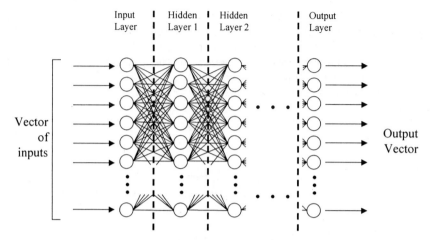

Figure 16.3: A typical structure of a neural network with hidden layers interposed between input and output layers.

It should be noted that if not used properly neural networks may tend to "memorize" the noise in training data. Various techniques that exist reduce this over-training problem. These techniques are discussed by Kreider et al. (1995). In general, however, NNs can be very flexible models which can approximate many kinds of input-output mappings.

A number of applications of NNs to estimate savings from energy conservation measures have been proposed and are discussed by Krarti et al. (1998). For instance, a NN-based approach to determine energy savings from building retrofits has been proposed by Cohen and Krarti (1995 and 1997). The retrofit savings were estimated using the difference between predicted pre-retrofit energy consumption extrapolated into the post period (using NNs) and the actual post-retrofit measured data. The following steps summarize the procedure used to determine retrofit savings with NNs:

- From the pre-retrofit building energy use data, a pre-retrofit NN model is developed to predict building energy consumption.

- Using the pre-retrofit NN model, prediction is made of what future (i.e., in the post-retrofit period) building energy use would be if the retrofit has not been implemented.

- Retrofit energy savings are estimated from the difference between the actual post-retrofit measured data and the energy use obtained from the previous step.

- A second NN model is developed to correlate selected input variables to the retrofit energy savings.

To illustrate the performance of the NN-based approach to estimate energy savings, Cohen and Krarti (1995) considered three energy conservation measures applied to an educational building:

1. High efficiency lighting retrofit (changing fluorescent lamps from F40-T12 to high efficiency T-8 lamps and electronic ballasts).

2. Variable air volume system retrofit to replace a dual-duct constant volume system (In particular, the supply fans are equipped with a variable-speed drive to allow the supply air flow to vary with load).

3. Combination variable-speed chiller drive retrofit and thermal storage system addition (the chiller is equipped with an adjustable-frequency drive on the compressor motor. In addition, a chilled water tank is installed. The chiller provides off-peak cooling and supplements the tank during the on-peak period if necessary).

To evaluate the performance of the trained NNs, a one week testing set was selected. The week used for testing was not part of the training set used to determine the weights of the NNs. Figure 16.4 illustrates the performance of the NNs in predicting the energy savings for the three retrofits when the educational building is assumed to be located in Chicago, IL. In particular, Figure 16.4(a) compares the predicted and the actual savings for the whole building chilled water consumption when lighting retrofit is implemented. The NN accurately estimates the hourly chilled water savings for the entire week. For the VAV retrofit, Figure 16.4(b) indicates that the NN model determines the hot water savings fairly well. Finally, for the ASD chiller drive/thermal energy storage (TES) retrofit, Figure 16.4(c) shows that the day-time peaks are well predicted by the NN model, but some of the evening charging peaks are not.

Currently, the NN-based approach to estimate savings has been applied to only a few building types. Generalization from one building to other similar buildings in other climates has not been totally successful (Krarti et al., 1998). Only nets trained on the specific building to be retrofitted are suggested at this time.

414 ENERGY AUDIT OF BUILDING SYSTEMS

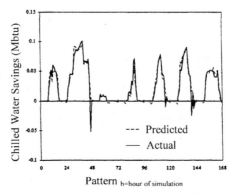

(a) Chilled water savings due to lighting retrofit

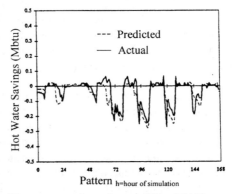

(b) Hot water savings due to CV to VAV retrofit

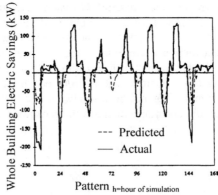

(c) Whole building electrical energy savings due to cooling plant retrofit

Figure 16.4: Energy savings estimation from a NN-based approach (Krarti et al., 1998)

16.3.4   Computer Simulation Models

Detailed computer simulation programs can be used to develop baseline models for building energy use. A brief discussion of the existing computer programs that can be applied to building energy simulation is provided in Chapter 4. The main feature specific to the application of computer simulation models to estimate retrofit energy savings is the calibration procedure to match the baseline model results with measured data. Typically, the calibration procedure of a computer simulation model can be time consuming and require significant efforts especially when daily or hourly measured data are used. Unfortunately, there is no consensus on a general calibration procedure that can be considered for any building type. To date, calibration of building simulation programs is rather an art form that relies on user knowledge and expertise. However, several authors have proposed graphical and statistical methods to aid in the calibration process and specifically to compare the simulated results with measured data. These comparative methods can then be used to adjust either manually or automatically certain input values until the simulated results match the measured data according to predefined accuracy level. Figure 16.5 illustrates a typical calibration procedure for building energy simulation models.

Calibration of computer simulation models is generally time-consuming especially for hourly predictions. Two methods have been used to compare the simulated results to the measured data and help reduce the efforts involved in the calibration procedure:

- Graphical techniques to efficiently view and compare the simulated results and the measured data. Recently, several graphical packages have been utilized or developed for energy computer simulation calibration including: 2-D time-series plots and x-y plots (for monthly and daily data), 3-D surface plots (for hourly data), BWM or box-whisker-mean plots (for weekly data). Haberl and Abbas (1998a, 1998b) and Haberl and Bou-Saada (1998) provide a detailed discussion of the most commonly used graphical tools to calibrate energy computer simulation models. Figure 16.6 illustrates 3-D surface plots for hourly energy use before, during, and after an energy retrofit project.

- Statistical indicators to evaluate the goodness-of-fit of the simulated results compared to the measured data. Several indicators have been considered including mean difference, mean bias error (MBE), root mean square error (RMSE), and coefficient of variance (COV). A discussion of these indicators can be found in Haberl and Bou-Saada (1998). To obain these indicators, classical statistical tools can be used such as SAS (1989).

Currently, ASHRAE is in the process of defining specific procedures that can be used to calibrate whole-building simulation models. These procedures use the basic methodology illustrated in Fig. 16.5.

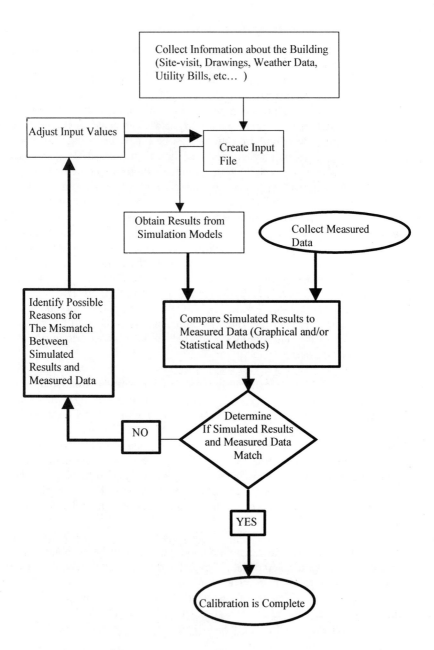

Figure 16.5: Typical calibration procedure for building energy simulation models (Note that the steps that are specific to the calibration process are highlighted in bold)

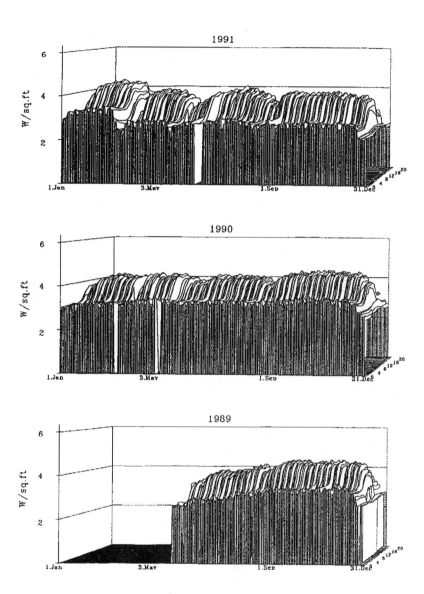

Figure 16.6: Typical 3-D graphs to show the hourly simulation results before and after a retrofit measure (Haberl and Abbas, 1998)

16.4 Applications

The analysis methods presented in this chapter are used to measure and verify savings from energy retrofit projects. These methods are recommended as part of the International Performance for Monitoring and Verification Protocol (IPMVP, 1997) which identifies four different options for energy savings estimation procedures depending on the type of measure, accuracy required, and the cost involved:

- *Option A* uses simplified calibration methods to estimate stipulated savings from specific energy end-uses (such as lighting and constant speed motors). This option may use spot or short-term measurements to verify specific parameters such as the electricity demand when lighting are on. Common applications of this option involve lighting retrofits (such as replacement with more energy efficient lighting systems and use of lighting controls)

- *Option B* typically involves long-term monitoring in attempt to estimate the energy savings from measured data (rather than stipulated energy consumption). Simplified estimation methods can be used to determine the savings from the monitored data. The energy savings obtained from motor replacements (i.e., use of more energy-efficient motor or installation of variable speed drives) are commonly estimated using option B.

- *Option C* is generally applied to estimate savings by monitoring whole-building energy use and by using regression analysis to establish baseline models. The regression models can be developed from monthly or daily measured data. Examples of energy retrofit projects for which option C procedure can be used to include conversion of constant-volume system to variable-air-volume system and chiller or boiler replacement.

- *Option D* is typically used when hourly savings need to be estimated. In addition to hourly monitoring, this option requires complex data analysis to establish baseline models. Dynamic models and calibrated simulation models are typically used to estimate hourly energy savings. The impact of energy management control systems (EMCS) is one of several applications for which option D can be considered.

Table 16.3 summarizes the requirements of each option including typical accuracy levels of savings estimation and the average cost (expressed as a percent of the total cost required to implement the energy retrofit project) needed for metering equipment and for data analysis.

Table 16.3: Basic characteristics of the various options for procedures to estimate energy savings (IPMVP, 1997)

	Models for Data Analysis	Accuracy		Metering requirements	Applications		
					End-Use	Load	Operation
A	Simplified Methods	20-100%	1-5%	Spot/short-term	Subsystem	Constant	Constant
B	Simplified Methods	10-20%	3-10%	Long-term	Subsystem	Constant/ Variable	Constant/ Variable
C	Regression Analysis	5-10%	1-10%	Long-term (daily or monthly)	Building	Variable	Variable
D	Dynamic Analysis/ Simulation	10-20%	3-10%	Long-term (hourly)	Subsystem/ Building	Variable	Variable

16.5 Summary

In this chapter, methods used to estimate energy savings from retrofit projects are briefly presented. The common applications of the presented methods are discussed with some indication of their typical expected accuracy and additional cost. As mentioned throughout the chapter, the presented measurement and verification protocols are applicable to estimate energy savings measured from energy conservation measures applied to existing buildings commercial buildings. However, similar procedures have been developed for other applications such as energy efficiency in new buildings, water efficiency, renewable technology, emissions trading, and indoor air quality (IPMVP, 1997).

REFERENCES

Akbari H., Huang J., and Davis S., Peak Power and Cooling Energy Savings of Shade Trees, *Energy and Buildings*, 25, 139-147, 1997.

Akbari H., *Monitoring Peak Power and Cooling Energy Savings of Shade Trees and White Surfaces in the Sacramento Municipal Utility District (SMUD) Service Area: Data Analysis, Simulations and Results*, LBNL Report # 34411, Berkeley, CA, 1993.

Akbari, H. and Sezgen O., *Case Studies of Thermal Energy Storage (TES) Systems: Evaluation and Verification of System Performance*. LBL Report # 30852. Lawrence Berkeley Laboratory, University of California, Energy and Environment Division. Berkeley, CA, 1992.

American Gas, Distributing Power in the Future, an article written by Kabous, *American Gas Journal*, July, 16, 1999.

Andreas, J., *Energy-Efficient Motors: Selection and Application*, Marcel Dekker, New York, 1992.

Anstett, M., and Kreider, J.F, Application of Artificial Neural Networks to Commercial Energy Use Prediction, *ASHRAE Transactions*, 99(1), 505, 1993.

ARI, American Refrigeration Institute, Statistics about US Refrigeration Equipment Shipment, Website: hhtp://www.ari.org. 1999.

ASHRAE, *Ventilation for Acceptable Indoor Air Quality, Standard 62-1989*, American Society of Heating, Refrigerating and Air-Conditioning Engineers, Inc., Atlanta, GA, 1989.

ASHRAE, *Handbook of HVAC Systems and Equipment*, American Society of Heating, Refrigerating and Air-Conditioning Engineers, Inc., Atlanta, GA, 1996.

ASHRAE, *Handbook of Fundamentals*, American Society of Heating, Refrigerating and Air-Conditioning Engineers, Inc., Atlanta, GA, 1997.

ASHRAE 14P, *Proposed Guideline 14P, Measurement of Energy and Demand Savings*, American Society of Heating, Refrigerating and Air-Conditioning Engineers, Inc., Atlanta, GA, 1997.

ASHRAE, *Handbook of HVAC Applications*, American Society of Heating, Refrigerating and Air-Conditioning Engineers, Inc., Atlanta, GA, 1999.

ASHRAE, *Ventilation for Acceptable Indoor Air Quality, Standard 62-1999*, American Society of Heating, Refrigerating and Air-Conditioning Engineers, Inc., Atlanta, GA, 1999.

ASHRAE, *Handbook of HVAC Systems and Equipment*, American Society of Heating, Refrigerating and Air-Conditioning Engineers, Inc., Atlanta, GA, 2000.

Avery, G., Updating the VAV Outside Air Economizer Controls, *ASHRAE Journal*, 31(4), 14, 1989.

Avery, G., The Instability of VAV Systems, *Heating, Piping, and Air Conditioning*, 64(2), 47, 1992.

AWWA, *Water Use Inside the Home*, Report of American Water Works Research Foundation, 1999.

Ayari, A., and Krarti, M., Evaluation of Design Ventilation Requirements for Enclosed Parking Garages, *ASHRAE Transactions*, 106(1), 2000.

Ayari, A., Grot, D., and Krarti, M., Field Evaluation of Ventilation System Performance in Enclosed Parking Garages, *ASHRAE Transactions*, 106(1), 2000.

Azebergi, R., Hunsberger, R., Zhou, N., *A Residential Building Energy Audit*, Report for Class Project CVEN5020, University of Colorado, 2000.

Biesemeyer, W.D., and Jowett, J., Facts and Fiction of HVAC Motor Measuring for Energy Savings, *Proceedings of the ACEEE 1996 Summer Study on Energy Efficiency in Buildings,* ACEEE, Washington DC, 1996.

BLAST: Building Load Analysis and System Thermodynamics, *User-Manual.* U.S. Army Construction Engineering Research Laboratory and University of Illinois, Urbana-Champaign, IL, 1994.

Bloomfield, D.P., and Fisk, D.J., The Optimization of Intermittent Heating, *Building and Environment*, 12, 43, 1977.

BNP, Florida Campus Cuts CFC Use, as well as Energy Consumption, *Air Conditioning, Heating and Refrigeration News*, 195 (11), 32, 1995.

Bourges, B., Calcul des Degrés Jours Mensuels à Température de Base Variable, Revue Chauffage Ventilation Conditionnement, 5, 1987.

BPA, *High Efficiency Motor Selection Handbook.* Bonneville Power Administration, Portland, OR, 1990.

Braun, J.E., Reducing Energy Costs and Peak Electrical Demands Through Optimal Control of Building Thermal Storage. *ASHRAE Transactions*, 96(2), 876-888, 1990.

Braun, J.E., A Comparison of Chiller-Priority, Storage-Priority, and Optimal Control of an Ice Storage System. *ASHRAE Transactions*, 98(1), 893, 1992.

Bushby, S.T., and Newman, H.M., The BACnet Communication Protocol for Building Automation Systems, *ASHRAE Journal*, 33(4), 14, 1991.

Carey, C. W., Mitchell, J. W. and Beckman, W.A., The Control of Ice-Storage Systems, *ASHRAE Transactions*, 104(1), 1345, 1995.

CEREN, *La Consommation d''Energie Dans les Regions Francaises*, Report from Centre d'Etudes et de Recherches Economiques sur l'Energie, 1997.

Claridge, D.E., Design Methods for Earth-Contact Heat Transfer. Progress in Solar Energy, Edited by K. Boer, American Solar Energy Society (ASES), Boulder, CO.

Claridge, D.E., Krarti, M., Bida, M., A Validation Study of Variable-Base Degree-Day Cooling Calculations, *ASHRAE Transactions*, 93(2), 90-104, 1987.

Claridge, D.E., Haberl, J., Turner W., O'Neal D., Heffington W., Tombari C., Roberts M., and Jaeger S., Improving Energy Conservation Retrofits With Measured Savings, *ASHRAE Journal*, 33(10), 14, 1991.

Cohen, D., and Krarti, M., A Neural Network Modeling Approach Applied To Energy Conservation Retrofits, *Building Simulation Fourth International Conference Proceedings*, Madison, WI, 423, 1995.

Cohen, D., and Krarti, M., Neural Network Modeling of Measured Data to Predict Building Energy System Retrofit Savings, *Proceedings of the ASME ISEC*, Washington, D.C., 27, 1997.

Cohen, B.M., and Kosar D.R., 2000, Humidity Issues in Bin Energy Analysis, *Heating/Piping/Air Conditioning*, 72(1), 65-78, 2000.

Conchilla M., *Interactions of Water and Energy Resources in Residential Buildings – A Modeling Study*, MS Thesis, University of Colorado, Boulder, CO, 1999.

Czarkowski, D., and A. Domijan, "Performance of Electric Power Meters and Analyzers in Adjustable Speed Drive applications", *ASHRAE Transactions*, 103(1), 1997.

DOE, US Department of Energy, Energy Efficiency for Buildings, Directory of Building Energy Software, Washington D.C., 2000. An updated list, Website: http://www.eren.doe.gov/buildings/tools_directory/.

DeMonsabert S., 1996, WATERGY: A Water and Energy Conservation Model for Federal Facilities, *CONSERV 96 Conference Proceedings*, 1996.

D'Albora, E.G., and Gillespie, K., Evaluating the Uncertainty in Cool Storage Inventory Using Energy Balance Method, *ASHRAE Transactions*, 108(2), 1999.

Dhar A., Reddy T.A., Claridge D.E., Modeling Hourly Energy Use in Commercial Buildings with Fourier Series Functional Forms, *ASME Solar Energy Engineering Journal*, 120(3), 217, 1998.

Diekerhoff, D.J., Grimsrud, D.T., and Lipschutz, R.D., Component Leakage Testing in Residential Buildings, *Proceedings of the American Council for an Energy Efficient Economy, 1982 Summer Study*, Santa Cruz, CA, 1982.

Domijan, A., Abu-Aisheh, and D. Czrakowski, Efficiency and Separation of Losses of an Induction Motor and its Adjustable Speed Drive at Different Loading/Speed Combinations, *ASHRAE Transactions*, 103(1), 1997.

Domijan, A., D. Czarkowski, A. Abu-Aisheh, and E. Embriz-Sander, *Measurements of Electrical Power Inputs to Variable Speed Motors and Their Solid State Power Converters: Phase I*, ASHRAE Final Report RP-770, 158, 1996.

Domijan, A., E. Embriz-Sander, A.J. Gilani, G. Lamer, C. Stiles, and C.W. Williams, Watthour Meter Accuracy under Controlled Unbalanced Harmonic Voltage and Current Conditions, *IEEE Transactions Power Delivery*. Winter Meeting of IEEE Power Engineering Society, 1995.

Drees, K., Wenger, J., and Janu, G., Ventilation Airflow Measurement for ASHRAE Standard 62-1989, *ASHRAE Journal*, 34(10), 40, 1992.

Drees, K.H., and J.E. Braun, Development and Evaluation of a Rule-Based Control Strategy for Ice Storage System. *HVAC&R Research*. 2(4), 312, 1996.

Drees, K. H., *Modeling and Control of Area Constrained Ice Storage Systems*, MS Thesis, Purdue University, 1994.

Drees, K.H., and Braun, J.E., Modeling of Area-Constrained Ice Storage Tanks, *International Journal of HVAC&R Research*, 1(2), 143, 1995.

EDF, Production d'electricite et courbes charges, Website: http://www.edf.fr, 1997.

Elovitz, D.M., Minimum Outside Air Control Methods for VAV Systems, *ASHRAE Transactions*, 101(2), 613, 1995.

EIA, Energy Information Agency, Energy Facts 1991, DOE/EIA-0469 (91), Washington, DC, 1991.

EIA, Energy Information Agency, Annual Review of Energy, DOE/EIA, Washington, DC, 1994.

EIA, Energy Information Agency, Nonresidential Buildings Energy Consumption Survey: Characteristics of Commercial Buildings 1995, DOE/EIA-0246(95), Washington, DC, 1997.

EIA, Annual Energy Review, Department of Energy, Energy Information Administration,, http://www.doe.eia.gov, 1998.

EPA, US Environmental Protection Agency, Community Water System Survey, Office of Ground Water and Drinking Water. Washington DC, 1995.

EPA, US Environmental Protection Agency, information provided in the Energy Star web site: http://www.epa.gov/appdstar/esoe/, US Environmental Protection Agency, 1999.

EPRI, Electric Power Research Institute, Website: http://www.epri.com, 1999.

Erbs, D.G., Klein, S.A., and Beckman, W.A., Estimation of Degree Days and Ambient Temperature Bin Data from Monthly-Average Temperatures, *ASHRAE Journal*, Vol. 25(6), 60-66, 1983.

Eto, J., On Using Degree-Days to Account for the Effects of Weather on Annual Energy Use in Office Buildings, *Energy and Buildings*, 12(2), 113, 1988.

Euromonitor, The World Market for White Goods, Euromonitor International Inc., 111 West Washington St., Chicago, IL, USA, 1994.

FEMP, Federal Energy Management Program, Energy Policy Act of 1992 Becomes Law, FEMP Focus Special Edition No. 2, 1992.

Fels M., Special Issue Devoted to Measuring Energy Savings: The Scorekeeping Approach, *Energy and Buildings*, 9(1 & 2), 1986.

Fels M., and Keatig K., Measurement of Energy Savings from Demand-Side Management Programs in US Electric Utilities, *Annual Review of Energy and Environment*, 18, 57, 1993.

Fels, J., Special Issue Devoted to Measuring Energy Savings: The Scorekeeping Approach, Energy and Buildings, 12, 2, 113, 1988.

FERC, Regulations Under Section 201 and 210 of the Public Utility Regulatory Policies Act of 1978 with Regard to Small Power Production of Cogeneration, 1978, 292.

Fuchs, E.F. and Fei R., A New Computer-Aided Method for the Efficiency Measurement of Low-Loss Transformers and Inductors under Nonsinusoidal Operation, *IEE Transactions on Power Delivery*, 11(1), 292-304, 1996.

GDF, Gas de France, "Rapport sur les Activites de GDF", website: http://www.gdf.fr, 1985.

GRI, *Electric and Gas Rates for the Residential, Commercial, and Industrial Sectors*, Volumes 1 and 2, by Whitem L.J., McVicker, C.M, and Stiles, E., Gas Research Institute, 1993.

Gibson, F.J. and Kraft, T.T, Electric Demand Prediction Using Artificial Neural Network Technology, *ASHRAE Journal*, 35(3), 60, 1993.

Gordon J.M., and Ng. K.C., Thermodynamic Modeling of Reciprocating Chillers, Journal of Applied Physics, 76(6), 2769, 1994.

Greely K., Harris J., and Hatcher A., *Measured Savings and Cost-Effectiveness of Conservation Retrofits in Commercial Buildings*, Lawrence Berkeley National Laboratory Report-27586, Berkeley, CA, 1990.

Green Seal, *Proposed Environmental Standards for Major Household Appliances,* Green Seal, Washington DC, 1993.

Haberl J.S., and Abbas M., Development of Graphical Indices for Viewing Building Energy Data: Part I, *ASME Solar Energy Engineering Journal*, 120(3), 156, 1998a.

Haberl, J.S., Abbas, M., Development of Graphical Indices for Viewing Building Energy Data: Part II, *ASME Solar Energy Engineering Journal*, Vol. 120(3), 162, 1998b.

Haberl J.S., and Bou-Saada T.E., Procedures for Calibrating Hourly Simulation Models to Measured Energy and Environmental Data, *ASME Solar Energy Engineering Journal*, 120(3), 193, 1998.

Haberl J.S., and Claridge D.E., An Expert System for Building Energy Consumption Analysis: Prototype Results, *ASHRAE Transactions*, 93(1), 979, 1987.

Haberl, J., and Thamilseran, S., Predicting Hourly Building Energy Use: The Great Energy Predictor Shootout II: Measuring Retrofit Savings: Overview and Discussion of Results, *ASHRAE Transactions*, 102(2), 1996.

Haines, R.W., Ventilation Air, The Economy Cycle, and VAV, *Heating Piping and Air Conditioning*, 66(10), 71, 1994.

Harrje, D.T., and Born, G.J., Cataloguing Air Leakage Components in Houses, *Proceedings of the American Council for an Energy Efficient Economy, 1982 Summer Study*, Santa Cruz, CA, 1982.

Harriman, L.G., et al., New Weather Data for Energy Calculations, *ASHRAE Journal*, 39(11), 37, 1999.

Henze, G.P., *Evaluation of Optimal Control for Ice Storage Systems*, PhD. Dissertation, University of Colorado, Boulder, CO, 1996.

Henze, G. P., Krarti, M., and Brandemuehl, M. J., A Simulation Environment for the Analysis of Ice Storage Control, *Int. J. HVAC & R Research*, 3, 128, 1997.

Henze, G.P., M. Krarti, and Brandemuehl, M.J., A simulation environment for the analysis of ice storage controls, *HVAC& R Research*, 3(2), 128, 1997a.

Henze, G.P, R.H. Dodier, and M. Krarti., Development of a Predictive Optimal Controller for Thermal Energy Storage Systems, *HVAC& R Research*, 3(3), 233, 1997b.

Henze, G.P., and Krarti, M., The Impact of Forecasting Uncertainty on the Performance of a Predictive Optimal Controller for Thermal Energy Storage Systems, *ASHRAE Transactions*, 105(1), 1999.

Herron, D.J., Understanding the basics of Compressed Air Systems, *Energy Engineering*, 96(2), 19, 1999.

Hoshide, R.K., Electric Motor Do's and Don'ts, *Energy Engineering*, 1(1), 6-24, 1994.

Howe B., and Scales B., Beyond Leaks: Demand-Side Strategies for Improving Compressed Air Efficiency, *Energy Engineering*, 95, 31, 1998.

428ENERGY AUDIT OF BUILDING SYSTEMS

Huang J., The Potential of Vegetation in Reducing Summer Cooling Load in Residential Buildings, *Journal of Climate and Applied Meteorology*, 26(9), 1103, 1987.

Huang J., The Energy and Comfort Performance of Evaporative Coolers for Residential Buildings in California Climates, *ASHRAE Transactions*, 97(2), 847-881, 1991.

Hunn, B.D., Peterson, J.L., Banks, J.A. Aanstoos, T.A., Srinivasan, R., *Energy Use in Texas State Facilities FY-1990 through FY-1993*, Conservation and Solar Research Report No. 13. Austin, TX: Center for Energy Studies, University of Texas, 1995.

ICARMA, International Council of Air-Conditioning and Refrigeration Manufacturers Association, Press Releases. Web-site: http://www.icarma.org, 1997.

IEEE, Guide for Harmonic Control and Reactive Compensation of Static Power Converters, IEEE 519-1992, 1992.

IPCC, Intergovernmental Panel on Climate Change, IPCC Technical Paper on Technologies, Policies and Measures for Mitigating Climate Change, IPCC, Geneva, 1996.

IPMVP, *International Performance Monitoring and Verification Protocol*, U.S. Department of Energy DOE/EE-0157. Washington, DC: US Government Printing Office, 1997.

Janu, G., Wenger, J.D., and Nesler, C.G., Strategies for Outdoor Airflow Control from a System Perspective, *ASHRAE Transactions*, 101(2), 631, 1995.

Jekel, T.B., Mitchell, J.W., and Klein, S.A., Modeling of Ice Storage Tanks, *ASHRAE Transactions*, 99(2), 1993.

Jobe, T., and Krarti, M., Field Implementation of Optimum Start Heating Controls, *Proceedings for ASME Solar Engineering*, 305, 1997.

Johnson B., and Zoi C., EPA Energy Star Computers: The next Generation of Office Equipment, *Proc. Conf. Of ACEEE 1992 Summer Study on Energy Efficiency in Buildings, Panel 6,* Pacific Grove, CA, 1992.

Katipamula S., Reddy T.A., and Claridge D.E., Development and Application of Regression Models to Predict Cooling Energy Use in Large Commercial Buildings, *Proceedings of the ASME/JSES/JSES International Solar Energy Conference,* San Francisco, CA, 307, 1994.

Katipamula S., Reddy T.A., and Claridge D.E., Effect of Time Resolution on Statistical Modeling of Cooling Energy Use in Large Commercial Buildings, *ASHRAE Transactions*, 101(2), 1995.

Katipamula S., Reddy T.A., and Claridge D.E., Multivariate Regression Modeling, ASME Solar Energy Engineering Journal, 120(3), 177, 1998.

Ke, Y., and Mumma, S.A., Using Carbon Dioxide Measurements to Determine Occupancy for Ventilation Controls, *ASHRAE Transactions*, 103(3), 1997.

Kersten, M.S., *Thermal Properties in Soils*, Bulletin from University of Minnesota, Institute of Technology and Engineering, Experimental Station Bulletin, No. 28, 1949.

Kettler, J.P., Field Problems Associated with Return Fans on VAV Systems, *ASHRAE Transactions*, 94(1), 1477, 1988.

Kettler, J.P., Minimum Ventilation Control for VAV Systems: Fan Tracking vs. Workable Solutions, *ASHRAE Transactions*, 101(2), 625, 1995.

Kettler, J.P., Controlling Minimum Ventilation Volume in VAV Systems, *ASHRAE Journal*, 40(5), 45, 1998.

Kiatreungwattan, K., *A Model for an Indirect Ice Storage Tank during Partial Charging and Discharging Cycles*, M.S. Thesis, University of Colorado, Boulder, Colorado, 1998.

Kim K., Yoon H., E. Lee E., S. Choi S., and Krarti M., Building Energy Performance Simulations to Evaluate Energy Conservation Measures, *ASME Solar Energy Engineering Conference Proceedings*, 45, 1998.

Kissock K., and Fels M., An Assessment of PRISM's Reliability for Commercial Buildings, *National Energy Program Evaluation Conference*, Chicago, IL, 1995.

Kissock K., Claridge D., Haberl J., and Reddy A., Measuring Retrofit Savings for the Texas LoanSTAR Program: Preliminary Methodology and Results, *Proceedings of the ASME/JSES/JSES International Solar Energy Conference*, Maui, Hawaii, 299, 1992.

Kissock K., Reddy T.A., and Claridge D.E., Ambient-Temperature Regression Analysis for Estimating Retrofit Savings in Commercial Buildings, *ASME J. Solar Energy Engineering*, 120(3), 168-176, 1998.

Knebel, D.E., *Simplified Energy Analysis Using the Modified Bin Method*, American Society of Heating, Refrigeration, and Air-conditioning Engineers, Atlanta, GA, 1983.

Koomey J.G., Dunham C., and Lutz J.D., *The Effect of Efficiency Standards on Water Use and Water Heating Energy Use in the US: A detailed End-Use Treatment*, LBNL-Report, LBL-35475, Lawrence Berkeley National Laboratory, Berkeley, CA, 1994.

Krarti, M., Foundation Heat Transfer, Chapter in *Advances in Solar Energy*, Edited by Y. Goswami and K. Boer, ASES, Boulder, CO, 1999.

Krarti M., Kreider J.F., Cohen D., and Curtiss P., Estimation of Energy Savings for Building Retrofits Using Neural Networks, *ASME Journal of Solar Energy Engineering*, Vol. 120(3), 211, 1998.

Krarti, M., Ayari, A., Grot, D., *Ventilation Requirements for Enclosed Vehicular Parking Garages,* Final Report for ASHRAE RP-945, American Society of Heating, Refrigerating, and Air Conditioning Engineering, Atlanta, GA, 1999a.

Krarti, M., Brandemuehl, M., Schroeder C., and Jeanette E., *Techniques for Measuring and Controlling Outside Air Intake Rates in Variable Air Volume Systems.* Final Report for ASHRAE RP-980, American Society of Heating, Refrigerating, and Air Conditioning Engineering, Atlanta, GA, 1999b.

Krarti, M., Henze, G.P., and Bell, D., Planning Horizon for a Predictive Optimal Controller for Thermal Energy Storage Systems, *ASHRAE Transactions*, 105(1), 1999c.

Krarti, M., and Chuangchid, P., *Cooler Floor Heat Gain in Refrigerated Structures*, Final Report for ASHRAE Project RP-953, American Society of Heating, Refrigerating, and Air Conditioning Engineering, Atlanta, GA, 1999.

Krarti, M., Schroeder, C., Jeanette, E., and Brandemuehl, M., Experimental Analysis of Measurement and Control Techniques of Outside Air Intake Rates in VAV Systems, *ASHRAE Transactions*, 106(2), 2000.

Kreider, J.F., *Heating, Ventilating, and Air Conditioning Control Systems*, Chapter in CRC Handbook of Energy Efficiency, Edited by F. Kreith and R. West, CRC Press, 337-378, 1997.

Kreider, J. F., and Rabl, A., *Heating and Cooling of Buildings*, McGraw-Hill, New York, 1994.

Kreider, J. F., Claridge, D. Z., Curtis, P., Dodier, R., Haberl, J. S., and Krarti, M., Building Energy Use Prediction and System Identification Using Recurrent Neural Networks, *J. Solar Energy Engineering*, 117, 161, 1995.

Kreider, J. F., S. L. Blanc, R. C. Kammerud, P. S. Curtiss, Operational Data as the Basis for Neural Network Prediction of Hourly Electrical Demand, *ASHRAE Transactions*, Vol. 103(2), 1997.

Kreider, J., F., and Haberl, J., Predicting Hourly Building Energy Usage: The Great Predictor Shootout -- Overview and Discussion of Results, *ASHRAE Transactions*, 100(2), 1104, 1994.

Kreider, J., F., and Wang, X.A., Improved Artificial Neural Networks for Commercial Building Energy Use Prediction, *Journal of ASME Solar Energy Engineering*, 361, 1992.

Kreider, J.F., Claridge D., Curtiss, P., Dodier, R., Haberl, J., and Krarti, M., Recurrent Neural Networks for Building Energy Use Prediction and System Identification: A Progress Report, *ASME Transactions, Journal of Solar Energy Engineering*, 117, 1995.

Kreider, J.F., Curtiss, P.S., Massie. D., and Jeanette, E., A Commercial-Scale University HVAC Laboratory, *ASHRAE Transactions*, 105, Part 1, 1999.

Lawrence Berkeley Laboratory, Radiance User Manual, Lawrence Berkeley Laboratory, Berkeley, CA, 1991.

LBL, DOE-2 User Guide, Version 2.1, LBL report No. LBL-8689 Rev. 2, Lawrence Berkeley Laboratory, Berkeley, California, 1980.

LBL, DOE-2 Engineers Manual, Lawrence Berkeley Laboratory Report LBL-11353, National Technical Information Services, Springfield, VA, 1982.

Levenhagen, J., Control Systems to Comply with ASHRAE Standard 62-1989, *ASHRAE Journal*, 34(9), 40, 1992.

Lobodovsky, K.K., Motor Efficiency Management, *Energy Engineering*, 1(2), 32-43, 1994.

Mayer, P.W.*, Residential Water Use and Conservation Effectiveness: A Process Approach*, MS Thesis, University of Colorado, 1995.

Mease, N.E., Cleaveland, W.G. Jr., Mattingly, G.E., and Hall, J.M., Air Speed Calibrations at the National Institute of Standards and Technology, *Proceedings of the 1992 Measurement Science Conference*, Anaheim, CA, 1992.

Meckler, M., Demand-Control Ventilation Strategies for Acceptable IAQ, *Heating Piping and Air Conditioning*, 66(5), 71, 1994.

Mont, J.A., and Turner, W.C., A Study on Real-Time Pricing Electric Tariffs, *Energy Engineering*, 96(5), 7, 1999.

Morris, F.B., Braun, J.E., and Treado, S.J, Experimental and Simulated Performance of Optimal Control of Building Thermal Storage, *ASHRAE Transactions*, 100(1), 402, 1994.

Mumma, S.A., and Wong, Y.M., Analytical Evaluation of Outdoor Airflow Rate Variation vs. Supply Airflow rate Variation in Variable-Air-Volume Systems When the Outdoor Air Damper Position is Fixed, *ASHRAE Transactions*, 96(1), 1197, 1990.

Nadel, S., M. Shepard, S. Greenberg, G. Katz, and A. de Almeida, Energy-Efficient Motor Systems: A handbook of Technology, Program, and Policy Opportunities, *American Coucil for Energy-Efficient Economy*, Washington D.C., 1991.

NAESCO, NAESCO Standard for Measurement of Energy Savings for Electric Utility Demand Side Management (DSM) Projects. Washington DC: National Association of Energy Services Companies, 1993.

NEC, National Electrical Code, published by the National Fire Protection Association, Quincy, MA, 1996.

NEMA, *Energy Management Guide for Selection and Use of Polyphase Motors, Standard MG-10-1994*, National Electrical Manufacturers Association, Rosslyn, VA, 1994.

Neto, J. H. M., and Krarti, M., Deterministic Model for an Internal Melt Ice-on-coil Thermal Storage Tank, *ASHRAE Transactions*, 106(1), 113, 1997a.

Neto, J. H. M., and Krarti, M., Experimental Validation of a Numerical Model for an Internal Melt Ice-on-coil Thermal Storage Tank, *ASHRAE Transactions*, 106(1), 125, 1997b.

Norford, L.K., S.L. Englander, and B.J. Wiseley, *Demonstration Knowledge Base to Aid Building Operators in Responding to Real-Time-Pricing Electricity rates*. Final Report for ASHRAE RP-833, American Society of Heating, Refrigerating, and Air Conditioning Engineering, Atlanta, GA, 1996.

NPLIP, Power Quality. Lighting Answers. *Newsletter by the National Lighting Product Information Program,* 2(2), 5, 1995.

OCDE, (also OECD: Organization for Economic Cooperation and Development), Economic Statistics, Website: http://www.ocde.org, 1999.

OmniComp, OmniComp Inc., State College, PA, 1984.

Orphelin M., *Methodes Pour la reconstitution de Courbes De Charge Agregees des Usages Thermiques de l'Electricite*, PhD Thesis, Ecole Nationale des Mines de Paris, 1999.

OTA, Office of Technology Administration, Internal Report, 1995.

Ower, E., and Pankhurst, R.C., *The Measurement of Airflow*, 5th Ed., Pergamon Press, New York, 1977.

Periera L.S., Evapotranspiration: Review of Concepts and Future Trends. Procedures: Evapotranspiration and Irrigation Schedules, *American Society of Agricultural Engineers Conference Proceedings*, San Antonio, TX, 109-115, 1996.

Persily, A., Ventilation, Carbon Dioxide, and ASHRAE Standard 62-1989, *ASHRAE Journal*, 35(7), 40, 1993.

Phelan J., Brandemuehl M.J., and Krarti M., Methodology Development to Measure In-Situ Chiller, Fan, and Pump Performance, Final Report for ASHRAE Project RP-827, American Society of Heating, Refrigerating, and Air Conditioning Engineering, Atlanta, GA, 1996.

Rabl A., Rialhe A., Energy Signature Models for Commercial Buildings: Test with Measured Data and Interpretation, *Energy and Buildings*, 1992.

Rabl A., Parameter Estimation in Buildings: Methods for Dynamic Analysis of Measured Energy Use, *ASME Solar Energy Engineering Journal*, 110(1), 52, 1988.

Reddy T.A., Application of Dynamic Building Inverse Models to Three Occupied Residences Monitored Non-Intrusively, *Proceedings of Thermal Performance of Exterior Envelopes of Buildings IV*, FL, 1989.

Reddy T.A., Kissock K., Katipamula S., Ruch D., and Claridge D., *An Overview of Measured Energy Retrofit Saving Methodologies Developed in the Texas LoanSTAR Program*, Energy Systems Laboratory Technical Report ESL-TR-94/03-04, Texas A&M University, 1994.

Ruch D., and Claridge D.E., A Four-Parameter Change-Point Model for Predicting Energy Consumption in Commercial Buildings, *ASME Journal of Solar Energy Engineering*, 104, 177, 1992.

SAS, *SAS Reference Manual*, SAS Institute Inc., SAS Circle, P.O. Box 8000, Cary, NC, 1989.

Seem, J.E., Armstrong, P.R., and Hancock, C.E., Comparison of Seven Methods for Forecasting the Time to return from Night Setback, *ASHRAE Transactions*, 95(2), 439, 1989.

Shaw S.R., Abler C.B., Lepard R.F., Luo D., Leeb S.B., and Norford L.K., Instrumentation for High Performance Non-intrusive Electrical Load Monitoring, ASME Journal of Solar Energy Engineering, 120(3), 224, 1998.

Sherman, M.H., and Grimsrud, D.T., Infiltration-Pressurization Correlation: Simplified Physical Modeling, *ASHRAE Transactions*, 86(2), 778-784, 1980.

Sherman, M.H., and Matson, N.E., Ventilation-Energy Liabilities in US Dwellings, *Proceedings of 14th AIVC Conference*, 23-41, Copenhagen, Denmark, 1993.

Solberg, D., Dougan, D., and Damiano, L., Measurement for the Control of Fresh Air Intakes, *ASHRAE Journal*, 32(1), 45, 1990.

Sonderegger, R.C., Thermal Modeling of Buildings as a Design Tool, *Proceedings of CHMA 2000*, Vol. 1, 1985.

Sterling, E.M., Collet, C.W., and Turner, S., Commissioning to Avoid Indoor Air Quality Problems, *ASHRAE Journal*, 34(10), 28, 1992.

Strand, R.K., Pederson, C. O., and Coleman, G. N., Development of Direct and Indirect Ice Storage Models for Energy Analysis Calculations, *ASHRAE Transactions*, 103(1), 1213, 1994.

Taha H., Urban Climates and Heat Islands: Albedo, Evapotranspiration, and Anthropogenic Heat, *Energy and Buildings*, 25, 99, 1997.

Tamblyn, R.T., Control Concepts for Thermal Storage, *ASHRAE Transactions*, 91(1b), 5, 1985.

Taylor, John R., *An Introduction to Error Analysis*, University Science Books, Mill Valley, 1982.

Terrell, R.E., Improving Compressed Air System Efficiency- Know What You Really Need, *Energy Engineering*, 96(1), 7, 1999.

Thumann, A., and Mehta, P., *Handbook of Energy Engineering*. The Fairmont Press Inc., Liburn, GA, 1997.

Tuluca A., and Steven Winter Associates, *Energy Efficient Design and Construction for Commercial Buildings*, Mc-Graw Hills, 1997.

Tumura, G.T., and Shaw, C.Y., Studies on Exterior Wall Air-tightness and Air Infiltration of Tall Buildings, *ASHRAE Transactions*, 82(1), 122-129, 1976.

Turiel, I., Present Status of Residential Appliance Energy Efficiency Standards, an International Review, *Energy and Buildings*, 26(1), 5, 1997.

UNEP, United Nations Environment Program, website: http://www.unep.org, 1999.

Verderber, R, Morse, C. Alling, R., Harmonics from Compact Fluorescent Lamps, *IEEE Transactions on Industry Applications*, 29(3), 670, 1993.

Vick, B., Nelson, D. J., and Yu, X., Model of an Ice-on-Pipe Brine Thermal Storage Component, *ASHRAE Transactions*, 105(1), 45, 1996a.

Vick, B., Nelson, D. J., and Yu, X., Validation of the Algorithm for Ice-on-Pipe Brine Thermal Storage Systems, *ASHRAE Transactions*, 105(1), 55, 1996b.

Yoon H., E. Lee E., Choi S., and Krarti M., Building Energy Audit: Samsung Building, Internal report, KIER Report, Korea, 1997.

Waide P., Lebot B., and Hinnells M., Appliance Energy Standards in Europe, *Energy and Buildings*, 26(1), 45, 1997.

West J., and Braun J.E., Modeling Partial Charging and Discharging of Area-Constrained Ice Storage Tanks, *HVAC&R Research*, 5(3), 209, 1999.

WSEO, MotorMaster Electric Motor Selection Software, Washington State Energy Office, Olympia, WA, 1990.

Appendix A. Conversion factors

Conversion Factors (Metric to English)

Area	1 m^2	$= 1550.0 \text{ in}^2$
		$= 10.764 \text{ ft}^2$
Energy	1 J	$= 9.4787 \times 10^{-4} \text{ Btu}$
Heat transfer rate	1 W	$= 3.4123 \text{ Btu/h}$
Heat flux	1 W/m^2	$= 0.3171 \text{ Btu/hr·ft}^2$
Heat generation rate	1 W/m^3	$= 0.09665 \text{ Btu/hr·ft}^3$
Heat transfer coefficient	$1 \text{ W/m}^2\text{·K}$	$= 0.17612 \text{ Btu/hr·ft}^2\text{·°F}$
Latent heat	1 J/kg	$= 4.2995 \times 10^{-4} \text{ Btu/lb}_m$
Length	1 m	$= 3.2808 \text{ ft}$
	1 km	$= 0.62137 \text{ mile}$
Mass	1 kg	$= 2.2046 \text{ lb}_m$
Mass density	1 kg/m3	$= 0.062428 \text{ lb}_m/\text{ft}^3$
Mass flow rate	1 kg/s	$= 7936.6 \text{ lb}_m/\text{h}$
Mass transfer coefficient	1 m/s	$= 1.1811 \times 10^4 \text{ ft/h}$
Pressure	1 N/m^2	$= 0.020886 \text{ lb}_f/\text{ft}^2$
		$= 1.4504 \times 10^{-4} \text{ lb}_f/\text{in.}^2$
		$= 4.015 \times 10^{-3} \text{ in. water}$
		$= 1 \times 10^{-5} \text{ bar}$
Power	1 kW	$= 1.340 \text{ HP}$
		$= 3,412 \text{ Btu/hr}$
Refrigeration capacity	1 kJ/hr	$= 94,782 \text{ Btu/hr}$
		$= 7.898 \times 10^{-5} \text{ ton}$
	1 kW	$= 0.2844 \text{ ton}$
Specific heat	1 J/kg·K	$= 2.3886 \times 10^{-4} \text{ Btu/lb}_m\text{·°F}$
Temperature	K	$= (5/9) \text{ °R}$
		$= (5/9)(\text{°F} + 459.67)$
		$= \text{°C} + 273.15$
Temperature difference	1 K	$= 1 \text{ °C}$
		$= (9/5) \text{ °R} = (9/5) \text{ °F}$
Thermal conductivity	1 W/m·K	$= 0.57782 \text{ Btu/ hr·ft·°F}$
Thermal resistance	1 K/W	$= 0.52750 \text{ °F/hr·Btu}$
Volume	1 m^3	$= 6.1023 \times 10^4 \text{ in}^3$
		$= 35.314 \text{ ft}^3$
		$= 264.17 \text{ gal}$
Volume flow rate	$1 \text{ m}^3/\text{s}$	$= 1.2713 \times 10^5 \text{ ft}^3/\text{hr}$
	1 L/s	$= 127.13 \text{ ft}^3/\text{hr}$
		$= 2.119 \text{ ft}^3/\text{min}$

Conversion Factors (English to Metric)

Area	1 in^2	$= 6.45 \times 10^{-4}$ m^2
	1 ft^2	$= 0.0929$ m^2
Energy	1 Btu	$= 1054.997$ J
Heat transfer rate	1 Btu/h	$= 0.2931$ W
Heat flux	1 Btu/hr·ft^2	$= 3.1536$ W/m^2
Heat generation rate	1 Btu/hr·ft^3	$= 10.3466$ W/m^3
Heat transfer coefficient	1 Btu/hr·ft^2·°F	$= 5.67795$ W/m^2·K
Latent heat	1 Btu/lb$_m$	$= 2325.852$ J/kg
Length	1 ft	$= 0.3048$ m
	1 mile	$= 1.60935$ km
Mass	1 lb$_m$	$= 0.4536$ kg
Mass density	1 lb$_m$/ft^3	$= 16.01845$ kg/m3
Mass flow rate	1 lb$_m$/h	$= 1.26 \times 10^{-4}$ kg/s
Mass transfer coefficient	1 ft/h	$= 8.4667 \times 10^{-5}$ m/s
Pressure	1 lb$_f$/ft^2	$= 47.8790$ N/m^2
	1 lb$_f$/in.^{2}	$= 6894.6498$ N/m^2
	1 in. water	$= 249.066$ N/m^2
	1 bar	$= 1 \times 10^5$ N/m^2
Power	1 HP	$= 0.7463$ kW
	1 Btu/hr	$= 2.931 \times 10^{-4}$ kW
Refrigeration capacity	1 Btu/hr	$= 1.0551 \times 10^{-5}$ kJ/hr
	1 ton	$= 1.2661 \times 10^4$ kJ/hr
		$= 3.5162$ kW
Specific heat	1 Btu/lb$_m$·°F	$= 4186.5528$ J/kg·K
Temperature	1 °R	$= (9/5)$ K
	1 °F	$= (9/5)$ K - 459.67
	1 °C	$=$ K - 273.15
Temperature difference	1 °R	$= (5/9)$ °K
	1 °F	$= (5/9)$ °K
	1 °C	$= 1$ K
Thermal conductivity	1 Btu/ hr·ft·°F	$= 1.7306$ W/m·K
Thermal resistance	1 °F/hr·Btu	$= 1.8957$ K/W
Volume	1 in^3	$= 1.6387 \times 10^{-5}$ m^3
	1 ft^3	$= 0.0283$ m^3
	1 gal	$= 3.785 \times 10^{-3}$ m^3
Volume flow rate	1 ft^3/hr	$= 7.8660 \times 10^{-6}$ m^3/s
	1 ft^3/hr	$= 7.8660 \times 10^{-3}$ L/s
	1 ft^3/min	$= 0.4719$ L/s

Appendix B. Weather Data

Table B-1 Monthly and Annual Degree Days (Base 65 °F) for Selected Locations in US and Canada (*Source:* ASHRAE, Handbook and Product Directory, 1993. With permission.)

State and Station	Average Winter Temperature °F	Average Winter Temperature °C	July	August	September	October	November	December	January	February	March	April	May	June	Yearly Total
Alabama, Birmingham	54.20	12.70	0	0	6	93	363	555	592	462	363	108	9	0	2551
Alaska, Anchorage	23.00	5.00	245	291	516	930	1284	1572	1631	1316	1293	879	592	315	10864
Arizona, Tucson	58.10	14.80	0	0	0	25	231	406	471	344	242	75	6	0	1800
Arkansas, Little Rock	50.50	10.60	0	0	9	127	465	716	736	577	434	126	9	0	3219
California, San Francisco	53.40	12.20	81	78	60	143	306	462	508	395	363	279	214	126	3015
Colorado, Denver	37.60	3.44	6	9	117	428	819	1035	1132	938	887	558	288	66	6283
Connecticut, Bridgeport	39.90	4.72	0	0	66	307	615	986	1079	966	853	510	208	27	5617
Delaware, Wilmington	42.50	6.17	0	0	51	270	588	927	980	874	735	387	112	6	4930
District of Columbia, Washington	45.70	7.94	0	0	33	217	519	834	871	762	626	288	74	0	4224
Florida, Tallahassee	60.10	15.90	0	0	0	28	198	360	375	286	202	86	0	0	1485
Georgia, Atlanta	51.70	11.28	0	0	18	124	417	648	636	518	428	147	25	0	2961
Hawaii, Honolulu	74.20	23.80	0	0	0	0	0	0	0	0	0	0	0	0	0
Idaho, Boise	39.70	4.61	0	0	132	415	792	1017	1113	854	722	438	245	81	5809
Illinois, Chicago	35.80	2.44	0	12	117	381	807	1166	1263	1086	939	534	260	72	6639
Indiana, Indianapolis	39.60	4.56	0	0	90	316	723	1051	1113	949	809	432	177	39	5699
Iowa, Soiux City	43.00	1.10	0	9	108	369	867	1240	1435	1198	989	483	214	39	6951
Kansas, Wichita	44.20	7.11	0	0	33	229	618	905	1023	804	645	270	87	6	4620
Kentucky, Louisville	44.00	6.70	0	0	54	248	609	890	930	818	682	315	105	9	4660
Louisiana, Shreveport	56.20	13.80	0	0	0	47	297	417	552	426	304	81	0	0	2184
Maine, Caribou	24.40	-3.89	78	115	336	682	1044	1535	1690	1470	1308	858	468	183	9767
Maryland, Baltimore	43.70	6.83	0	0	48	264	585	905	936	820	679	327	90	0	4654

State and Station	Average Winter Temperature °F	Average Winter Temperature °C	July	August	September	October	November	December	January	February	March	April	May	June	Yearly Total
Massachusetts, Boston	40.00	4.40	0	9	60	316	603	983	1088	972	846	513	208	36	5634
Michigan, Lancing	34.80	1.89	6	22	138	431	813	1163	1262	1142	1011	579	273	69	6909
Minnesota, Minneapolis	28.30	-1.72	22	31	189	505	1014	1454	1631	1380	1166	621	288	81	8382
Mississippi, Jackson	55.70	13.50	0	0	0	65	315	502	546	414	310	87	0	0	2239
Missouri, Kansas City	43.90	6.94	0	0	39	220	612	905	1032	818	682	294	109	0	4711
Montana, Billings	34.50	1.72	6	15	186	487	897	1135	1296	1100	970	570	285	102	7049
Nebraska, Lincoln	38.80	4.11	0	6	75	301	726	1066	1237	1016	834	402	171	30	5864
Nevada, Las Vegas	53.50	12.28	0	0	0	78	387	617	688	487	335	111	6	0	2709
New Hampshire, Concord	33.00	0.6	6	50	177	505	822	1240	1358	1184	1032	636	298	75	7383
New Jersey, Atlantic City	43.20	6.56	0	0	39	251	549	880	936	848	741	420	133	15	4812
New Mexico, Albuquerque	45.00	7.2	0	0	12	229	642	868	930	703	595	288	81	0	4348
New York, Syracuse	35.20	2.11	6	28	132	415	744	1153	1271	1140	1004	570	248	45	6756
North Carolina, Charlotte	50.40	10.56	0	0	6	124	438	691	691	582	481	156	22	0	3191
North Dakota, Bismarck	26.60	-2.67	34	28	222	577	1083	1463	1708	1442	1203	645	329	117	8851
Ohio, Cleveland	37.20	3.22	9	25	105	384	738	1088	1159	1047	918	552	260	66	6351
Oklahoma, Stillwater	48.30	9.39	0	0	15	164	498	766	868	664	527	189	34	0	3725
Oregon, Pendleton	42.60	6.22	0	0	111	350	711	884	1017	773	617	396	205	63	5127
Pennsylvania, Pittsburgh	38.40	3.89	0	9	105	375	726	1063	1119	1002	874	480	195	39	5987
Rhode Island, Providence	38.80	4.11	0	16	96	372	660	1023	1110	988	868	534	236	51	5954
South Carolina, Charleston	56.40	13.9	0	0	0	59	282	471	487	389	291	54	0	0	2033
South Dakota, Rapid City	33.40	1.11	22	12	165	481	897	1172	1333	1145	1051	615	326	126	7345
Tennessee, Memphis	50.50	10.6	0	0	18	130	447	698	729	585	456	147	22	0	3232
Texas, Dallas	55.30	13.3	0	0	0	62	321	524	601	440	319	90	6	0	2363
Utah, Salt Lake City	38.40	3.89	0	0	81	419	849	1082	1172	910	763	459	233	84	6052
Vermont, Burlington	29.40	-1.11	28	65	207	539	891	1349	1513	1333	1187	714	353	90	8269

State and Station	Average Winter Temperature °F	Average Winter Temperature °C	July	August	September	October	November	December	January	February	March	April	May	June	Yearly Total
Virginia, Norfolk	49.20	9.89	0	0	0	136	408	698	738	655	533	216	37	0	3421
Washington, Spokane	36.50	2.83	9	25	168	493	879	1082	1231	980	834	531	288	135	6655
West Virginia, Charleston	44.80	7.44	0	0	63	254	591	865	880	770	648	300	96	9	4476
Wisconsin, Milwaukee	32.60	0.667	43	47	174	471	876	1252	1376	1193	1054	642	372	135	7635
Wyoming, Casper	33.40	1.11	6	16	192	524	942	1169	1290	1084	1020	657	381	129	7410
Alberta, Calgary	–	–	109	186	402	719	1110	1389	1575	1379	1268	798	477	291	9703
British Columbia, Vancouver	–	–	81	87	219	456	657	787	862	723	676	501	310	156	5515
Manitoba, Winnipeg	–	–	38	71	322	683	1251	1757	2008	1719	1465	813	405	147	10679
New Brunswick, Fredericton	–	–	78	68	234	592	915	1392	1541	1379	1172	753	406	141	8671
Nova Scotia, Halifax	–	–	58	51	180	457	710	1074	1213	1122	1030	742	487	237	7361
Ontario, Ottawa	–	–	25	81	222	567	936	1469	1624	1441	1231	708	341	90	8735
Quebec, Montreal	–	–	9	43	165	521	882	1392	1566	1381	1175	684	316	69	8203
Saskatchewan, Regina	–	–	78	93	360	741	1284	1711	1965	1687	1473	804	409	201	10806

Table B-2 Weather Data for Selected Locations in the US and Canada (*Source:* ASHRAE, Handbook of Fundamentals, 1993. With permission.)

State and Station	Lat. °N	Long. °W	Elev. feet	Winter °F Design Dry-bulb 99%	97.5%	Summer °F Design Dry-bulb and Coincident Wet-bulb — Mean 1%	2.5%	5%	Mean daily Range	Design Wet-bulb 1%	2.5%	5%	Prevailing Wind Winter	Knots	Summer	Median of Annual Extr. Max	Min
ALABAMA																	
ALEXANDER CITY	32 57	85 57	660	18	22	96 / 77	93 / 76	91 / 76	21	79	78	78					
ANNISTON AP	33 35	85 51	599	18	22	97 / 77	94 / 76	92 / 76	21	79	78	78	SW	5	SW	98.4	12.4
AUBURN	32 36	85 30	652	18	22	96 / 77	93 / 76	91 / 76	21	79	78	78					
BIRMINGHAM AP	33 34	86 45	620	17	21	96 / 74	94 / 75	92 / 74	21	78	78	76	NNW	8	WNW	99.8	14.6
DECATUR	34 37	86 59	580	11	16	95 / 75	93 / 74	91 / 74	22	78	77	76					
DOTHAN AP	31 19	85 27	374	23	27	94 / 76	92 / 76	91 / 76	20	80	79	78				98.5	12.9
FLORENCE AP	34 48	87 40	581	17	21	97 / 74	94 / 74	92 / 74	22	78	79	76	NW	7	NW		
GADSDEN	34 01	86 00	554	16	20	96 / 75	94 / 75	92 / 74	22	78	77	76	NNW	8	WNW		
HUNTSVILLE AP	34 42	86 35	606	11	16	95 / 75	93 / 74	91 / 74	23	78	77	76	N	9	SW		
MOBILE AP	30 41	88 15	211	25	29	95 / 77	93 / 77	91 / 76	18	80	79	78	N	10	N		
MOBILE CO	30 40	88 15	211	25	29	95 / 77	93 / 77	91 / 76	16	80	79	78					
MONTGOMERY AP	32 23	86 22	169	22	25	96 / 76	95 / 77	93 / 76	21	79	79	78	NW	7	W	97.9	22.3
SELMA, CRAIG AFB	32 20	87 59	166	22	26	97 / 78	95 / 77	93 / 77	21	81	80	79	N	9	SW	98.9	18.2
TALLADEGA	33 27	86 06	565	18	22	97 / 77	94 / 76	92 / 76	21	79	78	78				100.1	17.6
TUSCALOOSA AP	33 13	87 37	169	20	23	98 / 75	96 / 76	94 / 76	22	79	78	77	N	5	WNW	99.6	11.2
ALASKA																	
ANCHORAGE AP	61 10	150 01	114	-23	-18	71 / 59	68 / 58	66 / 56	15	60	59	57	SE	3	WNW		
BARROW (S)	71 18	156 47	31	-45	-41	57 / 53	53 / 50	49 / 47	12	54	50	47	SW	8	SE		
FAIRBANKS AP (S)	64 49	147 52	436	-51	-47	82 / 62	78 / 60	75 / 59	24	64	62	60	N	5	S		
JUNEAU AP	58 22	134 35	12	-4	1	74 / 60	70 / 58	67 / 57	15	61	59	58	N	7	W		
KODIAK	57 45	152 29	73	10	13	69 / 58	65 / 56	62 / 55	10	60	58	56	WNW	14	NW		
NOME AP	64 30	165 26	13	-31	-27	66 / 57	62 / 55	59 / 54	10	58	56	55	N	4	W		
ARIZONA																	
DOUGLAS AP	31 27	109 36	4098	27	31	98 / 63	95 / 63	93 / 63	31	70	69	68				104.0	14.0
FLAGSTAFF AP	35 8	111 40	7006	-2	4	84 / 55	82 / 55	80 / 54	31	61	60	59	NE	5	SW	90.0	-11.6
FORT HUACHUCA AP (S)	31 35	110 20	4664	24	28	95 / 62	92 / 62	90 / 62	27	69	68	67					
KINGMAN AP	35 12	114 01	3539	18	25	100 / 64	100 / 64	97 / 64	30	70	69	69					
NOGALES	31 21	110 55	3800	28	32	99 / 64	96 / 64	94 / 64	31	71	69	69	SW	5	W		
PHOENIX AP (S)	33 26	112 01	1112	31	34	109 / 71	107 / 71	105 / 71	27	76	75	75	E	4	W	112.8	26.7
PRESCOTT AP	34 39	112 26	5010	4	9	96 / 61	94 / 60	92 / 60	30	66	65	64					
TUCSON AP (S)	32 07	110 56	2558	28	32	104 / 66	102 / 66	100 / 66	26	72	71	71	SE	6	WNW	108.9	19.3
WINSLOW AP	35 01	110 44	4895	5	10	97 / 61	95 / 60	93 / 60	32	66	65	64	SW	6	WSW	102.7	-0.4

State and Station	Lat. °	Lat. 'N	Long. °	Long. 'W	Elev. feet	Winter, °F Design Dry-bulb 99%	Winter 97.5%	Summer Design Dry-bulb and Coincident Wet-bulb — Mean 1%	2.5%	5%	Mean daily Range	Design Wet-bulb 1%	2.5%	5%	Prevailing Wind Winter	Knots	Prevailing Wind Summer	Temp °F Median of Annual Extr. Max.	Min.
YUMA AP	32	39	114	37	213	36	39	111 / 72	109 / 72	107 / 71	27	79	78	77	NNE	6	WSW	114.8	30.8
ARKANSAS																			
BLYTHEVILLE AFB	35	57	89	57	264	10	15	96 / 78	94 / 77	91 / 76	21	81	80	78	N	8	SSW		13.9
CAMDEN	33	36	92	49	116	18	23	98 / 76	96 / 76	94 / 76	21	80	79	78					
EL DORADO AP	33	13	92	49	277	18	23	98 / 76	96 / 76	94 / 76	21	80	79	78	S	6	SE	101.0	-0.3
FAYETTEVILLE AP	36	00	94	10	1251	7	12	97 / 72	94 / 73	92 / 73	23	77	76	75	NE	9	SSW	99.4	7.0
FORT SMITH AP	35	20	94	22	463	12	17	101 / 75	98 / 76	95 / 76	24	80	79	78	NW	8	SW	101.9	10.6
HOT SPRINGS	34	29	93	06	535	17	23	101 / 77	98 / 77	94 / 77	22	80	79	78	N	8	SW	103.0	7.3
JONESBORO	35	50	90	42	345	10	15	96 / 78	94 / 77	91 / 76	21	81	80	78				101.7	
LITTLE ROCK AP (S)	34	44	92	14	257	15	20	99 / 76	96 / 77	94 / 77	22	80	79	78	N	9	SSW	99.0	11.2
PINE BLUFF AP	34	18	92	05	241	16	22	100 / 78	97 / 77	95 / 78	22	81	80	80	N	7	SW	102.2	13.1
TEXARKANA AP	33	27	93	59	389	18	23	98 / 76	96 / 77	93 / 76	21	80	79	78	WNW	9	SSW	104.8	14.0
CALIFORNIA																			
BAKERSFIELD AP	35	25	119	03	475	30	32	104 / 70	101 / 69	98 / 68	32	73	71	70	ENE	5	WNW	109.8	25.3
BARSTOW AP	34	51	116	47	1927	26	29	106 / 68	104 / 68	102 / 67	37	73	71	70	WNW	7	W	110.4	17.4
BLYTHE AP	33	37	114	43	395	30	33	112 / 71	110 / 71	108 / 70	28	75	75	74	NW	3	S	116.8	24.1
BURBANK AP	34	12	118	21	775	37	39	95 / 68	91 / 68	88 / 67	25	71	70	69	NW	5	SSE		
CHICO	39	48	121	51	238	28	30	103 / 69	101 / 68	98 / 67	36	71	70	68	WNW	5	NW	109.0	22.6
CONCORD	37	58	121	59	200	24	27	100 / 69	97 / 68	94 / 67	32	71	71	70					
COVINA	34	05	117	52	575	32	35	98 / 69	95 / 68	92 / 67	31	73	71	70					
CRESCENT CITY AP	41	46	124	12	40	31	33	68 / 60	65 / 59	63 / 58	18	62	60	59					
DOWNEY	33	56	118	08	116	37	40	93 / 70	89 / 70	86 / 69	22	72	71	70					
EL CAJON	32	49	116	58	367	42	44	83 / 69	80 / 70	78 / 68	30	71	70	68					
EL CENTRO AP (S)	32	49	115	40	-43	35	38	112 / 74	110 / 74	108 / 74	34	81	80	78	W	6	SE		
ESCONDIDO	33	07	117	05	660	39	41	89 / 68	85 / 68	82 / 68	30	71	70	69					
EUREKA/ARCATA AP	40	59	124	06	218	31	33	68 / 60	65 / 59	63 / 58	11	62	60	59	E	5	NW	75.8	29.7
FAIRFIELD-TRAVIS AFB	38	16	121	56	62	29	32	99 / 68	95 / 67	91 / 66	34	70	68	67	N	5	WSW		
FRESNO AP (S)	36	46	119	43	328	28	30	102 / 70	100 / 69	97 / 68	34	72	71	70	E	4	WNW	108.7	25.8
HAMILTON AFB	38	04	122	30	3	30	32	89 / 68	84 / 66	80 / 65	28	72	69	67	N	4	SE		
LAGUNA BEACH	33	33	117	47	35	41	43	83 / 68	80 / 68	77 / 67	18	70	69	68					
LIVERMORE	37	42	121	57	545	24	27	100 / 69	97 / 68	93 / 67	24	71	70	67					
LOMPOC, VANDENBERG AFB	34	43	120	34	368	35	38	75 / 61	70 / 61	67 / 60	20	63	61	60	WNW	4	NW		
LONG BEACH AP	33	49	118	09	30	41	43	83 / 68	80 / 68	77 / 67	22	70	69	68	ESE	5	NW		
LOS ANGELES AP (S)	33	56	118	24	97	41	43	83 / 68	80 / 68	77 / 67	15	70	69	68	E	4	WNW		
LOS ANGELES CO (S)	34	03	118	14	270	37	40	93 / 70	89 / 70	86 / 69	20	72	71	70	NW	4	WSW	98.1	35.9

State and Station	Lat. ° N	Lat. ' N	Long. ° W	Long. ' W	Elev. feet	Winter °F Design Dry-bulb 99%	Winter °F Design Dry-bulb 97.5%	Summer Design Dry-bulb and Coincident Wet-bulb Mean 1%	2.5%	5%	Mean daily Range	Design Wet-bulb 1%	2.5%	5%	Prevailing Wind Winter	Knots	Summer	Temp °F Median of Annual Extr. Max.	Min.
MERCED-CASTLE AFB	37	23	120	34	188	29	31	102 / 70	99 / 69	96 / 68	36	72	71	70	ESE	4	NW	105.8	26.2
MODESTO	37	39	121	00	91	28	30	101 / 69	98 / 68	95 / 67	36	71	70	69			NW		
MONTEREY	36	36	121	54	39	35	38	75 / 63	71 / 61	68 / 61	20	64	62	61	SE	4			
NAPA	38	13	122	17	56	30	32	100 / 69	96 / 68	92 / 67	30	71	69	68				103.1	25.8
NEEDLES AP	34	46	114	37	913	30	33	112 / 71	110 / 71	108 / 70	27	75	75	74				116.4	26.7
OAKLAND AP	37	49	122	19	5	34	36	85 / 64	80 / 63	75 / 62	19	66	64	63	E	5	WNW		
OCEANSIDE	33	14	117	25	26	41	43	83 / 68	80 / 68	77 / 67	13	70	69	68				93.0	31.8
ONTARIO	34	03	117	36	952	31	33	102 / 70	99 / 69	96 / 67	36	74	72	71	E	4	WSW		
OXNARD	34	12	119	11	49	34	36	83 / 66	80 / 64	77 / 63	19	70	68	67					
PALMDALE AP	34	38	118	06	2542	18	22	103 / 65	101 / 65	98 / 64	35	69	67	66	SW	5	WSW		
PALM SPRINGS	33	49	116	32	411	33	35	112 / 71	110 / 70	108 / 70	35	76	74	73					
PASADENA	34	09	118	09	864	32	35	98 / 69	95 / 68	92 / 67	29	73	71	70				102.8	30.4
PETALUMA	38	14	122	38	16	26	29	94 / 68	90 / 66	87 / 65	31	72	70	68				102.0	24.2
POMONA CO	34	03	117	45	934	28	30	102 / 70	99 / 69	95 / 68	36	74	72	71	E	4	W	105.7	26.2
REDDING AP	40	31	122	18	495	29	31	105 / 68	102 / 67	100 / 66	32	71	69	68				109.2	26.0
REDLANDS	34	03	117	11	1318	31	33	102 / 70	99 / 69	96 / 68	33	74	72	71				106.7	27.1
RICHMOND	37	56	122	21	55	34	36	85 / 64	80 / 63	75 / 62	17	66	64	63					
RIVERSIDE-MARCH AFB (S)	33	54	117	15	1532	29	32	100 / 68	98 / 68	95 / 67	37	72	71	70	N	4	NW	107.6	26.6
SACRAMENTO AP	38	31	121	30	17	30	32	101 / 70	98 / 70	94 / 69	36	72	71	70	NNW	6	SW	105.1	27.6
SALINAS AP	36	40	121	36	75	30	32	74 / 61	70 / 60	67 / 59	24	62	61	59					
SAN BERNARDINO,NORTON AFB	34	08	117	16	1125	31	33	102 / 70	99 / 69	96 / 68	38	74	72	71	E	3	W	109.3	25.3
SAN DIEGO AP	32	44	117	10	13	42	44	83 / 69	80 / 69	78 / 68	12	71	70	68	NE	3	WNW	91.2	37.4
SAN FERNANDO	34	17	118	28	965	37	39	95 / 68	91 / 68	88 / 67	38	71	70	69					
SAN FRANCISCO AP	37	37	122	23	8	35	38	82 / 64	77 / 63	73 / 62	20	65	64	62	S	5	NW	91.3	35.9
SAN FRANCISCO CO	37	46	122	26	72	38	40	74 / 63	71 / 62	69 / 61	14	64	62	61	W	5	W	98.6	28.2
SAN JOSE AP	37	22	121	21	56	34	36	85 / 66	81 / 65	77 / 64	26	68	67	65	SE	4	NNW	99.8	29.3
SAN LUIS OBISPO	35	20	120	43	250	33	35	92 / 69	88 / 70	84 / 69	26	73	71	70	E	4	W	101.0	29.9
SANTA ANA AP	33	45	117	52	115	37	39	89 / 69	85 / 68	82 / 68	28	71	70	69	E	3	SW	97.1	31.7
SANTA BARBARA MAP	34	26	119	50	10	34	36	81 / 67	77 / 66	75 / 65	24	68	67	66	NE	3	SW	97.5	26.8
SANTA CRUZ	36	59	122	01	125	35	38	75 / 63	71 / 61	68 / 61	28	64	62	61					
SANTA MARIA AP (S)	34	54	120	27	236	31	33	81 / 64	76 / 63	73 / 62	23	65	64	63	E	4	WNW		
SANTA MONICA CO	34	01	118	29	64	41	43	83 / 68	80 / 68	77 / 67	16	70	69	68					
SANTA PAULA	34	21	119	05	263	33	35	90 / 68	86 / 67	84 / 66	36	71	69	68				102.5	23.4
SANTA ROSA	38	31	122	49	125	27	29	99 / 68	95 / 67	91 / 66	34	70	68	67	N	5	SE	104.1	24.5
STOCKTON AP	37	54	121	15	22	28	30	100 / 69	97 / 68	94 / 67	37	71	70	68	WNW	4	NW		

State and Station	Lat. °	′N	Long. °	′W	Elev. feet	Winter °F Design Dry-bulb 99%	97.5%	Summer Design Dry-bulb and Mean Coincident Wet-bulb 1%	2.5%	5%	Mean daily Range	Design Wet-bulb 1%	2.5%	5%	Prevailing Wind Winter	Knots	Summer	Temp Median of Annual Extr. Max	Min
UKIAH	39	09	123	12	623	27	29	99 / 69	95 / 68	91 / 67	40	70	68	67				108.1	21.6
VISALIA	36	20	119	18	325	28	30	102 / 70	100 / 69	97 / 68	38	72	71	70				108.4	25.1
YREKA	41	43	122	38	2625	13	17	95 / 65	92 / 64	89 / 63	38	67	65	64				102.8	7.1
YUBA CITY	39	08	121	36	80	29	31	104 / 68	101 / 67	99 / 66	36	71	69	68					
COLORADO																			
ALAMOSA AP	37	27	105	52	7537	-21	-16	84 / 57	82 / 57	80 / 57	35	62	61	60					-8.4
BOULDER	40	00	105	16	5445	2	8	93 / 59	91 / 59	89 / 59	27	64	63	62				96.0	-12.1
COLORADO SPRINGS AP	38	49	104	43	6145	-3	2	91 / 58	88 / 57	86 / 57	30	63	62	61	N 9		S	92.3	-10.4
DENVER AP	39	45	104	52	5283	-5	1	93 / 59	91 / 59	89 / 59	28	64	63	62	S 8		SE	96.8	-11.2
DURANGO	37	17	107	53	6550	-1	4	89 / 59	87 / 59	85 / 59	30	64	63	62				92.4	-18.1
FORT COLLINS	40	35	105	05	4999	-10	-4	93 / 59	91 / 59	89 / 59	28	64	63	62				95.2	
GRAND JUNCTION AP (S)	39	07	108	32	4843	2	7	96 / 59	94 / 59	92 / 59	29	64	63	62	ESE 5		WNW	99.9	-3.4
GREELEY	40	26	104	38	4648	-11	-5	96 / 60	94 / 60	92 / 60	29	65	64	63					
LAJUNTA AP	38	03	103	30	4160	-3	3	100 / 68	98 / 68	95 / 67	31	72	70	69	W 8		S		
LEADVILLE	39	15	106	18	10155	-8	-4	84 / 52	81 / 51	78 / 50	30	56	55	54				79.7	-17.8
PUEBLO AP	38	18	104	29	4641	-7	0	97 / 61	95 / 61	92 / 61	31	67	66	65	W 5		SE	100.5	-12.2
STERLING	40	37	103	12	3939	-7	-2	95 / 62	93 / 62	90 / 62	30	67	66	65				100.3	-15.4
TRINIDAD AP	37	15	104	20	5740	-2	3	93 / 61	91 / 61	89 / 61	32	66	65	64	W 7		WSW	96.8	-10.5
CONNECTICUT																			
BRIDGEPORT AP	41	11	73	11	25	6	9	86 / 73	84 / 71	81 / 70	18	75	74	73	NNW 13		WSW		
HARTFORD, BRAINARD FIELD	41	44	72	39	19	3	7	91 / 74	88 / 73	85 / 72	22	77	75	74	N 5		SSW	95.7	-4.4
NEW HAVEN AP	41	19	73	55	6	3	7	88 / 75	84 / 73	82 / 72	17	76	75	74	NNE 7		SW	93.0	-0.2
NEW LONDON	41	21	72	06	59	5	9	88 / 73	85 / 72	83 / 71	16	76	75	74					
NORWALK	41	07	73	25	37	6	9	86 / 73	84 / 71	81 / 70	19	75	74	73					
NORWICH	41	32	72	04	20	3	7	89 / 75	86 / 73	83 / 72	18	76	75	74					
WATERBURY	41	35	73	04	843	-4	2	88 / 73	85 / 71	82 / 70	21	75	74	72	N 8		SW		
WINDSOR LOCKS, BRADLEY FLD	41	56	72	41	169	0	4	91 / 74	88 / 72	85 / 71	22	76	75	73	N 8		SW		
DELAWARE																			
DOVER AFB	39	08	75	28	28	11	15	92 / 75	90 / 75	87 / 74	18	79	77	76	W 9		SW	97.0	7.0
WILMINGTON AP	39	40	75	36	74	10	14	92 / 74	89 / 74	87 / 73	20	77	76	75	WNW 9		WSW	95.4	4.9
DISTRICT OF COLUMBIA																			
ANDREWS AFB	38	05	76	05	279	10	14	92 / 75	90 / 74	87 / 73	18	78	76	75					
WASHINGTON, NATIONAL AP	38	51	77	02	14	14	17	93 / 75	91 / 74	89 / 74	18	78	77	76	WNW 11		S	97.6	7.4
FLORIDA																			
BELLE GLADE	26	39	80	39	16	41	44	92 / 76	91 / 76	89 / 76	16	79	78	78				94.7	30.9

State and Station	Lat. (° / N)	Long. (° / W)	Elev. feet	Winter, °F — Design Dry-bulb 99%	97.5%	Summer, °F — Design Dry-bulb and Mean Coincident Wet-bulb 1%	2.5%	5%	Mean daily Range	Design Wet-bulb 1%	2.5%	5%	Prevailing Wind Winter	Summer	Knots	Temp. °F — Median of Annual Extr. Max.	Min.
CAPE KENNEDY AP	28 29	80 34	16	35	38	90 / 78	88 / 78	87 / 78	15	80	79	79	NW	E	8		
DAYTONA BEACH AP	29 11	81 03	31	32	35	92 / 78	90 / 77	88 / 78	15	80	79	78	NW	ESE	9		
FORT LAUDERDALE	26 04	80 09	10	42	46	92 / 78	91 / 78	90 / 78	15	80	79	79	NNE	W	7		
FORT MYERS AP	26 35	81 52	15	41	44	93 / 78	92 / 78	91 / 77	18	80	79	79				94.9	34.9
FORT PIERCE	27 28	80 21	25	38	42	91 / 78	90 / 77	89 / 78	15	80	79	78				96.1	34.0
GAINESVILLE AP (S)	29 41	82 16	152	28	31	95 / 77	93 / 77	92 / 77	18	80	78	78	W	W	6	97.8	23.3
JACKSONVILLE AP	30 30	81 42	26	29	32	96 / 77	94 / 77	92 / 76	19	79	78	77	NW	SW	7	97.5	25.4
KEY WEST AP	24 33	81 45	4	55	57	90 / 78	90 / 78	89 / 78	9	80	79	79	NNE	SE	12	92.0	51.5
LAKELAND CO (S)	28 02	81 57	214	39	41	93 / 76	91 / 76	89 / 76	17	79	78	78	NNW	SSW	9		
MIAMI AP (S)	25 48	80 16	7	44	47	91 / 77	90 / 77	89 / 77	15	79	78	78	NNW	SE	8	92.5	39.0
MIAMI BEACH CO	25 47	80 17	10	45	48	90 / 77	89 / 77	88 / 77	10	79	79	78					
OCALA	29 11	82 08	89	31	34	95 / 77	93 / 77	92 / 76	18	80	79	78				98.6	24.8
ORLANDO AP	28 33	81 23	100	35	38	94 / 76	93 / 76	91 / 76	17	79	78	79	NNW	SSW	9		
PANAMA CITY,TYNDALL AFB	30 04	85 35	18	29	33	92 / 78	90 / 77	89 / 77	14	81	80	79	N	WSW	8	96.3	23.3
PENSACOLA CO	30 25	87 13	56	25	29	94 / 77	93 / 77	91 / 77	14	80	79	79	NNE	SW	7	97.6	25.8
ST. AUGUSTINE	29 58	81 20	10	31	35	92 / 78	89 / 78	87 / 78	16	79	79	78	NW	W	7		
ST. PETERSBURG	27 46	82 38	35	36	40	92 / 77	91 / 77	90 / 77	16	79	79	78	N	W	8	94.8	35.6
SANFORD	28 46	81 17	89	35	38	94 / 76	93 / 76	91 / 76	17	79	78	78					
SARASOTA	27 23	82 33	26	39	42	93 / 77	92 / 77	90 / 76	17	79	79	78					
TALLAHASSEE AP (S)	30 23	84 22	55	27	30	94 / 77	92 / 76	90 / 76	19	79	78	78	NW	NW	6	97.6	20.9
TAMPA AP (S)	27 58	82 32	19	36	40	92 / 77	91 / 77	90 / 76	17	79	79	78	N	W	8	95.0	31.5
WEST PALM BEACH AP	26 41	80 06	15	41	45	92 / 78	91 / 78	90 / 78	16	80	79	79	NW	ESE	9		
GEORGIA																	
ALBANY,TURNER AFB	31 36	84 05	223	25	29	97 / 77	95 / 76	93 / 76	20	80	79	78	N	W	7	100.6	19.9
AMERICUS	32 03	84 14	456	21	25	97 / 77	94 / 76	92 / 75	20	79	78	77				100.4	16.5
ATHENS	33 57	83 19	802	18	22	94 / 74	92 / 74	90 / 74	21	78	77	76	NW	WNW	9	98.7	13.5
ATLANTA AP (S)	33 39	84 26	1010	17	22	94 / 74	92 / 74	90 / 73	19	77	76	75	NW	NW	11	95.7	
AUGUSTA AP	33 22	81 58	145	20	23	97 / 77	95 / 76	93 / 76	19	80	79	78	W	WSW	4	99.0	17.5
BRUNSWICK	31 15	81 29	25	29	32	92 / 78	89 / 78	87 / 78	18	80	79	79				99.3	24.7
COLUMBUS,LAWSON AFB	32 31	84 56	242	21	24	95 / 76	93 / 76	91 / 75	21	79	78	77	NW	W	8		
DALTON	34 34	84 57	720	17	22	94 / 77	93 / 76	91 / 76	22	79	78	77				101.0	16.7
DUBLIN	32 20	82 54	215	21	25	96 / 77	93 / 76	91 / 75	20	79	78	77				97.1	10.5
GAINESVILLE	34 11	83 41	1269	16	21	93 / 74	91 / 74	89 / 73	21	77	76	75				97.1	
GRIFFIN (S)	33 15	84 16	981	18	22	93 / 76	91 / 75	88 / 74	21	78	77	76					
LA GRANGE	33 01	85 04	709	19	23	94 / 76	91 / 75	89 / 74	21	78	77	76				97.7	11.9

State and Station	Lat. (°N 'N)	Long. (° 'W)	Elev. feet	Winter Design Dry-bulb 99%	97.5%	Summer DB & Coincident WB 1%	2.5%	5%	Mean daily Range	Design Wet-bulb 1%	2.5%	5%	Prev. Wind Winter	Prev. Wind Summer	Temp. Max.	Temp. Min.
MACON AP	32 42	83 39	354	21	25	96/77	93/76	91/75	22	79	78	77	NW 8	WNW	99.6	16.9
MARIETTA,DOBBINS AFB	33 55	84 31	1068	17	21	94/74	92/74	90/74	21	78	77	76	NNW 12	NW		
MOULTRIE	31 08	83 42	292	27	30	97/77	95/77	92/76	20	80	79	78	NW 8	W	99.3	19.8
ROME AP	34 21	85 10	637	17	22	94/76	93/77	91/76	23	79	78	77	N 7	N	99.1	11.3
SAVANNAH-TRAVIS AP	32 08	81 12	50	24	27	96/77	93/77	92/77	20	80	79	78	WNW 7	SW	98.7	21.9
VALDOSTA-MOODY AFB	30 58	83 12	233	28	31	96/77	94/77	92/76	20	80	79	78	WNW 6	W		
WAYCROSS	31 15	82 24	148	26	29	96/77	94/77	91/76	20	80	79	78		W	100.0	19.5
HAWAII																
HILO AP (S)	19 43	155 05	36	61	62	84/73	83/72	82/72	15	75	74	74	SW 6	NE		
HONOLULU AP	21 20	157 55	13	62	63	87/73	86/73	85/72	12	76	75	74	ENE 12	ENE		
KANEOHE BAY MCAS	21 27	157 46	18	65	66	85/75	84/74	83/74	12	76	76	75	NNE 9	NE		
WAHIAWA	21 03	158 02	900	58	59	86/73	85/72	84/72	14	75	74	73	WNW 5	E		
IDAHO																
BOISE AP (S)	43 34	116 13	2838	3	10	96/65	94/64	91/64	31	68	66	65	SE 6	NW	103.2	0.6
BURLEY	42 32	113 46	4156	-3	2	99/62	95/61	92/66	35	64	63	61			98.6	-8.3
COEUR D'ALENE AP	47 46	116 49	2972	-8	-1	89/62	86/61	83/60	31	64	63	61			99.9	-4.5
IDAHO FALLS AP	43 31	112 04	4741	-11	-6	89/61	87/61	84/59	38	65	63	61	N 9	S	96.2	-16.0
LEWISTON AP	46 23	117 01	1413	-1	6	96/65	93/64	90/63	32	67	66	64	W 3	WNW	105.9	2.7
MOSCOW	46 44	116 58	2660	-7	0	90/63	87/62	84/61	32	65	64	62			98.0	-5.9
MOUNTAIN HOME AFB	43 02	115 54	2996	6	12	99/64	97/63	94/62	36	66	65	63	ESE 7	NW	103.2	-6.5
POCATELLO AP	42 55	112 36	4454	-8	-1	94/61	91/60	89/59	35	64	63	61	NE 5	W	97.9	-11.4
TWIN FALLS AP (S)	42 29	114 29	4150	-3	2	99/62	95/61	92/60	34	64	63	61	SE 6	NW	100.9	-5.1
ILLINOIS																
AURORA	41 45	88 20	744	-6	-1	93/76	91/76	88/75	20	79	78	76	WNW 8		96.7	-13.0
BELLEVILLE,SCOTT AFB	38 33	89 51	453	1	6	94/76	92/76	89/75	21	79	78	76		S		
BLOOMINGTON	40 29	88 57	876	-6	-2	92/75	90/74	88/73	21	78	76	75			98.4	-9.6
CARBONDALE	37 47	89 15	417	2	7	95/77	93/77	90/76	21	80	79	77			100.9	-0.8
CHAMPAIGN/URBANA	40 02	88 17	777	-3	2	95/75	92/74	90/73	21	78	77	75				
CHICAGO,MIDWAY AP	41 47	87 45	607	-5	0	94/74	91/73	88/72	20	77	75	74	NW 11	SW		
CHICAGO,O'HARE AP	41 59	87 54	658	-8	-4	91/74	89/74	86/72	20	77	76	74	WNW 9	SW		
CHICAGO CO	41 53	87 38	590	-3	2	94/75	91/74	89/73	15	79	77	75			96.1	-8.3
DANVILLE	40 12	87 36	695	-4	1	93/75	90/74	88/73	21	78	77	75	W 10	SSW	98.2	-8.4
DECATUR	39 50	88 52	679	-3	2	94/75	91/74	88/73	21	78	77	75	NW 10	SW	99.0	-8.1
DIXON	41 50	89 29	696	-7	-2	93/75	91/74	88/73	23	78	77	75				
ELGIN	42 02	88 16	758	-7	-2	91/75	88/74	86/73	21	78	77	75			97.5	-13.5

State and Station	Lat. °N (° ′)	Long. °W (° ′)	Elev. feet	Winter °F Design Dry-bulb 99%	97.5%	Summer °F Design Dry-bulb and Coincident Wet-bulb Mean 1%	2.5%	Wet-bulb 5%	Mean daily Range	Design Wet-bulb 1%	2.5%	5%	Prevailing Wind Winter	Knots	Summer	Temp. °F Median of Annual Extr. Max.	Min.
FREEPORT	42 18	89 37	780	-9	-4	91 / 74	89 / 73	87 / 72	24	77	76	74					
GALESBURG	40 56	90 26	764	-7	-2	93 / 75	91 / 75	88 / 74	22	78	77	75	WNW	8	SW		
GREENVILLE	38 53	89 24	563	-1	4	94 / 76	92 / 75	89 / 74	21	79	78	76					
JOLIET	41 31	88 10	582	-5	0	93 / 75	90 / 74	88 / 73	20	78	77	75	NW	11	SW		
KANKAKEE	41 05	87 55	625	-4	1	93 / 75	90 / 74	88 / 73	21	78	77	75					
LA SALLE/PERU	41 19	89 06	520	-7	-2	93 / 75	91 / 75	88 / 74	22	78	77	75					
MACOMB	40 28	90 40	702	-5	0	95 / 76	92 / 76	89 / 75	22	79	78	76					
MOLINE AP	41 27	90 31	582	-9	-4	93 / 75	91 / 75	88 / 74	23	79	77	75	WNW	8	SW	96.8	-12.7
MT VERNON	38 19	88 52	479	0	5	95 / 76	92 / 75	89 / 74	21	79	78	76				100.5	-2.9
PEORIA AP	40 40	89 41	652	-8	-4	91 / 75	89 / 74	87 / 73	22	78	76	75	WNW	8	SW	98.0	-10.9
QUINCY AP	39 57	91 12	769	-2	3	96 / 76	93 / 76	90 / 76	22	80	78	77	NW	11	SSW	101.1	-6.7
RANTOUL,CHANUTE AFB	40 18	88 08	753	-4	1	94 / 75	91 / 74	89 / 73	21	78	77	75	W	10	SSW		
ROCKFORD	42 21	89 03	741	-7	-4	91 / 74	89 / 73	87 / 72	24	77	76	74				97.4	-13.8
SPRINGFIELD AP	39 50	89 40	588	-3	2	94 / 75	92 / 74	89 / 74	21	79	77	76	NW	10	SW	98.1	-7.2
WAUKEGAN	42 21	87 53	700	-6	-3	92 / 76	89 / 74	87 / 73	21	78	76	75				96.5	-10.6
INDIANA																	
ANDERSON	40 06	85 37	919	0	6	95 / 76	92 / 75	89 / 74	22	79	78	76	W	9	SW	95.1	-6.0
BEDFORD	38 51	86 30	670	0	5	95 / 76	92 / 75	89 / 74	22	79	78	76				97.5	-4.4
BLOOMINGTON	39 08	86 37	847	0	5	95 / 76	92 / 75	89 / 74	22	79	78	76	W	9	SW	97.8	-4.6
COLUMBUS,BAKALAR AFB	39 16	85 54	651	3	7	95 / 76	92 / 75	90 / 74	22	79	78	76	W	9	SW	98.3	-6.4
CRAWFORDSVILLE	40 03	86 54	679	-2	3	94 / 75	91 / 74	88 / 73	22	79	77	76				98.4	-7.6
EVANSVILLE AP	38 03	87 32	381	4	9	95 / 76	93 / 75	91 / 75	22	79	78	77				98.2	0.2
FORT WAYNE AP	41 00	85 12	791	-4	1	92 / 73	89 / 72	87 / 72	24	77	75	74	NW	9	SW	96.8	-10.5
GOSHEN AP	41 32	85 48	827	-3	1	91 / 73	89 / 73	86 / 72	23	77	75	74	WSW	10	SW	98.5	-8.5
HOBART	41 32	87 15	600	-4	2	91 / 73	88 / 73	85 / 72	21	77	75	74				96.9	-8.1
HUNTINGTON	40 53	85 30	802	-4	1	92 / 73	89 / 72	87 / 72	23	77	75	74				95.6	-6.6
INDIANAPOLIS AP (S)	39 44	86 17	792	-2	2	92 / 74	90 / 74	87 / 73	22	78	76	75	WNW	10	SW	98.3	1.7
JEFFERSONVILLE	38 17	85 45	455	5	10	95 / 74	93 / 74	90 / 74	23	79	77	76				98.2	-7.5
KOKOMO	40 25	86 03	855	-4	0	91 / 74	90 / 73	88 / 73	22	77	75	74					
LAFAYETTE	40 02	86 05	600	-3	3	94 / 74	91 / 73	88 / 73	22	78	76	75					
LA PORTE	41 36	86 43	810	-3	3	93 / 74	90 / 74	87 / 73	22	77	76	75				98.1	-10.5
MARION	40 29	85 41	859	-4	0	91 / 74	90 / 73	88 / 73	23	77	75	74				97.0	-8.6
MUNCIE	40 11	85 21	957	-3	2	92 / 74	90 / 73	87 / 73	23	76	76	75					
PERU,GRISSOM AFB	40 39	86 09	813	-6	-1	90 / 74	88 / 73	86 / 73	22	77	75	74	W	10	SW		
RICHMOND AP	39 46	84 50	1141	-2	2	92 / 74	90 / 74	87 / 73	22	78	76	75				94.8	-8.5

State and Station	Lat. °N	Lat. ′N	Long. °W	Long. ′W	Elev. feet	Winter °F Design Dry-bulb 99%	97.5%	Summer Design Dry-bulb and Mean Coincident Wet-bulb 1%	2.5%	5%	Mean daily Range	Design Wet-bulb 1%	2.5%	5%	Prevailing Wind Winter (Knots)	Summer	Temp. °F Median of Annual Extr. Max.	Min.
SHELBYVILLE	39	31	85	47	750	-1	3	93 / 74	91 / 74	88 / 73	22	78	76	75	SW 11	SSW	97.7	-6.0
SOUTH BEND AP	41	42	86	19	773	-3	1	91 / 73	89 / 73	86 / 72	22	77	75	74			96.2	-9.2
TERRE HAUTE AP	39	27	87	18	585	-2	4	95 / 75	92 / 74	89 / 73	22	79	77	76	NNW 7	SSW	98.3	-4.9
VALPARAISO	41	31	87	02	801	-3	3	93 / 74	90 / 74	87 / 73	22	78	76	75			95.5	-11.0
VINCENNES	38	41	87	32	420	1	6	95 / 75	92 / 74	90 / 73	22	79	77	76			100.3	-2.8
IOWA																		
AMES (S)	42	02	93	48	1099	-11	-6	93 / 75	90 / 74	87 / 73	23	78	76	75		S	97.4	-17.8
BURLINGTON AP	40	47	91	07	692	-7	-3	94 / 74	91 / 75	88 / 73	22	78	77	75	NW 9	SSW	98.6	-11.0
CEDAR RAPIDS AP	41	53	91	42	863	-10	-5	91 / 76	88 / 75	86 / 74	23	78	77	75	NW 9	S	97.7	-15.6
CLINTON	41	50	90	13	595	-8	-3	92 / 75	90 / 75	87 / 74	23	78	77	75			97.5	-13.8
COUNCIL BLUFFS	41	20	95	49	1210	-8	-3	94 / 76	91 / 75	88 / 74	23	78	77	75				
DES MOINES AP	41	32	93	39	938	-10	-5	94 / 75	91 / 74	88 / 73	23	78	77	75	NW 11	S	98.2	-14.2
DUBUQUE	42	24	90	42	1056	-12	-7	90 / 74	88 / 73	86 / 72	22	77	75	74	N 10	SSW	95.2	-15.0
FORT DODGE	42	33	94	11	1162	-12	-7	91 / 74	88 / 74	86 / 72	23	78	77	75	NW 11	S	98.5	-19.1
IOWA CITY	41	38	91	33	661	-11	-6	92 / 76	89 / 76	87 / 74	22	80	78	76	NW 9	SSW	98.5	-15.2
KEOKUK	40	24	91	24	574	-5	0	95 / 75	92 / 75	89 / 74	22	79	77	76			98.4	-8.8
MARSHALLTOWN	42	04	92	56	898	-12	-7	92 / 76	90 / 75	88 / 74	23	78	77	75			98.5	-13.4
MASON CITY AP	43	09	93	20	1213	-15	-11	90 / 74	88 / 74	85 / 72	24	77	75	74	NW 11	S	96.5	-21.7
NEWTON	41	41	93	02	936	-10	-5	94 / 75	91 / 74	88 / 73	23	78	77	75			98.2	-14.7
OTTUMWA AP	41	06	92	27	840	-8	-4	94 / 75	91 / 74	88 / 73	22	78	77	75			99.1	-12.0
SIOUX CITY AP	42	24	96	23	1095	-11	-7	95 / 74	92 / 74	89 / 73	24	78	77	75	NNW 9	S	99.9	-17.7
WATERLOO	42	33	92	24	868	-15	-10	91 / 76	89 / 75	86 / 74	23	78	77	75	NW 9	S	97.7	-19.8
KANSAS																		
ATCHISON	39	34	95	07	945	-2	2	96 / 77	93 / 76	91 / 76	23	81	79	77	NNW 11	SSW	100.5	-8.8
CHANUTE AP	37	40	95	29	981	3	7	100 / 74	97 / 74	94 / 74	23	78	77	76			102.8	-2.8
DODGE CITY AP (S)	37	46	99	58	2582	0	5	100 / 69	97 / 69	95 / 69	25	74	73	71	N 12	SSW	102.9	-7.0
EL DORADO	37	49	96	50	1282	3	7	101 / 72	98 / 73	96 / 73	24	77	76	75			103.5	-5.0
EMPORIA	38	20	96	12	1210	1	5	100 / 74	97 / 74	94 / 73	25	78	76	75			102.4	-6.4
GARDEN CITY AP	37	56	100	44	2880	-1	4	99 / 69	96 / 69	94 / 69	28	74	73	71				
GOODLAND AP	39	22	101	42	3654	-5	0	99 / 66	96 / 65	93 / 66	31	71	70	68	WSW 10	S	103.2	-10.4
GREAT BEND	38	21	98	52	1889	0	4	101 / 73	98 / 73	95 / 73	28	78	76	75				
HUTCHINSON AP	38	04	97	52	1542	4	8	102 / 72	99 / 72	97 / 72	28	77	75	74	N 14			
LIBERAL	37	03	100	58	2870	2	7	99 / 68	96 / 68	94 / 68	28	73	72	71		S	105.3	-6.1
MANHATTAN, FORT RILEY (S)	39	03	96	46	1065	-1	3	99 / 75	95 / 75	92 / 74	24	78	77	76	NNE 8	S	105.8	-3.8
PARSONS	37	20	95	31	899	5	9	100 / 74	97 / 74	94 / 74	23	79	77	76	NNW 11	SSW	104.5	-8.6

State and Station	Lat. °	N ′	Long. °	′W	Elev. feet	Winter, °F Design Dry-bulb 99%	97.5%	Summer, °F Design Dry-bulb and Coincident Wet-bulb Mean 1%	2.5%	5%	Mean daily Range	Design Wet-bulb 1%	2.5%	5%	Prevailing Wind Winter	Summer	Knots	Temp. °F Median of Annual Extr. Max.	Min.
RUSSELL AP	38	52	98	49	1866	0	4	101 / 73	98 / 73	95 / 73	29	78	76	75	N	SSW	8		
SALINA	38	48	97	39	1272	0	5	103 / 74	100 / 74	97 / 73	26	78	77	75	NNW	S	10	101.8	-6.4
TOPEKA AP	39	04	95	38	877	0	4	99 / 75	96 / 75	93 / 74	24	79	78	76	NNW	SSW	12	102.5	-2.8
WICHITA AP	37	39	97	25	1321	3	7	101 / 72	98 / 73	96 / 73	23	77	76	75					
KENTUCKY																			
ASHLAND	38	33	82	44	546	5	10	94 / 76	91 / 74	89 / 73	22	78	77	75	W	SW	6	97.4	0.8
BOWLING GREEN AP	35	58	86	28	535	4	10	94 / 77	92 / 75	89 / 74	21	79	77	76				99.9	1.2
CORBIN AP	36	57	84	06	1174	4	9	94 / 73	92 / 73	89 / 72	23	77	76	75					
COVINGTON AP	39	03	84	40	869	1	6	92 / 73	90 / 72	88 / 72	22	77	75	74	W	SW	9		
HOPKINSVILLE,FT CAMPBELL	36	40	87	29	571	4	10	94 / 77	92 / 75	89 / 74	21	79	77	76	N	W	6	100.1	-0.4
LEXINGTON AP (S)	38	02	84	36	966	3	8	93 / 73	91 / 73	88 / 72	22	77	76	75	WNW	SW	9	95.3	-0.5
LOUISVILLE AP	38	11	85	44	477	5	10	95 / 74	93 / 74	90 / 74	23	79	77	76	NW	SW	8	97.4	1.2
MADISONVILLE	37	19	87	29	439	5	10	96 / 76	93 / 75	90 / 75	22	79	78	77					
OWENSBORO	37	45	87	10	407	5	10	97 / 76	94 / 75	91 / 75	23	79	78	77	NW	SW	9	98.0	-0.2
PADUCAH AP	37	04	88	46	413	7	12	98 / 76	95 / 75	92 / 75	20	79	78	77					
LOUISIANA																			
ALEXANDRIA AP	31	24	92	18	92	23	27	95 / 77	94 / 77	92 / 77	20	80	79	78	N	S	7	100.1	15.7
BATON ROUGE AP	30	32	91	09	64	25	29	95 / 77	93 / 77	92 / 77	19	80	80	79	ENE	W	8	98.0	21.4
BOGALUSA	30	47	89	52	103	24	28	95 / 77	93 / 77	92 / 77	19	80	80	79				99.3	20.2
HOUMA	29	31	90	40	13	31	35	95 / 78	93 / 78	92 / 77	15	81	80	79				97.2	22.5
LAFAYETTE AP	30	12	92	00	42	26	30	95 / 78	94 / 78	92 / 78	18	81	80	79	N	SW	8	98.2	22.6
LAKE CHARLES AP (S)	30	07	93	13	9	27	31	95 / 77	93 / 77	92 / 77	17	80	79	79	N	SSW	9	99.2	20.5
MINDEN	32	36	93	18	250	20	25	99 / 77	96 / 76	94 / 76	20	79	79	78				101.7	14.9
MONROE AP	32	31	92	02	79	20	25	99 / 77	96 / 76	94 / 76	20	79	79	78	N	S	9	101.1	15.9
NATCHITOCHES	31	46	93	05	130	22	26	97 / 77	95 / 77	93 / 77	20	80	79	78					
NEW ORLEANS AP	29	59	90	15	4	29	33	93 / 78	92 / 78	90 / 77	16	81	80	79	NNE	SSW	9	96.3	27.7
SHREVEPORT AP (S)	32	28	93	49	254	20	25	99 / 77	96 / 76	94 / 76	20	79	79	78	N	S	9		
MAINE																			
AUGUSTA AP	44	19	69	48	353	-7	-3	88 / 73	85 / 70	82 / 68	22	74	72	70	NNE	WNW	10		
BANGOR,DOW AFB	44	48	68	50	192	-11	-6	86 / 70	83 / 68	80 / 67	22	73	71	69	WNW	S	7		
CARIBOU AP (S)	46	52	68	01	624	-18	-13	84 / 69	81 / 67	78 / 66	21	71	69	67	WSW	SW	10	94.0	-13.7
LEWISTON	44	02	70	15	200	-7	-2	88 / 73	85 / 70	82 / 68	22	74	72	70					
MILLINOCKET AP	45	39	68	42	413	-13	-9	87 / 69	83 / 68	80 / 66	22	72	70	68	WNW	WNW	11	92.4	-23.0
PORTLAND (S)	43	39	70	19	43	-6	-1	87 / 72	84 / 71	81 / 69	22	74	72	70	W	S	7	93.5	-9.9
WATERVILLE	44	32	69	40	302	-8	-4	87 / 72	84 / 69	81 / 68	22	74	72	70					

State and Station	Lat. °N	Lat. '	Long. °W	Long. '	Elev. feet	Winter °F Design Dry-bulb 99%	97.5%	Summer °F Design Dry-bulb and Coincident Wet-bulb Mean 1%	2.5%	5%	Mean daily Range	Design Wet-bulb 1%	2.5%	5%	Prevailing Wind Winter	Summer	Knots	Temp. °F Median of Annual Extr. Max.	Min.
MARYLAND																			
BALTIMORE AP	39	11	76	40	148	10	13	94 / 75	91 / 75	89 / 74	21	78	77	76	W	WSW	9		
BALTIMORE CO	39	20	76	41	20	14	17	92 / 77	89 / 76	87 / 75	17	80	78	76	WNW	S	9	97.9	7.2
CUMBERLAND	39	37	78	46	790	6	10	92 / 75	89 / 74	87 / 74	22	77	76	75	WNW	W	10		
FREDERICK AP	39	27	77	25	313	8	12	94 / 76	91 / 75	88 / 74	22	78	77	76	N	WNW	9		
HAGERSTOWN	39	42	77	44	704	8	12	94 / 75	91 / 74	89 / 74	22	77	76	75	WNW	W	10		
SALISBURY (S)	38	20	75	30	59	12	16	93 / 75	91 / 75	88 / 74	18	79	77	76		W		96.8	7.4
MASSACHUSETTS																			
BOSTON AP (S)	42	22	71	02	15	6	9	91 / 73	88 / 71	85 / 70	16	75	74	72	WNW	SW	16	95.7	-1.2
CLINTON	42	24	71	41	398	-2	2	90 / 72	87 / 71	84 / 69	17	75	73	72				91.7	-8.5
FALL RIVER	41	43	71	08	190	5	9	87 / 72	84 / 71	81 / 69	18	75	73	72	NW	SW	10	92.1	-1.0
FRAMINGHAM	42	17	71	25	170	3	6	89 / 72	86 / 71	83 / 69	17	74	73	71				96.0	-7.7
GLOUCESTER	42	35	70	41	10	2	5	89 / 73	86 / 71	83 / 70	15	75	74	72					
GREENFIELD	42	03	72	04	205	-7	-2	88 / 72	85 / 71	82 / 69	23	76	73	71					
LAWRENCE	42	42	71	10	57	-6	0	90 / 73	87 / 72	84 / 70	22	76	74	73	NW	WSW	8	95.2	-9.0
LOWELL	42	39	71	19	90	-4	1	91 / 73	88 / 72	85 / 70	21	76	74	73				95.1	-8.5
NEW BEDFORD	41	41	70	58	79	5	9	85 / 72	82 / 71	80 / 69	19	74	73	72	NW	SW	10	91.4	2.2
PITTSFIELD AP	42	26	73	18	1194	-8	-3	87 / 71	84 / 70	81 / 68	23	73	72	70	NW	SW	12		
SPRINGFIELD, WESTOVER AFB	42	12	72	32	245	-5	0	90 / 72	87 / 71	84 / 69	19	75	73	72	N	SSW	8	95.7	-4.7
TAUNTON	41	54	71	04	20	5	9	89 / 73	86 / 72	83 / 70	18	75	74	73				92.9	-9.8
WORCESTER AP	42	16	71	52	986	0	4	87 / 71	84 / 70	81 / 68	18	73	72	70	W	W	14		
MICHIGAN																			
ADRIAN	41	55	84	01	754	-1	3	91 / 73	88 / 72	85 / 71	23	76	75	73				97.2	-7.0
ALPENA AP	45	04	83	26	610	-11	-6	89 / 70	85 / 70	83 / 69	27	73	72	70	W	SW	5	93.9	-14.8
BATTLE CREEK AP	42	19	85	15	941	1	5	92 / 74	88 / 72	85 / 70	23	76	74	73	SW	SW	8		
BENTON HARBOR AP	42	08	86	26	643	1	5	91 / 72	88 / 72	85 / 70	20	75	74	72	SSW	WSW	8		
DETROIT	42	25	83	01	619	3	6	91 / 73	88 / 72	86 / 71	20	76	74	73	W	SW	11	95.1	-2.6
ESCANABA	45	44	87	05	607	-11	-7	87 / 70	83 / 69	80 / 68	17	73	71	69				88.8	-16.1
FLINT AP	42	58	83	44	771	-4	1	90 / 73	87 / 72	85 / 71	25	76	74	72	SW	SW	8	95.3	-9.9
GRAND RAPIDS AP	42	53	85	31	784	1	5	91 / 72	88 / 72	85 / 71	24	75	74	72	WNW	WSW	8	95.4	-5.6
HOLLAND	42	42	86	06	678	2	6	88 / 72	86 / 71	83 / 70	22	75	73	72				94.1	-6.8
JACKSON AP	42	16	84	28	1020	1	5	92 / 74	88 / 72	85 / 70	23	76	74	73				96.5	-7.8
KALAMAZOO	42	17	85	36	955	1	5	92 / 74	88 / 72	85 / 70	23	76	74	73				95.9	-6.7
LANSING AP	42	47	84	36	873	-3	1	90 / 73	87 / 72	84 / 70	24	75	74	72	SW	W	12	94.6	-11.0
MARQUETTE CO	46	34	87	24	735	-12	-8	84 / 70	81 / 69	77 / 66	18	72	70	68		W		94.5	-11.8

State and Station	Lat. °	′N	Long. °	′W	Elev. feet	Winter Design Dry-bulb 99%	97.5%	Summer DB & Coincident WB 1%	2.5%	5%	Mean daily Range	Design Wet-bulb 1%	2.5%	5%	Wind Winter	Wind Summer	Knots	Temp Max.	Min.
MT PLEASANT	43	35	84	46	796	0	4	91 / 73	87 / 72	84 / 71	24	76	74	72				95.4	-11.1
MUSKEGON AP	43	10	86	14	625	2	6	86 / 72	84 / 70	82 / 70	21	75	73	72	E	SW	8	95.0	-6.8
PONTIAC	42	40	83	25	981	0	4	90 / 73	87 / 72	85 / 71	21	76	74	73					
PORT HURON	42	59	82	25	586	0	4	90 / 73	87 / 72	83 / 71	21	76	74	73	W	S	8		
SAGINAW AP	43	32	84	05	667	0	4	91 / 73	87 / 72	84 / 71	23	76	74	72	WSW	SW	7	96.1	-7.6
SAULT STE. MARIE AP (S)	46	28	84	22	721	-12	-8	84 / 70	81 / 69	77 / 66	23	72	70	68	E	SW	7	89.8	-21.0
TRAVERSE CITY AP	44	45	85	35	624	-3	1	89 / 72	86 / 71	83 / 69	22	75	73	71	SSW	SW	9	95.4	-10.7
YPSILANTI	42	14	83	32	716	1	5	92 / 72	89 / 71	86 / 70	22	75	74	72	SW	SW	10		
MINNESOTA																			
ALBERT LEA	43	39	93	21	1220	-17	-12	90 / 74	87 / 72	84 / 71	24	77	75	73				95.1	-28.0
ALEXANDRIA AP	45	52	95	23	1430	-22	-16	91 / 72	88 / 72	85 / 70	24	76	74	72		S			
BEMIDJI AP	47	31	94	56	1389	-31	-26	88 / 69	85 / 69	81 / 67	24	73	71	69	N		8	94.5	-36.9
BRAINERD	46	24	94	08	1227	-20	-16	90 / 73	87 / 71	84 / 69	24	75	73	71					
DULUTH AP	46	50	92	11	1428	-21	-16	85 / 70	82 / 68	79 / 66	22	72	70	68	WNW	WSW	12	90.9	-27.4
FAIRBAULT	44	18	93	16	940	-17	-12	91 / 74	88 / 72	85 / 71	24	77	75	73				95.8	-24.3
FERGUS FALLS	46	16	96	04	1210	-21	-17	91 / 72	88 / 72	85 / 70	24	76	74	72				96.9	-27.8
INTERNATIONAL FALLS AP	48	34	93	23	1179	-29	-25	85 / 68	83 / 68	80 / 66	26	71	70	68				93.4	-36.5
MANKATO	44	09	93	59	1004	-17	-12	91 / 72	88 / 72	85 / 70	24	77	75	73					
MINNEAPOLIS/ST. PAUL AP	44	53	93	13	834	-16	-12	92 / 75	89 / 73	86 / 71	22	77	75	73	N	S	9	96.5	-22.0
ROCHESTER AP	43	55	92	30	1297	-17	-12	90 / 74	87 / 72	84 / 71	24	77	75	73	NW	S	8		
ST. CLOUD AP (S)	45	35	94	11	1043	-15	-11	91 / 74	88 / 72	85 / 71	24	76	74	72	NW	SSW	9		
VIRGINIA	47	30	92	33	1435	-25	-21	85 / 69	83 / 68	80 / 66	23	71	70	68				92.6	-33.0
WILLMAR	45	07	95	05	1128	-15	-11	91 / 74	88 / 72	85 / 71	24	76	74	72				96.8	-24.3
WINONA	44	03	91	38	652	-14	-10	91 / 75	88 / 73	85 / 72	24	77	75	74					
MISSISSIPPI																			
BILOXI,KEESLER AFB	30	25	88	55	26	28	31	94 / 79	92 / 79	90 / 78	16	82	81	80	N	S	8	98.0	23.0
CLARKSDALE	34	12	90	34	178	14	19	96 / 77	94 / 77	92 / 76	21	80	79	78				100.9	13.2
COLUMBUS AFB	33	39	88	27	219	15	20	95 / 77	93 / 77	91 / 76	22	80	79	78	N	W	7	101.6	12.7
GREENVILLE AFB	33	29	90	59	138	15	20	95 / 77	93 / 77	91 / 76	21	80	79	78				99.5	14.9
GREENWOOD	33	30	90	05	148	15	20	95 / 77	93 / 77	91 / 76	21	80	80	79				100.6	15.3
HATTIESBURG	31	16	89	15	148	24	27	96 / 78	94 / 77	92 / 77	21	81	80	79				99.9	18.2
JACKSON AP	32	19	90	05	310	21	25	97 / 76	95 / 76	93 / 76	21	79	78	78	NNW	NW	6	99.8	16.0
LAUREL	31	40	89	10	236	24	27	96 / 78	94 / 77	92 / 77	21	81	80	79				99.7	17.8
MCCOMB AP	31	15	90	28	469	21	26	96 / 77	94 / 76	92 / 77	18	80	79	78					
MERIDIAN AP	32	20	88	45	290	19	23	97 / 77	95 / 76	93 / 76	22	80	79	78	N	WSW	6	98.3	15.7

State and Station	Lat. °N	Lat. '	Long. °W	Long. '	Elev. feet	Winter 99%	Winter 97.5%	Summer DB/Coinc. WB 1%	2.5%	5%	Mean daily Range	Design Wet-bulb 1%	2.5%	5%	Wind Winter	Knots	Wind Summer	Median Annual Extr. Max	Min
NATCHEZ	31	33	91	23	195	23	27	96 / 78	94 / 78	92 / 77	21	81	80	79				98.4	18.4
TUPELO	34	16	88	46	361	14	19	96 / 77	94 / 77	92 / 76	22	80	79	78				100.7	11.8
VICKSBURG CO	32	24	90	47	262	22	26	97 / 78	95 / 78	93 / 77	21	81	80	79				96.9	18.0
MISSOURI																			
CAPE GIRARDEAU	37	14	89	35	351	8	13	98 / 76	95 / 75	92 / 75	21	79	78	77					
COLUMBIA AP (S)	38	58	92	22	778	-1	4	97 / 74	94 / 74	91 / 73	22	78	77	76	WNW	9	WSW	99.5	-6.2
FARMINGTON AP	37	46	90	24	928	3	8	96 / 76	93 / 75	90 / 74	22	78	77	75				99.9	-2.1
HANNIBAL	39	42	91	21	489	-2	3	96 / 76	93 / 76	90 / 76	22	80	78	77	NNW	11	SSW	98.4	-7.6
JEFFERSON CITY	38	34	92	11	640	2	7	98 / 75	95 / 74	92 / 74	23	78	77	76				101.2	-6.1
JOPLIN AP	37	09	94	30	980	6	10	100 / 73	97 / 73	94 / 73	24	78	77	76	NNW	12	SSW		
KANSAS CITY AP	39	07	94	35	791	2	6	99 / 75	96 / 74	93 / 74	20	78	77	76	NW	9	S	100.2	-4.3
KIRKSVILLE AP	40	06	92	33	964	-5	0	96 / 74	93 / 74	90 / 73	24	78	77	76				98.3	-10.8
MEXICO	39	11	92	54	775	-1	4	97 / 74	94 / 74	91 / 73	22	78	77	76				101.2	-8.0
MOBERLY	39	24	92	26	850	-2	3	97 / 74	94 / 74	91 / 73	23	78	77	76					
POPLAR BLUFF	36	46	90	25	380	11	16	98 / 78	95 / 76	92 / 76	22	81	79	78					
ROLLA	37	59	91	43	1204	3	9	94 / 77	91 / 75	89 / 74	22	78	77	76				99.4	-3.1
ST. JOSEPH AP	39	46	94	55	825	-3	2	96 / 77	93 / 76	91 / 76	23	81	79	77	NNW	9	S	100.6	-8.0
ST. LOUIS AP	38	45	90	23	535	2	6	97 / 75	94 / 75	91 / 74	21	78	77	76	NW	9	WSW		
ST. LOUIS CO	38	39	90	38	462	3	8	98 / 75	94 / 75	91 / 74	18	78	77	76	NW	6	S	99.1	-2.7
SIKESTON	36	53	89	36	325	9	15	98 / 77	95 / 76	92 / 75	21	80	78	77					
SEDALIA,WHITEMAN AFB	38	43	93	33	869	-1	4	95 / 76	92 / 76	90 / 75	22	79	78	76	NNW	7	SSW	100.0	-5.1
SIKESTON	36	53	89	36	325	9	15	98 / 77	95 / 76	92 / 75	21	80	78	77					
SPRINGFIELD AP	37	14	93	23	1268	3	9	96 / 73	93 / 74	91 / 74	23	78	77	75	NNW	10	S	97.2	-2.4
MONTANA																			
BILLINGS AP	45	48	108	32	3567	-15	-10	94 / 64	91 / 64	88 / 63	31	67	66	64	NE	9	SW	100.5	-19.1
BOZEMAN	45	47	111	09	4448	-20	-14	90 / 61	87 / 60	84 / 59	32	63	62	60				92.2	-23.2
BUTTE AP	45	57	112	30	5553	-24	-17	86 / 58	83 / 56	80 / 56	35	60	58	57	S	5	NW	91.8	-26.3
CUT BANK AP	48	37	112	22	3838	-25	-20	88 / 61	85 / 61	82 / 60	35	64	62	61				94.7	-30.9
GLASGOW AP (S)	48	25	106	32	2760	-22	-18	92 / 64	89 / 63	85 / 62	29	68	66	64					
GLENDIVE	47	08	104	48	2476	-18	-13	95 / 66	92 / 64	89 / 62	29	69	67	65	E	8	S	103.3	-29.8
GREAT FALLS AP (S)	47	29	111	22	3662	-21	-15	91 / 60	88 / 60	85 / 59	28	64	62	60	SW	7	S	98.0	-25.1
HAVRE	48	34	109	40	2492	-18	-11	94 / 65	90 / 64	87 / 63	33	68	66	65				99.7	-31.3
HELENA AP	46	36	112	00	3828	-21	-16	91 / 60	88 / 60	85 / 59	32	64	62	61	N	12	WNW	95.6	-23.7
KALISPELL AP	48	18	114	16	2974	-14	-7	91 / 62	87 / 61	84 / 60	34	65	63	62				94.4	-16.8
LEWISTON AP	47	04	109	27	4122	-22	-16	90 / 62	87 / 61	83 / 60	30	65	63	62	NW	9	NW	96.2	-27.7

State and Station	Lat °	Lat '	Long °	Long '	Elev, feet	Winter 99%	Winter 97.5%	Summer Dry-bulb/Mean Coincident Wet-bulb 1%	2.5%	5%	Mean daily Range	Design Wet-bulb 1%	2.5%	5%	Wind Winter	Knots	Wind Summer	Temp Max	Temp Min
LIVINGSTOWN AP	45	42	110	26	4618	-20	-14	90 / 61	87 / 60	84 / 59	32	63	62	60	NW	7	SE	97.2	-21.2
MILES CITY AP	46	26	105	52	2634	-20	-15	98 / 66	95 / 66	92 / 65	30	70	68	67	ESE	7	NW	103.6	-27.7
MISSOULA AP	46	55	114	05	3190	-13	-6	92 / 62	88 / 61	85 / 60	36	65	63	62				98.6	-13.9
NEBRASKA																			
BEATRICE	40	16	96	45	1235	-5	-2	99 / 75	95 / 74	92 / 74	24	78	77	76				103.1	-11.3
CHADRON AP	42	50	103	05	3313	-8	-3	97 / 66	94 / 65	91 / 65	30	71	69	68					
COLUMBUS	41	28	97	20	1450	-6	-2	98 / 74	95 / 73	92 / 73	25	77	76	75					
FREMONT	41	26	96	29	1200	-6	-2	98 / 75	95 / 74	92 / 74	22	78	77	76					
GRAND ISLAND AP	40	59	98	19	1860	-8	-3	97 / 72	94 / 71	91 / 71	28	75	74	73	NNW	10	S	103.3	-14.2
HASTINGS	40	36	98	26	1954	-7	-3	97 / 72	94 / 71	91 / 71	27	75	74	73	NNW	10	S	103.5	-10.7
KEARNEY	40	44	99	01	2132	-9	-4	96 / 71	93 / 70	90 / 70	28	74	73	72				102.9	-13.7
LINCOLN CO (S)	40	51	96	45	1180	-6	-2	99 / 75	95 / 74	92 / 74	24	78	77	76	N	8	S	102.0	-12.4
MCCOOK	40	12	100	38	2768	-6	-2	98 / 69	95 / 69	91 / 69	28	74	72	71					
NORFOLK	41	59	97	26	1551	-8	-4	97 / 74	93 / 74	90 / 73	30	78	77	75				102.0	-20.0
NORTH PLATTE AP (S)	41	08	100	41	2779	-8	-4	97 / 69	94 / 69	90 / 69	28	74	72	71	NW	9	SSE	100.8	-15.8
OMAHA AP	41	18	95	54	977	-8	-3	94 / 76	91 / 75	88 / 74	22	78	77	75	NW	8	S	100.2	-13.2
SCOTTSBLUFF AP	41	52	103	36	3958	-8	-3	95 / 65	92 / 65	90 / 64	31	70	68	67	NW	9	SE	101.6	-18.9
SIDNEY AP	41	13	103	06	4399	-8	-3	95 / 65	92 / 65	90 / 64	31	70	68	67					
NEVADA																			
CARSON CITY	39	10	119	46	4675	4	9	94 / 60	91 / 59	89 / 58	42	63	61	60	SSW	3	WNW	99.2	-5.0
ELKO AP	40	50	115	47	5050	-8	-2	94 / 59	92 / 59	90 / 58	42	63	62	60	E	4	SW		
ELY AP (S)	39	17	114	51	6253	-10	-4	89 / 57	87 / 56	85 / 55	39	60	59	58	S	9	SSW		
LAS VEGAS AP (S)	36	05	115	10	2178	25	28	108 / 66	106 / 65	104 / 65	30	71	70	69	ENE	7	SW	103.0	-1.0
LOVELOCK AP	40	04	118	33	3903	8	12	98 / 63	96 / 63	93 / 62	42	66	65	64					
RENO AP (S)	39	30	119	47	4404	5	10	95 / 61	92 / 60	90 / 59	45	64	62	61	SSW	3	WNW	98.9	0.2
RENO CO	39	30	119	47	4408	6	11	96 / 61	93 / 60	91 / 59	45	64	62	61	N	8	S		
TONOPAH AP	38	04	117	05	5426	5	10	94 / 60	92 / 59	90 / 58	40	64	62	61					
WINNEMUCCA AP	40	54	117	48	4301	-1	3	96 / 60	94 / 60	92 / 60	42	64	62	61	SE	10	W	100.1	-8.1
NEW HAMPSHIRE																			
BERLIN	44	03	71	01	1110	-14	-9	87 / 71	84 / 69	81 / 68	22	73	71	70				93.2	-24.7
CLAREMONT	43	02	72	02	420	-9	-4	89 / 72	86 / 70	83 / 69	24	74	73	71					
CONCORD AP	43	12	71	30	342	-8	-3	90 / 72	87 / 70	84 / 69	26	74	73	71	NW	7	SW	94.8	-16.0
KEENE	42	55	72	17	490	-12	-7	90 / 72	87 / 70	83 / 69	24	74	73	71				94.6	-18.9
LACONIA	43	03	71	03	505	-10	-5	89 / 72	86 / 70	83 / 69	25	74	73	71					
MANCHESTER, GRENIER AFB	42	56	71	26	233	-8	-3	91 / 72	88 / 71	85 / 70	24	75	74	72	N	11	SW	93.7	-12.6

State and Station	Lat. °	'N	Long. °	'W	Elev. feet	Winter °F Design Dry-bulb 99%	97.5%	Summer Design Dry-bulb and Coincident Wet-bulb Mean 1%	2.5%	5%	Mean daily Range	Design Wet-bulb 1%	2.5%	5%	Prevailing Wind Winter	Knots	Summer	Knots	Temp °F Median of Annual Extr. Max.	Min.
PORTSMOUTH,PEASE AFB	43	04	70	49	101	-2	2	89 / 73	85 / 71	83 / 70	22	75	74	72	W	8	W			
NEW JERSEY																				
ATLANTIC CITY CO	39	23	74	26	11	10	13	92 / 74	89 / 74	86 / 72	18	78	77	75	NW	11	WSW		93.0	7.5
LONG BRANCH	40	19	74	01	15	10	13	93 / 74	90 / 73	87 / 72	18	78	77	75					95.9	4.3
NEWARK AP	40	42	74	10	7	10	14	94 / 74	91 / 73	88 / 72	20	77	76	75	WNW	11	WSW			
NEW BRUNSWICK	40	29	74	26	125	6	10	92 / 74	89 / 73	86 / 72	19	77	76	75						
PATERSON	40	54	74	09	100	6	10	94 / 74	91 / 73	88 / 72	21	76	75	74						
PHILLIPSBURG	40	41	75	11	180	1	6	92 / 73	89 / 72	86 / 71	21	76	75	74					97.4	-0.7
TRENTON CO	40	13	74	46	56	11	14	91 / 75	88 / 74	85 / 73	19	78	76	75	W	9	SW		96.2	4.2
VINELAND	39	29	75	00	112	8	11	91 / 75	89 / 74	86 / 73	19	78	76	75						
NEW MEXICO																				
ALAMAGORDO,HOLLOMAN AFB	32	51	106	06	4093	14	19	98 / 64	96 / 64	94 / 64	30	69	68	67						
ALBUQUERQUE AP (S)	35	03	106	37	5311	12	16	96 / 61	94 / 61	92 / 61	27	66	65	64					98.1	5.1
ARTESIA	32	46	104	23	3320	13	19	103 / 67	100 / 67	97 / 67	30	72	71	70	N	7			105.5	3.7
CARLSBAD AP	32	20	104	16	3293	13	19	103 / 67	100 / 67	97 / 67	28	72	71	70	N	6	SSE			
CLOVIS AP	34	23	103	19	4294	8	13	95 / 65	93 / 65	91 / 65	28	69	68	67					102.0	2.5
FARMINGTON AP	36	44	108	14	5503	1	6	95 / 63	93 / 62	91 / 61	30	67	65	64						
GALLUP	35	31	108	47	6465	0	5	90 / 59	89 / 58	86 / 58	32	64	62	61	ENE	5	SW			
GRANTS	35	10	107	54	6524	-4	4	89 / 59	88 / 58	85 / 57	32	64	62	61						
HOBBS AP	32	45	103	13	3690	13	18	101 / 66	99 / 66	97 / 66	29	71	70	69						
LAS CRUCES	32	18	106	55	4544	15	20	99 / 64	96 / 64	94 / 64	30	69	68	67	SE	5	SE		89.8	-2.3
LOS ALAMOS	35	52	106	19	7410	5	9	89 / 60	87 / 60	85 / 60	32	62	61	60						
RATON AP	36	45	104	30	6373	-4	1	91 / 60	89 / 60	87 / 60	34	65	64	63						
ROSWELL,WALKER AFB	33	18	104	32	3676	13	18	100 / 66	98 / 66	96 / 66	33	71	70	69	N	6	SSE		103.0	2.7
SANTA FE CO	35	37	106	05	6307	6	10	90 / 61	88 / 61	86 / 61	28	63	62	61					90.1	-1.2
SILVER CITY AP	32	38	108	10	5442	5	10	95 / 61	94 / 60	91 / 60	30	66	64	63						
SOCORRO AP	34	03	106	53	4624	13	17	97 / 62	95 / 62	93 / 62	30	67	66	65						
TUCUMCARI AP	35	11	103	36	4039	8	13	99 / 66	97 / 66	95 / 65	28	70	69	68	NE	8	SW		102.7	1.1
NEW YORK																				
ALBANY AP (S)	42	45	73	48	275	-6	-1	91 / 73	88 / 72	85 / 70	23	75	74	72	WNW	8	S		95.2	-11.4
ALBANY CO	42	39	73	45	19	-4	1	91 / 73	88 / 72	85 / 70	20	75	74	72					92.4	-9.5
AUBURN	42	54	76	32	715	-3	2	90 / 73	87 / 71	84 / 70	22	75	73	72					92.2	-7.5
BATAVIA	43	00	78	11	922	1	5	90 / 72	87 / 71	84 / 70	22	75	73	72						
BINGHAMTON AP	42	13	75	59	1590	-2	1	86 / 71	83 / 69	81 / 68	20	73	72	70	WSW	10	WSW		92.9	-9.3
BUFFALO AP	42	56	78	44	705	2	6	88 / 71	85 / 70	83 / 69	21	74	73	72	W	10	SW		90.0	-3.2

State and Station	Lat ° N	Lat ′	Long ° W	Long ′	Elev. feet	Winter Design Dry-bulb 99%	97.5%	Summer DB/Coinc. WB 1%	2.5%	5%	Mean daily Range	Design Wet-bulb 1%	2.5%	5%	Prev. Wind Winter	Summer	Knots	Median Annual Extr. Max.	Min.
CORTLAND	42	36	76	11	1129	-5	0	88 / 71	85 / 71	82 / 70	23	74	73	71				93.8	-11.2
DUNKIRK	42	29	79	16	692	4	9	88 / 73	85 / 72	83 / 71	18	75	74	72	SSW	WSW	10	96.2	-6.7
ELMIRA AP	42	10	76	54	955	-4	1	89 / 71	86 / 71	83 / 70	24	74	73	71				96.1	-6.5
GENEVA (S)	42	45	76	54	613	-3	2	90 / 73	87 / 71	84 / 70	22	75	73	72					
GLENS FALLS	43	20	73	37	328	-11	-5	88 / 72	85 / 71	82 / 69	23	75	73	71	NNW	S	6	93.2	-14.6
GLOVERSVILLE	43	02	74	21	760	-8	-2	89 / 72	86 / 71	83 / 69	24	75	74	72					
HORNELL	42	21	77	42	1325	-4	0	88 / 71	85 / 71	82 / 69	24	74	73	72					
ITHACA (S)	42	27	76	29	928	-5	0	88 / 71	85 / 71	82 / 70	24	74	73	71	W	SW	6		
JAMESTOWN	42	07	79	14	1390	-1	3	88 / 70	86 / 70	83 / 69	20	74	72	71	WSW	WSW	9		
KINGSTON	41	56	74	00	279	-3	2	91 / 73	88 / 72	85 / 70	22	76	74	73					
LOCKPORT	43	09	79	15	638	4	7	89 / 74	86 / 72	84 / 71	21	76	74	73	N	SW	9	92.2	-4.8
MASSENA AP	44	56	74	51	207	-13	-8	86 / 70	83 / 69	80 / 68	20	73	72	70					
NEWBURGH,STEWART AFB	41	30	74	06	471	-1	4	90 / 73	88 / 72	85 / 70	21	76	74	73	W	W	10		
NYC-CENTRAL PARK (S)	40	47	73	58	157	11	15	92 / 74	89 / 73	87 / 72	17	76	75	74				94.9	3.8
NYC-KENNEDY AP	40	39	73	47	13	12	15	90 / 73	87 / 72	84 / 71	16	76	75	74					
NYC-LA GUARDIA AP	40	46	73	54	11	11	15	92 / 74	89 / 73	87 / 72	16	76	75	74					
NIAGARA FALLS AP	43	06	79	57	590	4	7	89 / 74	86 / 72	84 / 71	20	76	74	73	W	SW	9		
OLEAN	42	14	78	22	2119	-2	2	87 / 71	84 / 71	81 / 70	23	74	73	71					
ONEONTA	42	31	75	04	1775	-7	-4	86 / 71	83 / 69	80 / 68	24	73	72	70				91.3	-7.4
OSWEGO CO	43	28	76	33	300	1	7	86 / 73	83 / 71	80 / 70	20	75	73	72	E	WSW	7		
PLATTSBURG AFB	44	39	73	28	235	-13	-8	86 / 70	83 / 69	80 / 68	22	73	72	70	NW	SE	6		
POUGHKEEPSIE	41	38	73	55	165	0	6	92 / 74	89 / 74	86 / 72	21	77	75	74	NNE	SSW	6		
ROCHESTER AP	43	07	77	40	547	1	5	91 / 73	88 / 71	85 / 70	22	75	73	72	WSW	WSW	11	98.1	-5.6
ROME,GRIFFISS AFB	43	14	75	25	514	-11	-5	88 / 71	85 / 70	83 / 69	22	75	74	72	NW	W	5		
SCHENECTADY (S)	42	51	73	57	377	-4	1	90 / 73	87 / 72	84 / 70	22	75	74	72	WNW	S	8		
SUFFOLK COUNTY AFB	40	51	72	38	67	7	10	86 / 72	83 / 71	80 / 70	16	76	74	73	NW	S	9		
SYRACUSE AP	43	07	76	07	410	-3	2	90 / 73	87 / 71	84 / 70	20	75	73	72	N	WNW	7	93.0	-10.0
UTICA	43	09	75	23	714	-12	-6	88 / 73	85 / 71	82 / 70	22	75	73	71	NW	W	12		
WATERTOWN	43	59	76	01	325	-11	-6	86 / 73	83 / 71	81 / 70	20	75	73	72	E	WSW	7	91.7	-19.6
NORTH CAROLINA																			
ASHEVILLE AP	35	26	82	32	2140	10	14	89 / 73	87 / 72	85 / 71	21	75	74	72	NNW	NNW	12	91.9	5.8
CHARLOTTE AP	35	13	80	56	736	18	22	95 / 74	93 / 74	91 / 74	20	77	76	76	NNW	SW	6	97.8	12.6
DURHAM	35	52	78	47	434	16	20	94 / 75	92 / 75	90 / 75	20	78	77	76				98.9	9.6
ELIZABETH CITY AP	36	16	76	11	12	12	19	93 / 78	91 / 77	89 / 76	18	80	78	78	NW	SW	8		
FAYETTEVILLE,POPE AFB	35	10	79	01	218	17	20	95 / 76	92 / 76	90 / 75	20	79	78	77	N	SSW	6	99.1	13.1

State and Station	Lat. °	Lat. 'N	Long. °	Long. 'W	Elev. feet	Winter 99%	Winter 97.5%	Summer Dry-bulb/Coinc. WB Mean 1%	2.5%	5%	Mean daily Range	Design Wet-bulb 1%	2.5%	5%	Wind Winter (Knots)	Wind Summer	Temp Max.	Temp Min.
GOLDSBORO,SEYMOUR-JOHNSON	35	20	77	58	109	18	21	94 / 77	91 / 76	89 / 75	18	79	78	77	N 8	SW	99.8	13.0
GREENSBORO AP (S)	36	05	79	57	897	14	18	93 / 74	91 / 73	89 / 73	21	77	76	75	NE 7	SW 7	97.7	9.7
GREENVILLE	35	37	77	25	75	18	21	93 / 77	91 / 76	89 / 75	19	79	78	77				
HENDERSON	36	22	78	25	480	12	15	95 / 77	92 / 76	90 / 76	20	79	78	77				
HICKORY	35	45	81	23	1187	14	18	92 / 73	90 / 72	88 / 72	21	75	74	73			96.5	9.6
JACKSONVILLE	34	50	77	37	95	20	24	92 / 78	90 / 78	88 / 77	18	80	79	78				
LUMBERTON	34	37	79	04	129	18	21	95 / 76	92 / 76	90 / 75	20	79	78	77				
NEW BERN AP	35	05	77	03	20	20	24	92 / 78	90 / 78	88 / 77	18	80	79	78				
RALEIGH/DURHAM AP (S)	35	52	78	47	434	16	20	94 / 75	92 / 75	90 / 75	20	78	77	76	N 7	SW	98.2	15.1
ROCKY MOUNT	35	58	77	48	121	18	21	94 / 77	91 / 76	90 / 75	19	79	78	77			97.7	12.2
WILMINGTON AP	34	16	77	55	28	23	26	93 / 79	91 / 78	89 / 77	18	81	80	79	N 8	SW	96.9	18.2
WINSTON-SALEM AP	36	08	80	13	969	16	20	94 / 74	91 / 73	89 / 73	20	76	75	74	NW 8	WSW		
NORTH DAKOTA																		
BISMARCK AP (S)	46	46	100	45	1647	-23	-19	95 / 68	91 / 68	88 / 67	27	73	71	70	WNW 7	S	100.3	-31.5
DEVILS LAKE	48	07	98	54	1450	-25	-21	91 / 69	88 / 68	85 / 66	25	73	71	69			97.5	-30.4
DICKINSON AP	46	48	102	48	2585	-21	-17	94 / 68	90 / 66	87 / 65	25	71	69	68	WNW 12	SSE	101.0	-31.3
FARGO AP	46	54	96	48	896	-22	-18	92 / 73	89 / 71	85 / 69	25	76	74	72	SSE 11	S	97.3	-29.7
GRAND FORKS AP	47	57	97	24	911	-26	-22	91 / 70	87 / 70	84 / 68	25	74	72	70	N 8	S	97.6	-29.0
JAMESTOWN AP	46	55	98	41	1492	-22	-18	94 / 70	90 / 69	87 / 68	26	74	74	71			101.3	-27.9
MINOT AP	48	25	101	21	1668	-24	-20	92 / 68	89 / 67	86 / 65	25	72	70	68				
WILLISTON	48	09	103	35	1876	-25	-21	91 / 68	88 / 67	85 / 65	25	76	75	68	WSW 10	S	99.7	-32.9
OHIO																		
AKRON-CANTON AP	40	55	81	26	1208	1	6	89 / 72	86 / 71	84 / 70	21	75	73	72	SW 9	SW	94.4	-4.6
ASHTABULA	41	51	80	48	690	4	9	88 / 73	85 / 72	83 / 70	18	75	74	72				
ATHENS	39	20	82	06	700	0	6	95 / 75	92 / 74	90 / 73	22	78	76	74				
BOWLING GREEN	41	23	83	38	675	-2	2	92 / 73	89 / 73	86 / 71	23	76	75	74			96.7	-7.3
CAMBRIDGE	40	04	81	35	807	1	7	93 / 75	90 / 74	87 / 73	23	78	76	75				
CHILLICOTHE	39	21	83	00	640	0	6	95 / 75	92 / 74	90 / 73	22	78	76	74				
CINCINNATI CO	39	09	84	31	758	1	6	92 / 73	90 / 72	88 / 72	21	77	75	74	W 8	WSW	98.2	-2.1
CLEVELAND AP (S)	41	24	81	51	777	1	5	91 / 73	88 / 72	86 / 71	22	76	74	73	W 9	SW	97.2	-0.2
COLUMBUS AP (S)	40	00	82	53	812	0	5	92 / 73	90 / 73	87 / 72	24	77	75	74	SW 12	N	94.7	-3.1
DAYTON AP	39	54	84	13	1002	-1	4	91 / 73	89 / 72	86 / 71	20	76	75	74	W 8	SSW	96.0	-3.4
DEFIANCE	41	17	84	23	700	-1	4	94 / 74	91 / 73	88 / 72	24	77	76	74	WNW 11	SW	96.6	-4.5
FINDLAY AP	41	01	83	40	804	2	3	92 / 74	90 / 73	87 / 72	24	77	76	74			97.4	-7.4

State and Station	Lat. °	Lat. ′	Long. °	Long. ′	Elev. feet	Winter Design Dry-bulb 99%	Winter 97.5%	Summer Design Dry-bulb and Coincident Wet-bulb Mean 1%	2.5%	5%	Mean daily Range	Design Wet-bulb 1%	2.5%	5%	Prevailing Wind Winter	Knots	Summer	Median of Annual Extr. Max.	Min.
FREMONT	41	20	83	07	600	-3	1	90 / 73	88 / 73	85 / 71	24	76	75	73					
HAMILTON	39	24	84	35	650	0	5	92 / 73	90 / 72	87 / 71	22	76	75	73				98.2	-2.8
LANCASTER	39	44	82	38	860	0	5	93 / 74	91 / 73	88 / 72	23	77	75	74					
LIMA	40	42	84	02	975	-1	4	94 / 74	91 / 73	88 / 72	24	77	76	74	WNW	11	SW	96.0	-6.5
MANSFIELD AP	40	49	82	31	1295	0	5	90 / 73	87 / 72	85 / 72	22	76	74	73	W	8	SW	93.8	-10.7
MARION	40	36	83	10	920	0	5	93 / 74	91 / 73	88 / 72	23	77	76	74					
MIDDLETOWN	39	31	84	25	635	0	5	92 / 73	90 / 72	87 / 71	22	76	75	73					
NEWARK	40	01	82	28	880	-1	5	94 / 73	92 / 73	89 / 72	23	77	75	74	W	8	SSW	95.8	-6.8
NORWALK	41	16	82	37	670	-3	1	90 / 73	88 / 73	85 / 71	22	76	75	73				97.3	-8.3
PORTSMOUTH	38	45	82	55	540	5	10	95 / 76	92 / 74	89 / 73	22	78	77	75	W	8	SW	97.9	1.0
SANDUSKY CO	41	27	82	43	606	1	6	93 / 73	91 / 72	88 / 71	21	76	74	73				96.7	-1.9
SPRINGFIELD	39	50	83	50	1052	-1	3	91 / 74	89 / 73	87 / 72	21	77	76	74	W	7	W		
STEUBENVILLE	40	23	80	38	992	1	5	89 / 72	86 / 71	84 / 70	22	74	73	72					
TOLEDO AP	41	36	83	48	669	-3	1	90 / 73	88 / 73	85 / 71	25	76	75	73	WSW	8	SW	95.4	-5.2
WARREN	41	20	80	51	928	0	5	89 / 71	87 / 71	85 / 70	23	74	73	71					
WOOSTER	40	47	81	55	1020	1	6	89 / 72	86 / 71	84 / 70	22	75	73	72				94.0	-7.7
YOUNGSTOWN AP	41	16	80	40	1178	-1	4	88 / 71	86 / 71	84 / 70	23	74	73	71	SW	10	SW		
ZANESVILLE AP	39	57	81	54	900	1	7	93 / 75	90 / 74	87 / 73	23	78	76	75	W	6	WSW		
OKLAHOMA																			
ADA	34	47	96	41	1015	10	14	100 / 74	97 / 74	95 / 74	23	77	76	75					
ALTUS AFB	34	39	99	16	1378	11	16	102 / 73	100 / 73	98 / 73	25	77	76	75	N	10	S		
ARDMORE	34	18	97	01	771	13	17	100 / 74	98 / 74	95 / 74	23	77	77	76					
BARTLESVILLE	36	45	96	00	715	6	10	101 / 73	98 / 74	95 / 74	23	77	77	76					
CHICKASHA	35	03	97	55	1085	10	14	101 / 74	98 / 74	95 / 74	24	78	77	76					
ENID,VANCE AFB	36	21	97	55	1307	9	13	103 / 74	100 / 74	97 / 74	24	79	77	76					
LAWTON AFB	34	34	98	25	1096	12	16	101 / 74	99 / 74	96 / 74	24	78	77	76	N	10	S		
MCALESTER	34	50	95	55	776	14	19	99 / 74	96 / 74	93 / 74	23	77	76	75					
MUSKOGEE AP	35	40	95	22	610	10	15	101 / 74	98 / 75	95 / 75	23	79	78	77			S		
NORMAN	35	15	97	29	1181	9	13	99 / 74	96 / 74	94 / 74	24	77	76	75	N	10	S		
OKLAHOMA CITY AP (S)	35	24	97	36	1285	9	13	100 / 74	97 / 74	95 / 73	23	78	77	76	N	10	S		
PONCA CITY	36	44	97	06	997	5	9	100 / 74	97 / 74	94 / 74	24	77	76	76	N	14	SSW		
SEMINOLE	35	14	96	40	865	11	15	99 / 74	96 / 74	94 / 73	23	77	76	75					
STILLWATER (S)	36	10	97	05	984	8	13	100 / 74	96 / 74	93 / 74	24	77	76	75	N	12	SSW	103.7	1.6
TULSA AP	36	12	95	54	650	8	13	101 / 74	98 / 75	95 / 75	22	79	78	77	N	11	SSW		
WOODWARD	36	36	99	31	2165	6	10	100 / 73	97 / 73	94 / 73	26	78	76	75				107.1	-1.3

State and Station	Lat. °	Lat. ′N	Long. °	Long. ′W	Elev. feet	Winter °F 99%	Winter °F 97.5%	Summer Design Dry-bulb & Coincident Wet-bulb Mean 1%	2.5%	5%	Mean daily Range	Design Wet-bulb 1%	2.5%	5%	Prevailing Wind Winter	Knots	Prevailing Wind Summer	Temp °F Median Annual Extr. Max	Min
OREGON																			
ALBANY	44	38	123	07	230	18	22	92 / 67	89 / 66	86 / 65	31	69	67	66	ESE	7	NNW	97.5	16.6
ASTORIA AP (S)	46	09	123	53	8	25	29	75 / 65	71 / 62	68 / 61	16	65	63	61					
BAKER AP	44	50	117	49	3372	−1	6	92 / 63	89 / 61	86 / 60	30	65	63	61				97.5	−6.8
BEND	44	04	121	19	3595	−3	4	90 / 62	87 / 60	84 / 59	33	64	62	60				96.4	−5.8
CORVALLIS (S)	44	30	123	17	246	18	22	92 / 67	89 / 66	86 / 65	31	69	67	66	N	6	N	98.5	17.1
EUGENE AP	44	07	123	13	359	17	22	92 / 67	89 / 66	86 / 65	31	69	67	66	N	7	N		
GRANTS PASS	42	26	123	19	925	20	24	99 / 69	96 / 68	93 / 67	33	71	69	68	N	5	N	103.6	16.4
KLAMATH FALLS AP	42	09	121	44	4092	4	9	90 / 61	87 / 60	84 / 59	36	63	61	60	N	4	W	96.3	0.9
MEDFORD AP (S)	42	22	122	52	1298	19	23	98 / 68	94 / 67	91 / 66	35	70	68	67	S		WMW	103.8	15.0
PENDLETON AP	45	41	118	51	1482	−2	5	97 / 65	93 / 64	90 / 62	29	66	65	63	NNW	6	WNW		
PORTLAND AP	45	36	122	36	21	17	23	89 / 68	85 / 67	81 / 65	23	69	67	66	ESE	12	NW	96.6	18.3
PORTLAND CO	45	32	122	40	75	18	24	90 / 68	86 / 67	82 / 65	21	69	67	66				97.6	20.5
ROSEBURG AP	43	14	123	22	525	18	23	93 / 67	90 / 66	87 / 65	30	69	68	66				99.6	19.5
SALEM AP	44	55	123	01	196	18	23	92 / 68	88 / 66	84 / 65	31	69	68	66	N	6	N	98.9	15.9
THE DALLES	45	36	121	12	100	13	19	93 / 69	89 / 68	85 / 66	28	70	68	67				105.1	7.9
PENNSYLVANIA																			
ALLENTOWN AP	40	39	75	26	387	4	9	92 / 73	88 / 72	86 / 72	22	76	75	73	W	11	SW		
ALTOONA CO	40	18	78	19	1504	0	5	90 / 72	87 / 71	84 / 70	23	74	73	72	WMW	11	WSW	93.7	−5.2
BUTLER	40	52	79	54	1100	1	6	90 / 73	87 / 72	85 / 71	22	75	74	73					
CHAMBERSBURG	39	56	77	38	640	4	8	93 / 75	90 / 74	87 / 73	23	77	76	75					
ERIE AP	42	05	80	11	731	4	9	88 / 73	85 / 72	83 / 71	18	75	74	72	SSW	10	WSW	97.4	−0.3
HARRISBURG AP	40	12	76	46	308	7	11	94 / 75	91 / 74	88 / 73	21	77	76	75	NW	11	WSW	91.3	−2.2
JOHNSTOWN	40	19	78	50	2284	−3	2	86 / 70	83 / 70	80 / 68	23	72	71	70	WNW	8	WSW	96.5	3.7
LANCASTER	40	07	76	18	403	4	8	93 / 75	90 / 74	87 / 73	22	77	76	75	NW	11	WSW	96.4	−1.8
MEADVILLE	41	38	80	10	1065	0	4	88 / 71	85 / 70	83 / 69	21	73	72	71					
NEW CASTLE	41	01	80	22	825	2	7	91 / 73	88 / 72	86 / 71	23	75	74	73	WSW	10	WSW	93.2	−8.5
PHILADELPHIA AP	39	53	75	15	5	10	14	93 / 75	90 / 74	87 / 72	21	77	76	75	WNW	10	WSW	94.7	−6.4
PITTSBURGH AP	40	30	80	13	1137	1	5	89 / 72	86 / 71	84 / 70	22	74	73	72	WSW	10	WSW	96.4	5.9
PITTSBURGH CO	40	27	80	00	1017	3	7	91 / 72	88 / 71	86 / 70	19	74	73	72					
READING CO	40	20	75	38	266	9	13	92 / 73	89 / 72	86 / 70	19	76	75	73	W	11	SW	94.6	−1.1
SCRANTON/WILKES-BARRE	41	20	75	44	930	1	5	90 / 72	87 / 71	84 / 70	19	74	73	72	SW	8	WSW	97.0	3.6
STATE COLLEGE (S)	40	48	77	52	1175	3	7	90 / 72	87 / 71	84 / 70	23	74	73	72	NNW	8	WSW	94.8	−2.2
SUNBURY	40	53	76	46	446	2	7	92 / 73	89 / 72	86 / 70	22	75	74	73				93.2	−3.6
UNIONTOWN	39	55	79	43	956	5	9	91 / 74	88 / 73	85 / 72	22	76	75	74				93.9	−2.5

State and Station	Lat. ° N	Lat. ' N	Long. ° W	Long. ' W	Elev. feet	Winter Design Dry-bulb 99%	Winter Design Dry-bulb 97.5%	Summer Design Dry-bulb and Coincident Wet-bulb Mean 1%	2.5%	5%	Mean daily Range	Design Wet-bulb 1%	2.5%	5%	Prevailing Wind Winter	Knots	Prevailing Wind Summer	Temp Median of Annual Extr. Max.	Min.
WARREN	41	51	79	08	1280	-2	4	89 / 71	86 / 71	83 / 70	24	74	73	72				93.3	-10.7
WEST CHESTER	39	58	75	38	450	9	13	92 / 75	89 / 74	86 / 72	20	77	76	75				95.5	-4.6
WILLIAMSPORT AP	41	15	76	55	524	2	7	92 / 73	89 / 72	86 / 70	23	75	74	73	W	9	WSW	97.0	-2.4
YORK	39	55	76	45	390	8	12	94 / 75	91 / 74	88 / 73	22	77	76	75					
RHODE ISLAND																			
NEWPORT (S)	41	30	71	20	10	5	9	88 / 73	85 / 72	82 / 70	16	76	75	73	WNW	10	SW		
PROVIDENCE AP	41	44	71	26	51	5	9	89 / 73	86 / 72	83 / 70	19	75	74	73	WNW	11	SW	94.6	-0.5
SOUTH CAROLINA																			
ANDERSON	34	30	82	43	774	19	23	94 / 74	92 / 74	90 / 74	21	77	76	75				99.5	13.3
CHARLESTON AFB (S)	32	54	80	02	45	24	27	93 / 78	91 / 78	89 / 77	18	81	80	79	NNE	8	SW		
CHARLESTON CO	32	54	79	58	3	25	28	94 / 78	92 / 78	90 / 77	13	81	80	79				97.8	21.4
COLUMBIA AP	33	57	81	07	213	20	24	97 / 76	95 / 75	93 / 75	22	79	78	77	W	6	SW	100.6	16.2
FLORENCE AP	34	11	79	43	147	22	25	94 / 77	92 / 77	90 / 76	21	80	79	78	N	7	SW	99.5	16.5
GEORGETOWN	33	23	79	17	14	23	26	92 / 79	90 / 78	88 / 77	18	81	80	79	N	7	SSW	98.2	19.1
GREENVILLE AP	34	54	82	13	957	18	22	93 / 74	91 / 74	89 / 74	21	77	76	75	NW	8	SW	97.3	12.6
GREENWOOD	34	10	82	07	620	18	22	95 / 75	93 / 74	91 / 74	21	78	77	76				99.5	14.1
ORANGEBURG	33	30	80	52	260	20	24	97 / 76	95 / 75	93 / 75	21	79	78	77				101.2	18.0
ROCK HILL	34	59	80	58	470	19	23	96 / 75	94 / 74	92 / 74	20	78	77	76					
SPARTANBURG AP	34	58	82	00	823	18	22	93 / 74	91 / 74	89 / 74	20	77	76	75				99.5	13.9
SUMTER,SHAW AFB	33	54	80	22	169	22	25	95 / 77	92 / 76	90 / 75	21	79	78	77	NNE	6	W	100.0	15.4
SOUTH DAKOTA																			
ABERDEEN AP	45	27	98	26	1296	-19	-15	94 / 73	91 / 72	88 / 70	27	77	75	73	NNW	8	S	102.3	-28.1
BROOKINGS	44	18	96	48	1637	-17	-13	95 / 73	92 / 72	89 / 71	25	77	75	73				97.8	-26.5
HURON AP	44	23	98	13	1281	-18	-14	96 / 73	93 / 72	90 / 71	28	77	75	73	NNW	8	S	101.5	-25.8
MITCHELL	43	41	98	01	1346	-15	-10	96 / 72	93 / 71	90 / 70	28	76	75	73				103.0	-22.7
PIERRE AP	44	23	100	17	1742	-15	-10	99 / 71	95 / 71	92 / 69	29	75	74	72	NW	11	SSE	105.7	-20.6
RAPID CITY AP (S)	44	03	103	04	3162	-11	-7	95 / 66	92 / 65	89 / 65	28	71	69	67	NNW	10	SSE	100.9	-19.0
SIOUX FALLS AP	43	34	96	44	1418	-15	-11	94 / 73	91 / 72	88 / 71	24	76	75	73	NW	8	S		
WATERTOWN AP	44	55	97	09	1738	-19	-15	94 / 73	91 / 72	88 / 71	26	76	75	73				97.8	-26.5
YANKTON	42	55	97	23	1302	-13	-7	94 / 73	91 / 72	88 / 71	25	77	76	74				100.8	-19.1
TENNESSEE																			
ATHENS	35	26	84	35	940	13	18	95 / 74	92 / 73	90 / 73	22	77	76	75					
BRISTOL-TRI CITY AP	36	29	82	24	1507	9	14	91 / 72	89 / 72	87 / 71	22	75	75	73	WNW	6	SW		§
CHATTANOOGA AP	35	02	85	12	665	13	18	96 / 75	93 / 74	91 / 74	22	78	77	76	NNW	8	WSW	97.2	9.8
CLARKSVILLE	36	33	87	22	382	6	12	95 / 76	93 / 74	90 / 74	21	78	77	76				99.8	3.7

State and Station	Lat °	Lat ′N	Long °	Long ′W	Elev (feet)	Winter 99%	Winter 97.5%	Summer Design Dry-bulb / Mean Coincident Wet-bulb 1%	2.5%	5%	Mean daily Range	Design Wet-bulb 1%	Design Wet-bulb 2.5%	Design Wet-bulb 5%	Wind Winter	Knots	Wind Summer	Median Annual Extr. Max.	Min.
COLUMBIA	35	38	87	02	690	10	15	97 / 75	94 / 74	91 / 74	21	78	77	76					
DYERSBURG	36	01	89	24	344	10	15	96 / 78	94 / 77	91 / 76	21	81	80	78					
GREENVILLE	36	04	82	50	1319	11	16	92 / 73	90 / 72	88 / 72	22	76	75	74				95.6	0.8
JACKSON AP	35	36	88	55	423	11	16	98 / 76	95 / 75	92 / 75	21	79	78	77				99.2	6.6
KNOXVILLE AP	35	49	83	59	980	13	19	94 / 74	92 / 73	90 / 73	21	77	76	75	NE	8	W	96.0	7.0
MEMPHIS AP	35	03	90	00	258	13	18	98 / 77	95 / 76	93 / 76	21	80	79	78	N	10	SW	97.9	10.4
MURFREESBORO	34	55	86	28	600	9	14	97 / 75	94 / 74	91 / 74	22	78	77	76				97.7	4.5
NASHVILLE AP (S)	36	07	86	41	590	9	14	97 / 75	94 / 74	91 / 74	21	78	77	76	NW	8	WSW		
TULLAHOMA	35	23	86	05	1067	8	13	96 / 74	93 / 73	91 / 73	22	77	76	75	NW	9	WSW	96.7	3.7
TEXAS																			
ABILENE AP	32	25	99	41	1784	15	20	101 / 71	99 / 71	97 / 71	22	75	74	74	N	12	SSE	103.6	10.4
ALICE AP	27	44	98	02	180	31	34	100 / 78	98 / 77	95 / 77	20	82	81	79	N	11	S	104.9	24.8
AMARILLO AP	35	14	101	42	3604	6	11	98 / 67	95 / 67	93 / 67	26	71	70	70	N	11	S	100.8	0.9
AUSTIN AP	30	18	97	42	597	24	28	100 / 74	98 / 74	97 / 74	22	78	77	77	N	11	S	101.6	19.7
BAY CITY	29	00	95	58	50	29	33	96 / 77	94 / 77	92 / 77	16	80	79	79					
BEAUMONT	29	57	94	01	16	27	31	95 / 79	93 / 78	91 / 78	19	81	80	80				99.7	23.5
BEEVILLE	28	22	97	40	190	30	33	99 / 78	97 / 77	95 / 77	18	82	81	79	N	9	SSE	103.1	22.5
BIG SPRING AP (S)	32	18	101	27	2598	16	20	100 / 69	97 / 69	95 / 69	26	74	73	72				105.3	10.7
BROWNSVILLE AP (S)	25	54	97	26	19	35	39	94 / 77	93 / 77	92 / 77	18	80	79	79	NNW	13	SE	98.1	30.1
BROWNWOOD	31	48	98	57	1386	18	22	101 / 73	99 / 73	96 / 73	22	77	76	75	N	9	S	105.3	13.0
BRYAN AP	30	40	96	33	276	24	29	98 / 76	96 / 76	94 / 76	20	79	78	78					
CORPUS CHRISTI AP	27	46	97	30	41	31	35	95 / 78	94 / 78	92 / 78	19	80	80	79	N	12	SSE	97.0	27.2
CORSICANA	32	05	96	28	425	20	25	100 / 75	98 / 75	96 / 75	21	79	78	77				104.2	15.2
DALLAS AP	32	51	96	51	481	18	22	102 / 75	100 / 75	97 / 75	20	78	78	77	N	11	S		
DEL RIO,LAUGHLIN AFB	29	22	100	47	1081	26	31	100 / 73	98 / 73	97 / 73	24	79	77	76				103.8	23.0
DENTON	33	12	97	06	630	17	22	101 / 74	99 / 74	97 / 74	22	78	77	76				104.5	11.8
EAGLE PASS	28	52	100	32	884	27	32	101 / 73	99 / 73	98 / 73	24	78	78	77	NNW	9	ESE	107.7	22.1
EL PASO AP (S)	31	48	106	24	3918	20	24	100 / 64	98 / 64	96 / 64	27	69	68	68	N	7	S	103.0	15.7
FORT WORTH AP (S)	32	50	97	03	537	17	22	101 / 74	99 / 74	97 / 74	22	78	77	76	NW	11	S	103.2	13.5
GALVESTON AP	29	18	94	48	7	31	36	90 / 79	89 / 79	88 / 78	10	81	80	80	N	15	S	93.9	27.5
GREENVILLE	33	04	96	03	535	17	22	101 / 74	99 / 74	97 / 74	21	78	77	76				103.6	11.7
HARLINGEN	26	14	97	39	35	35	39	96 / 77	94 / 77	93 / 77	19	80	79	79	NNW	10	SSE	102.3	29.3
HOUSTON AP	29	58	95	21	96	27	32	96 / 77	94 / 77	92 / 77	18	80	79	79	NNW	11	S		
HOUSTON CO	29	59	95	22	108	28	33	97 / 77	95 / 77	93 / 77	18	80	79	79				99.0	23.5
HUNTSVILLE	30	43	95	33	494	22	27	100 / 75	98 / 75	96 / 75	20	78	78	77				100.8	18.7

State and Station	Lat °	Lat ′N	Long °	Long ′W	Elev. feet	Winter, °F Design Dry-bulb 99%	97.5%	Summer Design Dry-bulb and Coincident Wet-bulb (Mean) 1%	2.5%	5%	Mean daily Range	Design Wet-bulb 1%	2.5%	5%	Prevailing Wind Winter	Knots	Summer	Temp. Median of Annual Extr. Max.	Min.
KILLEEN,ROBERT GRAY AAF	31	05	97	41	850	20	25	99 / 73	97 / 73	95 / 73	22	77	76	75					
LAMESA	32	42	101	56	2965	13	17	99 / 69	96 / 69	94 / 69	26	73	72	71				105.5	8.9
LAREDO AFB	27	32	99	27	512	32	36	102 / 73	101 / 73	99 / 74	23	78	78	77	N	8	SE		
LONGVIEW	32	28	94	44	330	19	24	99 / 76	97 / 76	95 / 76	20	80	79	78					
LUBBOCK AP	33	39	101	49	3254	10	15	98 / 69	96 / 69	94 / 69	26	73	72	71	NNE	10	SSE		
LUFKIN AP	31	25	94	48	277	25	29	99 / 76	97 / 76	94 / 76	20	80	79	78	NNW	12	S		
MCALLEN	26	12	98	13	122	35	39	97 / 77	95 / 77	94 / 77	21	80	79	79					
MIDLAND AP (S)	31	57	102	11	2851	16	21	100 / 69	98 / 69	96 / 69	26	73	72	71	NE	9	SSE	103.6	10.8
MINERAL WELLS AP	32	47	98	04	930	17	22	101 / 74	99 / 74	97 / 74	22	78	77	76					
PALESTINE CO	31	47	95	38	600	23	27	100 / 76	98 / 76	96 / 76	20	79	79	78				101.2	16.3
PAMPA	35	32	100	59	3250	7	12	99 / 67	96 / 67	94 / 67	26	71	70	70					
PECOS	31	25	103	30	2610	16	21	100 / 69	98 / 69	96 / 69	27	73	72	71					
PLAINVIEW	34	11	101	42	3370	8	13	98 / 68	96 / 68	94 / 68	26	72	71	70				102.7	3.1
PORT ARTHUR AP	29	57	94	01	16	27	31	95 / 79	93 / 78	91 / 78	19	81	80	80	N	9	S	97.7	24.0
SAN ANGELO,GOODFELLOW AFB	31	26	100	24	1877	18	22	101 / 71	99 / 71	97 / 70	24	75	74	73	NNE	10	SSE		
SAN ANTONIO AP (S)	29	32	98	28	788	25	30	99 / 72	97 / 73	96 / 73	19	77	76	76	N	8	SSE	101.3	21.1
SHERMAN,PERRIN AFB	33	43	96	40	763	15	20	100 / 75	98 / 75	95 / 74	22	78	77	76	N	10	SSE		
SNYDER	32	43	100	55	2325	13	18	100 / 70	98 / 70	96 / 70	26	74	73	72	S	10	S	103.0	11.9
TEMPLE	31	06	97	21	700	22	27	100 / 74	98 / 74	97 / 74	22	78	77	77					
TYLER AP	32	21	95	16	530	19	24	99 / 76	97 / 76	95 / 76	21	80	79	78	NNE	23	S		
VERNON	34	10	99	18	1212	13	17	102 / 73	100 / 73	97 / 73	24	77	76	75					
VICTORIA AP	28	51	96	55	104	29	32	98 / 78	96 / 77	94 / 77	18	82	81	79				101.4	23.4
WACO AP	31	37	97	13	500	21	26	101 / 75	99 / 75	97 / 75	22	78	78	77					
WICHITA FALLS AP	33	58	98	29	994	14	18	103 / 73	101 / 73	98 / 73	24	77	76	75	NNW	12	S		
UTAH																			
CEDAR CITY AP	37	42	113	06	5617	-2	5	93 / 60	91 / 60	89 / 59	32	65	63	62	SE	5	SW	95.5	-7.8
LOGAN	41	45	111	49	4785	-3	2	93 / 62	91 / 61	88 / 60	33	65	64	63					
MOAB	38	36	109	36	3965	6	11	100 / 60	98 / 60	96 / 60	30	65	64	63					
OGDEN AP	41	12	112	01	4455	1	5	93 / 63	91 / 61	88 / 61	33	66	65	64	S	6	SW	99.5	-3.9
PRICE	39	37	110	50	5580	-2	5	93 / 60	91 / 60	89 / 59	33	65	63	62					
PROVO	40	13	111	43	4448	1	6	98 / 62	96 / 62	94 / 61	32	66	65	64					
RICHFIELD	38	46	112	05	5270	-2	5	93 / 60	91 / 60	89 / 59	34	65	63	62	SE	5	SW	98.1	-10.5
ST GEORGE CO	37	02	113	31	2900	14	21	103 / 65	101 / 65	99 / 64	33	70	68	67				109.3	11.1
SALT LAKE CITY AP (S)	40	46	111	58	4220	3	8	97 / 62	95 / 62	92 / 61	32	66	65	64	SSE	6	N	99.4	-0.1
VERNAL AP	40	27	109	31	5280	-5	0	91 / 61	89 / 60	86 / 59	32	64	63	62					

Climatic design data. Columns: Latitude/Longitude, Elevation; Winter °F (Design Dry-bulb 99% / 97.5%); Summer °F (Design Dry-bulb and Coincident Wet-bulb — Mean / Wet-bulb — at 1% / 2.5% / 5%); Mean daily Range; Design Wet-bulb (1% / 2.5% / 5%); Prevailing Wind (Winter, Summer, Knots); Temp. °F Median of Annual Extremes (Max., Min.).

State and Station	Lat. °	N	Long. °	W	Elev. feet	Winter 99%	Winter 97.5%	Summer DB/WB 1%	2.5%	5%	Mean daily Range	Design WB 1%	2.5%	5%	Wind Winter	Wind Summer	Knots	Temp Max.	Temp Min.
VERMONT																			
BARRE	44	12	72	31	600	-16	-11	84 / 71	81 / 69	78 / 68	23	73	71	70					
BURLINGTON AP (S)	44	28	73	09	332	-12	-7	88 / 72	85 / 70	82 / 69	23	74	72	71	E	SSW	7	92.4	-16.9
RUTLAND	43	36	72	58	620	-13	-8	87 / 72	84 / 70	81 / 69	23	74	72	71				92.5	-17.5
VIRGINIA																			
CHARLOTTESVILLE	38	02	78	31	870	14	18	94 / 74	91 / 74	88 / 73	23	77	76	75	NE	SW	7	97.4	8.0
DANVILLE AP	36	34	79	20	590	14	16	94 / 74	92 / 73	90 / 73	21	77	76	75				100.1	9.2
FREDERICKSBURG	38	18	77	28	100	10	14	96 / 76	93 / 75	90 / 74	21	78	77	76					
HARRISONBURG	38	27	78	54	1370	12	16	93 / 72	91 / 72	88 / 71	23	75	74	73					
LYNCHBURG AP	37	20	79	12	916	12	16	93 / 74	90 / 74	88 / 73	21	77	76	75	NE	SW	7	97.2	7.6
NORFOLK AP	36	54	76	12	22	20	22	93 / 77	91 / 76	89 / 76	18	79	78	77	NW	SW	10	97.2	15.3
PETERSBURG	37	11	77	31	194	14	17	95 / 76	92 / 76	90 / 75	20	79	78	77					
RICHMOND AP	37	30	77	20	164	14	17	95 / 76	92 / 76	90 / 75	21	79	78	77	N	SW	6	97.9	9.6
ROANOKE AP	37	19	79	58	1193	12	16	93 / 72	91 / 72	88 / 71	23	75	74	73	NW	SW	9	97.9	
STAUNTON	38	16	78	54	1201	12	16	93 / 72	91 / 72	88 / 71	23	75	74	73	NW	SW	9	95.9	2.5
WINCHESTER	39	12	78	10	760	6	10	93 / 75	90 / 74	88 / 74	21	77	76	75				97.3	3.7
WASHINGTON																			
ABERDEEN	46	59	123	49	12	25	28	80 / 65	77 / 62	73 / 61	16	65	63	62	ESE	NNW	6	91.9	19.3
BELLINGHAM AP	48	48	122	32	158	10	15	81 / 67	77 / 65	74 / 63	19	68	65	63	NNE	WSW	15	87.4	10.3
BREMERTON	47	34	120	40	162	21	25	82 / 65	78 / 64	75 / 62	20	66	64	63	E	N	8		
ELLENSBURG AP	47	02	120	31	1735	2	6	94 / 65	91 / 64	87 / 62	34	66	65	63					
EVERETT,PAINE AFB	47	55	122	17	596	21	25	80 / 65	76 / 64	73 / 62	20	67	64	63	ESE	NNW	6	84.9	15.2
KENNEWICK	46	13	119	08	392	5	11	99 / 68	96 / 67	92 / 66	30	70	68	67		NW		103.4	2.0
LONGVIEW	46	10	122	56	12	19	24	88 / 68	85 / 67	81 / 65	30	69	67	66	ESE	SSW	9	96.0	14.8
MOSES LAKE,LARSON AFB	47	12	119	19	1185	1	7	97 / 66	94 / 65	90 / 63	32	67	66	64	N	NE	8		
OLYMPIA AP	46	58	122	54	215	16	22	87 / 66	83 / 65	79 / 64	32	67	66	64	NE		4		
PORT ANGELES	48	07	123	26	99	24	27	72 / 62	69 / 61	67 / 60	18	64	62	61				83.5	19.4
SEATTLE-BOEING FIELD	47	32	122	18	23	21	26	84 / 68	81 / 66	77 / 65	24	69	67	65					
SEATTLE CO (S)	47	39	122	18	20	22	27	85 / 68	82 / 66	78 / 65	19	69	67	65	N	N	7	90.2	22.0
SEATTLE-TACOMA AP (S)	47	27	122	18	400	21	26	84 / 65	80 / 64	76 / 62	22	66	64	63	E	N	9	90.1	19.9
SPOKANE AP (S)	47	38	117	31	2357	-6	2	93 / 64	90 / 63	87 / 62	28	65	64	62	NE	SW	6	98.8	-4.9
TACOMA,MCCHORD AFB	47	15	122	30	100	19	24	86 / 66	82 / 65	79 / 63	22	68	66	64	S	NNE	5	89.4	18.8
WALLA WALLA AP	46	06	118	17	1206	0	7	97 / 67	94 / 66	90 / 64	27	69	67	66	W	W	5	103.0	3.8
WENATCHEE	47	25	120	19	632	7	11	99 / 67	96 / 66	92 / 64	32	68	67	65				101.1	1.0
YAKIMA AP	46	34	120	32	1052	-2	5	96 / 65	93 / 65	89 / 63	36	68	66	65	W	NW	5		

State and Station	Lat. °N	Lat. ′N	Long. °W	Long. ′W	Elev. feet	Winter °F Design Dry-bulb 99%	Winter 97.5%	Summer Design Dry-bulb and Coincident Wet-bulb 1% (Mean)	2.5%	5%	Mean daily Range	Design Wet-bulb 1%	2.5%	5%	Prevailing Wind Winter (Knots)	Summer	Temp °F Median of Annual Extr. Max.	Min.
WEST VIRGINIA																		
BECKLEY	37	47	81	07	2504	-2	4	83 / 71	81 / 69	79 / 69	22	73	71	70	WNW 9	WNW		
BLUEFIELD AP	37	18	81	13	2867	-2	4	83 / 71	81 / 69	79 / 69	22	73	71	70				
CHARNESDON	38	22	81	36	939	7	11	92 / 74	90 / 73	87 / 72	20	76	75	74	SW 8	SW	97.2	2.9
CLARKSBURG	39	16	80	21	977	6	10	92 / 74	90 / 73	87 / 73	21	76	75	74				
ELKINS AP	38	53	79	51	1948	1	6	86 / 72	84 / 70	82 / 70	22	74	72	71	WNW 9	WNW	90.6	-7.3
HUNTINGTON CO	38	25	82	30	565	5	10	94 / 76	91 / 74	89 / 73	22	78	77	75	W 6	SW	97.1	2.1
MARTINSBURG AP	39	24	77	59	556	6	10	93 / 75	90 / 74	88 / 74	21	77	76	75	WNW 10	W	99.0	1.1
MORGANTOWN AP	39	39	79	55	1240	4	8	90 / 74	87 / 73	85 / 73	22	76	75	74			95.9	0.7
PARKERSBURG CO	39	16	81	34	615	7	11	93 / 75	90 / 74	88 / 73	21	77	76	75	WSW 7	WSW		
WHEELING	40	07	80	42	665	1	5	89 / 72	86 / 71	84 / 70	21	74	73	72	WSW 10	WSW	97.5	-0.6
WISCONSIN																		
APPLETON	44	15	88	23	730	-14	-9	89 / 74	86 / 72	83 / 71	23	76	74	72			94.6	-16.2
ASHLAND	46	34	90	58	650	-21	-16	85 / 70	82 / 68	79 / 66	23	72	70	68			94.1	-26.8
BELOIT	42	30	89	02	780	-7	-3	92 / 75	90 / 75	88 / 74	24	78	77	75				
EAU CLAIRE AP	44	52	91	29	888	-15	-11	92 / 75	89 / 73	86 / 71	23	77	75	73				
FOND DU LAC	43	48	88	27	760	-12	-8	89 / 74	86 / 72	84 / 71	23	76	74	72				
GREEN BAY AP	44	29	88	08	682	-13	-9	88 / 74	85 / 72	83 / 71	23	76	74	72	W 8	SW	96.0	-17.7
LA CROSSE AP	43	52	91	15	651	-13	-9	91 / 75	88 / 73	85 / 72	22	77	75	74	NW 10	S	94.3	-17.9
MADISON AP (S)	43	08	89	20	858	-11	-7	91 / 74	88 / 73	85 / 71	22	77	75	73	NW 8	SW	95.7	-21.3
MANITOWOC	44	06	87	41	660	-11	-7	89 / 74	86 / 72	83 / 71	21	76	74	72			93.6	-16.8
MARINETTE	45	06	87	38	605	-15	-11	87 / 73	84 / 71	82 / 70	20	75	73	71			94.1	-13.7
MILWAUKEE AP	42	57	87	54	672	-8	-4	90 / 74	87 / 73	84 / 71	21	76	74	73	WNW 10	SSW	95.9	-15.8
RACINE	42	43	87	51	730	-6	-2	91 / 75	88 / 73	85 / 72	21	77	75	74				
SHEBOYGAN	43	45	87	43	648	-10	-6	89 / 75	86 / 73	83 / 71	20	77	75	73			97.0	-12.4
STEVENS POINT	44	30	89	34	1079	-15	-11	92 / 75	89 / 73	86 / 71	23	77	75	73			95.3	-24.1
WAUKESHA	43	01	88	14	860	-9	-5	90 / 74	87 / 73	84 / 71	22	76	74	73			95.7	-14.3
WAUSAU AP	44	55	89	37	1196	-16	-12	91 / 74	88 / 72	85 / 70	23	76	74	72				
WYOMING																		
CASPER AP	42	55	106	28	5338	-11	-5	92 / 58	90 / 57	87 / 57	31	63	61	60	NE 10	SW	97.3	-20.9
CHEYENNE	41	09	104	49	6126	-9	-1	89 / 58	86 / 58	84 / 57	30	63	62	60	N 11	WNW	92.5	-15.9
CODY AP	44	33	109	04	4990	-19	-13	89 / 60	86 / 60	83 / 59	32	64	63	61			97.4	-21.9
EVANSTON	41	16	110	57	6780	-9	-3	86 / 55	84 / 55	82 / 54	32	59	58	57			89.2	-21.2
LANDER AP (S)	42	49	108	44	5563	-16	-11	91 / 61	88 / 61	85 / 60	32	64	63	61	E 5	NW	94.9	-22.6
LARAMIE AP (S)	41	19	105	41	7266	-14	-6	84 / 56	81 / 56	79 / 55	28	61	60	59				

State and Station	Lat. °	N	Long. °	W	Elev. feet	Winter °F Design Dry-bulb 99%	97.5%	Summer °F Design Dry-bulb and Coincident Wet-bulb Mean 1%	2.5%	5%	Mean daily Range	Design Wet-bulb 1%	2.5%	5%	Prevailing Wind Winter	Knots	Summer	Temp °F Median of Annual Extr. Max.	Min.
NEWCASTLE	43	51	104	13	4265	-17	-12	91 / 64	87 / 63	84 / 63	30	69	68	66				99.4	-19.0
RAWLINS	41	48	107	12	6740	-12	-4	86 / 57	83 / 57	81 / 56	40	62	61	60					
ROCK SPRINGS AP	41	36	109	04	6745	-9	-3	86 / 55	84 / 55	82 / 54	32	59	58	57	WSW	10	W		
SHERIDAN AP	44	46	106	58	3964	-14	-8	94 / 62	91 / 62	88 / 61	32	66	65	63	NW	7	N	99.8	-23.6
TORRINGTON	42	05	104	13	4098	-14	-8	94 / 62	91 / 62	88 / 61	30	66	65	63				101.1	-20.7
ALBERTA																			
CALGARY AP	51	06	114	01	3540	-27	-23	84 / 63	81 / 61	79 / 60	25	65	63	62	NNW	8	SE		
EDMONTON AP	53	34	113	31	2219	-29	-25	85 / 66	82 / 65	79 / 63	23	68	66	65	E	9	SE		
GRANDE PRAIRIE AP	55	11	118	53	2190	-39	-33	83 / 64	80 / 63	78 / 61	23	66	64	62					
JASPER	52	53	118	04	3480	-31	-26	83 / 64	80 / 62	77 / 61	28	66	64	63					
LETHBRIDGE AP (S)	49	38	112	48	3018	-27	-22	90 / 65	87 / 64	84 / 63	28	68	66	65					
MCMURRAY AP	56	39	111	13	1216	-41	-38	86 / 67	82 / 65	79 / 64	26	69	67	65					
MEDICINE HAT AP	50	01	110	43	2365	-29	-24	93 / 66	90 / 65	87 / 64	28	70	68	66					
RED DEER AP	52	11	113	54	2965	-31	-26	84 / 65	81 / 64	78 / 62	25	67	66	64					
BRITISH COLUMBIA																			
DAWSON CREEK	55	44	120	11	2164	-37	-33	82 / 64	79 / 63	76 / 61	26	66	64	62					
FORT NELSON AP (S)	58	50	122	35	1230	-43	-40	84 / 64	81 / 63	78 / 62	23	67	65	64					
KAMLOOPS CO	50	43	120	25	1133	-21	-15	94 / 66	91 / 65	88 / 64	29	68	66	65					
NANAIMO (S)	49	11	123	58	230	16	20	83 / 67	80 / 65	77 / 64	21	68	66	65					
NEW WESTMINSTER	49	13	122	54	50	14	18	84 / 68	81 / 67	78 / 66	19	69	68	66					
PENTICTON AP	49	28	119	36	1121	0	4	92 / 68	89 / 67	87 / 66	31	70	68	67					
PRINCE GEORGE AP (S)	53	53	122	41	2218	-33	-28	84 / 64	80 / 62	77 / 61	26	66	64	62	N	11	N		
PRINCE RUPERT CO	54	17	130	23	170	-2	2	64 / 59	63 / 57	61 / 56	12	60	58	57					
TRAIL	49	08	117	44	1400	-5	0	92 / 66	89 / 65	86 / 64	33	68	67	65					
VANCOUVER AP (S)	49	11	123	10	16	15	19	79 / 67	77 / 66	74 / 65	17	68	67	66	E	6	WNW		
VICTORIA CO	48	25	123	19	228	20	23	77 / 64	73 / 62	70 / 60	16	64	62	60					
MANITOBA																			
BRANDON	49	52	99	59	1200	-30	-27	89 / 72	86 / 70	83 / 68	25	74	72	70					
CHURCHILL AP (S)	58	45	94	04	155	-41	-39	81 / 66	77 / 64	74 / 62	18	67	65	63	SE	11	S		
DAUPHIN AP	51	06	100	03	999	-31	-28	87 / 71	84 / 70	81 / 68	23	74	72	70					
FLIN FLON	54	46	101	51	1098	-41	-37	84 / 68	81 / 66	79 / 65	19	70	68	67					
PORTAGE LA PRAIRIE AP	49	54	98	16	867	-28	-24	88 / 73	86 / 72	83 / 70	22	76	74	71					
THE PAS AP (S)	53	58	101	06	894	-37	-33	85 / 68	82 / 67	79 / 66	20	71	69	68	W	8	W		
WINNIPEG AP (S)	49	54	97	14	786	-30	-27	89 / 73	86 / 71	84 / 70	22	75	73	71	W	8	S		

State and Station	Lat. °	Lat. '	Long. °	Long. '	Elev. feet	Winter °F Design Dry-bulb 99%	97.5%	Summer °F Design Dry-bulb and Coincident Wet-bulb Mean 1%	2.5%	5%	Mean daily Range	Design Wet-bulb 1%	2.5%	5%	Prevailing Wind Winter	Knots	Summer	Temp °F Median of Annual Extr. Max.	Min.
NEW BRUNSWICK																			
CAMPBELLTON CO	48	00	66	40	25	-18	-14	85 / 68	82 / 67	79 / 66	21	72	70	68					
CHATHAM AP	47	01	65	27	112	-15	-10	89 / 69	85 / 68	82 / 67	22	72	71	69					
EDMUNDSTON CO	47	22	68	20	500	-21	-16	87 / 70	83 / 68	80 / 67	21	73	71	69					
FREDERICTON AP (S)	45	52	66	32	74	-16	-11	89 / 71	85 / 69	82 / 68	23	73	71	70					
MONCTON AP (S)	46	07	64	41	248	-12	-8	85 / 70	82 / 69	79 / 67	23	72	71	69					
SAINT JOHN AP	45	19	65	53	352	-12	-8	80 / 67	77 / 65	75 / 64	19	70	68	66					
NEWFOUNDLAND																			
CORNER BROOK	48	58	57	57	15	-5	0	76 / 64	73 / 63	71 / 62	17	67	66	65					
GANDER AP	48	57	54	34	482	-5	-1	82 / 66	79 / 65	77 / 64	19	69	67	66	WNW	11	SW		
GOOSE BAY AP (S)	53	19	60	25	144	-27	-24	85 / 66	81 / 64	77 / 63	19	68	66	64	N	9	SW		
ST JOHN'S AP (S)	47	37	52	45	463	3	7	77 / 66	75 / 65	73 / 64	18	69	67	66	N	20	WSW		
STEPHENVILLE AP	48	32	58	33	44	-3	4	76 / 65	74 / 64	71 / 63	14	67	66	65	WNW	10	S		
NORTHWEST TERRITORIES																			
FORT SMITH AP(S)	60	01	111	58	665	-49	-45	85 / 66	81 / 64	78 / 63	24	68	66	65	NW	4	S		
FROBISHER AP (S)	63	45	68	33	68	-43	-41	66 / 53	63 / 51	59 / 50	14	54	52	51	NNW	9	NW		
INUVIK (S)	68	18	133	29	200	-56	-53	79 / 62	77 / 60	75 / 59	21	64	62	61					
RESOLUTE AP (S)	74	43	94	59	209	-50	-47	57 / 48	54 / 46	51 / 45	10	50	48	46					
YELLOWKNIFE AP	62	28	114	27	682	-49	-46	79 / 62	77 / 61	74 / 60	16	64	63	62	SSE	7	S		
NOVA SCOTIA																			
AMHERST	45	49	64	13	65	-11	-6	84 / 69	81 / 68	79 / 67	21	72	70	68					
HALIFAX AP (S)	44	39	63	34	83	1	5	79 / 66	76 / 65	74 / 64	16	69	67	66					
KENTVILLE (S)	45	03	64	36	40	-3	1	85 / 69	83 / 68	80 / 67	22	72	71	69					
NEW GLASGOW	45	37	62	37	317	-9	-5	81 / 69	79 / 68	77 / 67	20	72	70	69					
SYDNEY AP	46	10	60	03	197	-1	3	82 / 69	80 / 68	77 / 66	19	71	70	68					
TRURO CO	45	22	63	16	131	-8	-5	82 / 70	80 / 69	78 / 68	22	73	71	70					
YARMOUTH AP	43	50	66	05	136	5	9	74 / 65	72 / 64	70 / 63	15	68	66	65	NW	11	S		
ONTARIO																			
BELLEVILLE	44	09	77	24	250	-11	-7	86 / 73	84 / 72	82 / 71	20	75	74	73					
CHATHAM	42	24	82	12	600	0	3	89 / 74	87 / 73	85 / 72	19	76	75	74					
CORNWALL	45	01	74	45	210	-13	-9	89 / 73	87 / 72	84 / 71	21	75	74	72					
HAMILTON	43	16	79	54	303	-3	1	88 / 73	86 / 72	83 / 71	21	76	74	73					
KAPUSKASING AP (S)	49	25	82	28	752	-31	-28	86 / 70	83 / 69	80 / 67	23	72	70	69					
KENORA AP	49	48	94	22	1345	-32	-28	84 / 70	82 / 69	80 / 68	19	73	71	70					
KINGSTON	44	16	76	30	300	-11	-7	87 / 73	84 / 72	82 / 71	20	75	74	73					

State and Station	Lat. °N	Lat. 'N	Long. °W	Long. 'W	Elev. feet	Winter °F Design Dry-bulb 99%	Winter °F Design Dry-bulb 97.5%	Summer °F Design Dry-bulb and Coincident Wet-bulb Mean 1%	2.5%	5%	Mean daily Range	Design Wet-bulb 1%	2.5%	5%	Prevailing Wind Winter (Knots)	Summer	Temp °F Median of Annual Extr. Max.	Min.
KITCHENER	43	26	80	30	1125	-6	-2	88 / 73	85 / 72	83 / 71	23	75	74	72				
LONDON AP	43	02	81	09	912	-4	0	87 / 74	85 / 73	83 / 72	21	76	74	73				
NORTH BAY AP	46	22	79	25	1210	-22	-18	84 / 68	81 / 67	79 / 66	20	71	70	68				
OSHAWA	43	54	78	52	370	-6	-3	88 / 73	86 / 72	84 / 71	20	75	74	73				
OTTAWA AP (S)	45	19	75	40	413	-17	-13	90 / 72	87 / 71	84 / 70	21	75	73	72				
OWEN SOUND	44	34	80	55	597	-6	-2	84 / 71	82 / 70	80 / 69	21	73	72	70				
PETERBOROUGH	44	17	78	19	635	-13	-9	87 / 72	85 / 71	83 / 70	21	75	73	72				
ST CATHARINES	43	11	79	14	325	-1	3	87 / 73	85 / 72	83 / 71	20	76	74	73				
SARNIA	42	58	82	22	625	0	3	88 / 73	86 / 72	84 / 71	19	76	74	73				
SAULT STE MARIE AP	46	32	84	30	675	-17	-13	85 / 71	82 / 69	79 / 68	22	73	71	70				
SUDBURY AP	46	37	80	48	1121	-22	-19	86 / 69	83 / 67	81 / 66	22	72	70	68				
THUNDER BAY AP	48	22	89	19	644	-27	-24	85 / 70	83 / 68	80 / 67	24	72	70	68	W 8	W		
TIMMINS AP	48	34	81	22	965	-33	-29	87 / 69	84 / 68	81 / 66	25	72	70	68				
TORONTO AP (S)	43	41	79	38	578	-5	-1	90 / 73	87 / 72	85 / 71	20	75	74	73	N 10	SW		
WINDSOR AP	42	16	82	58	637	0	4	90 / 74	88 / 73	86 / 72	20	77	75	74				
PRINCE EDWARD ISLAND																		
CHARLOTTETOWN AP (S)	46	17	63	08	186	-7	-4	80 / 69	78 / 68	76 / 67	16	71	70	68				
SUMMERSIDE AP	46	26	63	50	78	-8	-4	81 / 69	79 / 68	77 / 67	16	72	70	68				
QUEBEC																		
BAGOTVILLE AP	48	20	71	00	536	-28	-23	87 / 70	83 / 68	80 / 67	21	72	70	68				
CHICOUTIMI	48	25	71	05	150	-26	-22	86 / 70	83 / 68	80 / 67	20	72	70	68				
DRUMMONDVILLE	45	53	72	29	270	-18	-14	88 / 72	85 / 71	82 / 69	21	75	73	71				
GRANBY	45	23	72	42	550	-19	-14	88 / 72	85 / 71	83 / 70	21	75	73	72				
HULL	45	26	75	44	200	-18	-14	90 / 72	87 / 71	84 / 70	21	75	73	72				
MEGANTIC AP	45	35	70	52	1362	-20	-16	86 / 71	83 / 70	81 / 69	20	74	72	71				
MONTREAL AP (S)	45	28	73	45	98	-16	-10	88 / 73	85 / 72	83 / 71	17	75	74	72				
QUEBEC AP	46	48	71	23	245	-19	-14	87 / 72	84 / 70	81 / 68	20	74	72	70				
RIMOUSKI	48	27	68	32	117	-16	-12	83 / 68	79 / 66	76 / 65	18	71	69	67				
ST JEAN	45	18	73	16	129	-15	-11	88 / 73	86 / 72	84 / 71	20	75	74	72				
ST JEROME	45	48	74	01	556	-17	-13	88 / 72	86 / 71	83 / 70	23	75	73	72				
SEPT. ILES AP (S)	50	13	66	16	190	-26	-21	76 / 63	73 / 61	70 / 60	17	67	65	63				
SHAWINIGAN	46	34	72	43	306	-18	-14	86 / 72	84 / 70	82 / 69	21	74	72	71				
SHERBROOKE CO	45	24	71	54	595	-25	-21	86 / 72	84 / 71	81 / 69	20	74	73	71				
THETFORD MINES	46	04	71	19	1020	-19	-14	87 / 71	84 / 70	81 / 69	21	74	72	71				
TROIS RIVIERES	46	21	72	35	50	-17	-13	88 / 72	85 / 70	82 / 69	23	74	72	71				

State and Station	Lat. °	Lat. 'N	Long. °	Long. 'W	Elev. feet	Winter, °F Design Dry-bulb 99%	97.5%	Summer, °F Design Dry-bulb and Coincident Wet-bulb Mean 1%	2.5%	5%	Mean daily Range	Design Wet-bulb 1%	2.5%	5%	Prevailing Wind Winter Knots	Summer	Temp. °F Median of Annual Extr. Max.	Min.
VAL D'OR AP	48	03	77	47	1108	-32	-27	85 / 70	83 / 68	80 / 67	22	72	70	68				
VALLEYFIELD	45	16	74	06	150	-14	-10	89 / 73	86 / 72	84 / 71	20	75	74	72				
SASKATCHEWAN																		
ESTEVAN AP	49	04	103	00	1884	-30	-25	92 / 70	89 / 68	86 / 67	26	72	70	69				
MOOSE JAW AP	50	20	105	33	1857	-29	-25	93 / 69	89 / 67	86 / 66	27	71	69	68				
NORTH BATTLEFORD AP	52	46	108	15	1796	-33	-30	88 / 67	85 / 66	82 / 65	23	69	68	66				
PRINCE ALBERT AP	53	13	105	41	1414	-42	-35	87 / 67	84 / 66	81 / 65	25	70	68	67				
REGINA AP	50	26	104	40	1884	-33	-29	91 / 69	88 / 68	84 / 67	26	72	70	68				
SASKATOON AP (S)	52	10	106	41	1645	-35	-31	89 / 68	86 / 66	83 / 65	26	70	68	67				
SWIFT CURRENT AP (S)	50	17	107	41	2677	-28	-25	93 / 68	90 / 66	87 / 65	25	70	69	67				
YORKTON AP	51	16	102	28	1653	-35	-30	87 / 69	84 / 68	80 / 66	23	72	70	68				
YUKON TERRITORY																		
WHITEHORSE AP (S)	60	43	135	04	2289	-46	-43	80 / 59	77 / 58	74 / 56	22	61	59	58	NW 5	SE		

Table B-3 Weather Data for Other Countries (*Source:* ASHRAE, *Handbook of Fundamentals*, 1993. With permission.)

State and Station	Lat. °	Lat. 'N	Long. °	Long. 'W	Elev. feet	Winter, °F Mean of Annual Extremes	99%	97.5%	Summer Dry-bulb 1%	2.5%	5%	Mean Daily Range	Wet-bulb 1%	2.5%	5%	Wind Winter	Summer	Knots
AFGHANISTAN																		
KABUL	34	35N	69	12E	5955	2	6	9	98	96	93	32	66	65	64	N	N	4
ALGERIA																		
ALGIERS	36	46N	3	03E	194	38	43	45	95	92	89	14	77	76	75			
ARGENTINA																		
BUENOS AIRES	34	35S	58	29W	89	27	32	34	91	89	86	22	77	76	75	SW	NNE	9
CORDOBA	31	22S	64	15W	1388	21	28	32	100	96	93	27	76	75	74			
TUCUMAN	26	50S	65	10W	1401	24	32	36	102	99	96	23	76	75	74			
AUSTRALIA																		
ADELAIDE	34	56S	138	35E	140	36	38	40	98	94	91	25	72	70	68	NE	NW	5
ALICE SPRINGS	23	48S	133	53E	1795	28	34	37	104	102	100	27	75	74	72	N	SE	6
BRISBANE	27	28S	153	02E	137	39	44	47	91	88	86	18	77	76	75	N	NNE	7
DARWIN	12	28S	130	51E	88	60	64	66	94	93	91	16	82	81	81	E	WNW	10
MELBOURNE	37	49S	144	58E	114	31	35	38	95	91	86	21	71	69	68			
PERTH	31	57S	115	51E	210	38	40	42	100	96	93	22	76	74	73	N	E	6
SYDNEY	33	52S	151	12E	138	38	40	42	89	84	80	13	74	73	72	N	NE	8
AUSTRIA																		
VIENNA	48	15N	16	22E	644	-2	6	11	88	86	83	16	71	69	67	W	SSE	13
AZORES																		
LAJES (TERCEIRA)	38	45N	27	05W	170	42	46	49	80	78	77	11	73	72	71	W	NW	9
BAHAMAS																		
NASSAU	25	05N	77	21W	11	55	61	63	90	89	88	13	80	80	79			
BANGLADESH																		
CHITTAGONG	22	21N	91	50E	87	48	52	54	93	91	89	20	82	81	81			

State and Station	Lat. °	'N	Long. °	'W	Elev. feet	Winter, °F Mean of Annual Extremes	99%	97.5%	Summer, °F Design Dry-bulb 1%	2.5%	5%	Mean Daily Range	Design Wet-bulb 1%	2.5%	5%	Prevailing Wind Winter	Summer	Knots Winter	Summer
BELGIUM																			
BRUSSELS	50	48N	4	21E	328	13	15	19	83	79	77	19	70	68	67	NE	ENE	8	
BERMUDA																			
KINDLEY AFB	33	22N	64	41W	129	47	53	55	87	86	85	12	79	78	78	NW	S	16	
BOLIVIA																			
LA PAZ	16	30S	68	09W	12001	28	31	33	71	69	68	24	58	57	56				
BRAZIL																			
BELEM	1	27S	48	29W	42	67	70	71	90	89	87	19	80	79	78	SE	E	5	
BELO HORIZONTE	19	56S	43	57W	3002	42	47	50	86	84	83	18	76	75	75				
BRASILIA	15	53S	47	56W	3481	46	52	54	86	84	84	23	71	70	70	W	E	1	
CAMPINAS	23	01S	47	08W	2169	--	52	54	91	90	88	23	76	75	74	ESE	N	7	
CONGONHAS	23	38S	46	39W	2635	--	46	49	88	86	84	21	73	72	71	S	NNW	3	
CURITIBA	25	25S	49	17W	3114	28	34	37	86	84	82	21	75	74	74				
FORTALEZA	3	46S	38	33W	89	66	69	70	91	90	89	17	79	78	78				
GALEAO	22	50S	43	15W	20	--	61	63	97	95	92	21	80	79	78	NW	SSE	2	
PORTO ALEGRE	30	02S	51	13W	33	32	37	40	95	92	89	20	76	76	75				
RECIFE	8	04S	34	53W	97	67	69	70	88	87	86	10	78	77	77	S	ESE	7	
RIO DE JANEIRO	22	55S	43	12W	201	56	58	60	94	92	90	11	80	79	78	N	S	5	
SALVADOR	13	00S	38	30W	154	65	67	68	88	87	86	12	79	79	78				
SAO PAULO	23	31S	46	37W	2598	36	47	50	89	87	85	14	74	73	72	E	NW	6	
BRITISH HONDURAS																			
BELIZE	17	31N	88	11W	17	55	60	62	90	90	89	13	82	82	81				
BULGARIA																			
SOFIA	42	42N	23	20E	1805	-2	3	8	89	86	84	26	71	70	69				
BURMA																			
MANDALAY	21	59N	96	06E	252	50	54	56	104	102	101	30	81	80	80				
RANGOON	16	47N	96	09E	18	59	62	63	100	98	95	25	83	82	82	W	W	6	

State and Station	Lat. °	Lat. 'N	Long. °	Long. 'W	Elev. feet	Winter, °F Mean of Annual Extremes	99%	97.5%	Summer, °F Design Dry-bulb 1%	2.5%	5%	Mean Daily Range	Design Wet-bulb 1%	2.5%	5%	Prevailing Wind Winter	Summer	Knots
CAMBODIA																		
PHNOM PENH	11	33N	104	51E	36	62	66	68	98	96	94	19	83	82	82	N	W	4
CHILE																		
CONCEPCION	36	47S	73	04W	30	--	41	41	75	73	72	22	64	63	62	S	SW	4
PUNTA ARENAS	53	10S	70	54W	26	22	25	27	68	66	64	14	56	55	54			
SANTIAGO	33	24S	70	47W	1555	27	30	32	90	88	86	37	68	67	66	NE	SW	1
VALPARAISO	33	01S	71	38W	135	39	43	46	81	79	77	16	67	66	65			
CHINA																		
CHUNGKING	29	33N	106	33E	755	34	37	39	99	97	95	18	81	80	79			
SHANGHAI	31	12N	121	26E	23	16	23	26	94	92	90	16	81	81	80	WNW	S	6
COLOMBIA																		
BARANQUILLA	10	59N	74	48W	44	66	70	72	95	94	93	17	83	82	82			
BOGOTA	4	36N	74	05W	8406	42	45	46	72	70	69	19	60	59	58	E	E	8
CALI	3	25N	76	30W	3189	53	57	58	84	82	79	15	70	69	68			
MEDELLIN	6	13N	75	36W	4650	48	53	55	87	85	84	25	73	72	72			
CONGO																		
BRAZZAVILLE	4	15S	15	15E	1043	54	60	62	93	92	91	21	81	81	80			
CUBA																		
GUANTANAMO BAY	19	54N	75	09W	21	60	64	66	94	93	92	16	82	81	80	N	ESE	6
HAVANA	23	08N	82	21W	80	54	59	62	92	91	89	14	81	81	80	N	E	11
CYPRUS																		
AKROTIRI	34	36N	32	59E	75	--	41	43	91	90	87	17	78	77	76	NNW	SW	4
LAINACA	34	53N	33	39E	7	--	37	41	93	91	90	21	78	77	76	NW	SSW	6
PAPHOS	34	43N	32	30E	26	--	42	45	88	86	86	13	79	79	78	NE	W	7
CZECHOSLOVAKIA																		
PRAGUE	50	05N	14	25E	662	3	4	9	88	85	83	16	66	65	64			
DENMARK																		
COPENHAGEN	55	41N	12	33E	43	11	16	19	79	76	74	17	68	66	64	NE	N	11

State and Station	Lat. °	'N	Long. °	'W	Elev. feet	Winter, °F — Mean of Annual Extremes	99%	97.5%	Summer, °F — Design Dry-bulb 1%	2.5%	5%	Mean Daily Range	Design Wet-bulb 1%	2.5%	5%	Prevailing Wind Winter	Summer	Knots Winter	Knots Summer
DOMINICAN REPUBLIC																			
SANTO DOMINGO	18	29N	69	54W	57	61	63	65	92	90	88	16	81	80	80	NNE	SE	6	
EQUADOR																			
GUAYAQUIL	2	10S	79	53W	20	61	64	65	92	91	89	20	80	80	79		N		3
QUITO	0	13S	78	32W	9446	30	36	39	73	72	71	32	63	62	62	N	N		3
EGYPT																			
CAIRO	29	52N	31	20E	381	39	45	46	102	100	98	26	76	75	74	N	NNW	9	
LUXOR	25	40N	32	43E	289	--	38	41	109	108	106	31	76	74	73	E	N	1	
EL SALVADOR																			
SAN SALVADOR	13	42N	89	13W	2238	51	54	56	98	96	95	32	77	76	75	N	S	7	
ETHIOPIA																			
ADDIS ABABA	9	02N	38	45E	7753	35	39	41	84	82	81	28	66	65	64	E	S	10	
ASMARA	15	17N	38	55E	7628	36	40	42	83	81	80	27	65	64	63	E	WNW	9	
FINLAND																			
HELSINKI	60	10N	24	57E	30	-11	-7	-1	77	74	72	14	66	65	63	E	S	4	
FRANCE																			
LYON	45	42N	4	47E	938	-1	10	14	91	89	86	23	71	70	69	N	S	7	
MARSEILLES	43	18N	5	23E	246	23	25	28	90	87	84	22	72	71	69	SE	W	14	
NANTES	47	15N	1	34W	121	17	22	26	86	83	80	21	70	69	67	NNE	E	6	
NICE	43	42N	7	16E	39	31	34	37	87	85	83	15	73	72	72		E		
PARIS	48	49N	2	29E	164	16	22	25	89	86	83	21	70	68	67	NE	E	7	
STRASBOURG	48	35N	7	46E	465	9	11	16	86	83	80	20	70	69	67				
FRENCH GUIANA																			
CAYENNE	4	56N	52	27W	20	69	71	72	92	91	90	17	83	83	82	ENE	E	5	
GERMANY																			
BERLIN(WEST)	52	27N	13	18E	187	6	7	12	84	81	78	19	68	67	66	E	E	6	
HAMBURG	53	33N	9	58E	66	10	12	16	80	76	73	13	68	66	65	E	E	6	
HANNOVER	52	24N	9	40E	561	7	16	20	82	78	75	17	68	67	65	E	E	8	

State and Station	Lat. °	Lat. 'N	Long. °	Long. 'W	Elev. feet	Winter, °F Mean of Annual Extremes	99%	97.5%	Summer, °F Design Dry-bulb 1%	2.5%	5%	Mean Daily Range	Design Wet-bulb 1%	2.5%	5%	Prevailing Wind Winter	Summer	Knots Winter	Knots Summer
MANNHEIM	49	34N	8	28E	359	2	8	11	87	85	82	18	71	69	68	N	S	5	5
MUNICH	48	09N	11	34E	1729	-1	5	9	86	83	80	18	68	66	64	S	N	4	4
GHANA																			
ACCRA	5	33N	0	12W	88	65	68	69	91	90	89	13	80	79	79	WSW	SW	5	5
GIBRALTAR																			
GIBRALTAR	36	09N	5	22W	11	38	42	45	92	89	86	14	76	75	74				
GREECE																			
ATHENS	37	58N	23	43E	351	29	33	36	96	93	91	18	72	71	71	N	NNE	9	9
SOUDA	35	32N	24	09E	479	--	38	41	95	91	90	21	76	74	72	NNW	WNW	4	4
THESSALONIKI	40	37N	22	57E	78	23	28	32	95	93	91	20	77	76	75				
GREENLAND																			
NARSARSSUAQ	61	11N	45	25W	85	-23	-12	-8	66	63	61	20	56	54	52				
GUATEMALA																			
GUATEMALA CITY	14	37N	90	31W	4855	45	48	51	83	82	81	24	69	68	67	N	S	9	9
GUYANA																			
GEORGETOWN	6	50N	58	12W	6	70	72	73	89	88	87	11	80	79	79				
HAITI																			
PORT AU PRINCE	18	33N	72	20W	121	63	65	67	97	95	93	20	82	81	80	N	ESE	6	6
HONDURAS																			
TEGUCIGALPA	14	06N	87	13W	3094	44	47	50	89	87	85	28	73	72	71	N	E	8	8
HONG KONG																			
HONG KONG	22	18N	114	10E	109	43	48	50	92	91	90	10	81	80	80	N	W	9	9
HUNGARY																			
BUDAPEST	47	31N	19	02E	394	8	10	14	90	86	84	21	72	71	70	N	S	5	5
ICELAND																			
REYKJAVIK	64	08N	21	56E	59	8	14	17	59	58	56	16	54	53	53	E	E	12	12
INDIA																			
AHMENABAD	23	02N	72	35E	163	49	53	56	109	107	105	28	80	79	78				

State and Station	Lat. °	Lat. 'N	Long. °	Long. 'W	Elev. feet	Winter, °F Mean of Annual Extremes	Winter 99%	Winter 97.5%	Design Dry-bulb 1%	Design Dry-bulb 2.5%	Design Dry-bulb 5%	Summer Mean Daily Range	Design Wet-bulb 1%	Design Wet-bulb 2.5%	Design Wet-bulb 5%	Prevailing Wind Winter	Prevailing Wind Summer	Knots Winter	Knots Summer
BANGALORE	12	57N	77	37E	3021	53	56	58	96	94	93	26	75	74	74	NW	NW	6	
BOMBAY	18	54N	72	49E	37	62	65	67	96	94	92	13	82	81	81	N	S	4	
CALCUTTA	22	32N	88	20E	21	49	52	54	98	97	96	22	83	82	82	W	W	3	
MADRAS	13	04N	80	15E	51	61	64	66	104	102	101	19	84	83	83				
NAGPUR	21	09N	79	07E	1017	45	51	54	110	108	107	30	79	79	78	N	NW	6	
NEW DELHI	28	35N	77	12E	703	35	39	41	110	107	105	26	83	82	82	N	N		
INDONESIA																			
DJAKARTA	6	11S	106	50E	26	69	71	72	90	89	88	14	80	79	78	N	N	11	
KUPANG	10	10S	123	34E	148	63	66	68	94	93	92	20	81	80	80				
MAKASSAR	5	08S	119	28E	61	64	66	68	90	89	88	17	80	80	79				
MEDAN	3	35N	98	41E	77	66	69	71	92	91	90	17	81	80	79				
PALEMBANG	3	00S	104	46E	20	67	70	71	92	91	90	17	80	79	79				
SURABAYA	7	13S	112	43E	10	64	66	68	91	90	89	18	80	79	79				
IRAN																			
ABADAN	30	21N	48	16E	7	32	39	41	116	113	110	32	82	81	81	W	WNW	6	
MESHED	36	17N	59	36E	3104	3	10	14	99	96	93	29	68	67	66				
TEHRAN	35	41N	51	25E	4002	15	20	24	102	100	98	27	75	74	73	W	SE	5	
IRAQ																			
BAGHDAD	33	20N	44	24E	111	27	32	35	113	111	108	34	73	72	72	WNW	WNW	5	
MOSUL	36	19N	43	09E	730	23	29	32	114	112	110	40	73	72	72				
IRELAND																			
DUBLIN	53	22N	6	21W	155	19	24	27	74	72	70	16	65	64	62	W	SW	9	
SHANNON	52	41N	8	55W	8	19	25	28	76	73	71	14	65	64	63	SE	W	4	
IRIAN BARAT																			
MANOKWARI	0	52S	134	05E	62	70	71	72	89	88	87	12	82	81	81				
ISRAEL																			
JERUSALEM	31	47N	35	13E	2485	31	36	38	95	94	92	24	70	69	69	W	NW	12	
TEL AVIV	32	06N	34	47E	36	33	39	41	96	93	91	16	74	73	72	N	W	8	

State and Station	Lat. °	Lat. 'N	Long. °	Long. 'W	Elev. feet	Winter, °F Mean of Annual Extremes	Winter 99%	Winter 97.5%	Summer Design Dry-bulb 1%	2.5%	5%	Mean Daily Range	Summer Design Wet-bulb 1%	2.5%	5%	Prevailing Wind Winter	Prevailing Wind Summer	Knots
ITALY																		
MILAN	45	27N	9	17E	341	12	18	22	89	87	84	20	76	75	74	W	SW	4
NAPLES	40	53N	14	18E	220	28	34	36	91	88	86	19	74	73	72	N	SSW	6
ROME	41	48N	12	36E	377	25	30	33	94	92	89	24	74	73	72	E	WSW	6
IVORY COAST																		
ABIDJAN	5	19N	4	01W	65	64	67	69	91	90	88	15	83	82	81	WSW	SW	5
JAMAICA																		
KINGSTON	17	56N	76	47W	46	--	71	72	93	91	89	10	81	80	80	N	ESE	10
MONTEGO	18	30N	77	55W	10	--	68	69	89	89	88	11	80	79	79	ESE	ENE	10
JAPAN																		
FUKUOKA	33	35N	130	27E	22	26	29	31	92	90	89	20	82	80	79			
SAPPORO	43	04N	141	21E	56	-7	1	5	86	83	80	20	76	74	72	SE	SE	3
TOKYO	35	41N	139	46E	19	21	26	28	91	89	87	14	81	80	79	SW	S	10
JOHNSTON ISLAND	16	44N	169	31W	16	--	72	73	87	86	85	6	80	80	79	NE	E	15
JORDAN																		
AMMAN	31	57N	35	57E	2548	29	33	36	97	94	92	25	70	69	68	N	NNW	6
KENYA																		
NAIROBI	1	16S	36	48E	5971	45	48	50	81	80	78	24	66	65	65	E	ENE	13
KOREA																		
PYONGYANG	39	02N	125	41E	186	-10	-2	3	89	87	85	21	77	76	76			
SEOUL	37	34N	126	58E	285	-1	7	9	91	89	87	16	81	79	78	NW	W	7
LEBANON																		
BEIRUT	33	54N	35	28E	111	40	42	45	93	91	90	15	78	77	76	N	SW	7
LIBERIA																		
MONROVIA	6	18N	10	48W	75	64	68	69	90	89	88	19	82	82	81	E	WSW	3
LIBYA																		
BENGHAZI	32	06N	20	04E	82	41	46	48	97	94	91	13	77	76	75	SSE	S	8

State and Station	Lat. °	Lat. 'N	Long. °	Long. 'W	Elev. feet	Winter Mean of Annual Extremes	Winter 99%	Winter 97.5%	Summer Design Dry-bulb 1%	2.5%	5%	Mean Daily Range	Design Wet-bulb 1%	2.5%	5%	Prevailing Wind Winter	Knots	Prevailing Wind Summer
MADAGASCAR																		
TANANARIVE	18	55S	47	33E	4531	39	43	46	86	84	83	23	73	72	71			
MALAYSIA																		
KUALA LUMPUR	3	07N	101	42E	127	67	70	71	94	93	92	20	82	82	81	N	4	W
PENANG	5	25N	100	19E	17	69	72	73	93	93	92	18	82	81	80			
MARTINIQUE																		
FORT DE FRANCE	14	37N	61	05W	13	62	64	66	90	89	88	14	81	81	80			
MEXICO																		
GUADALAJARA	20	41N	103	20W	5105	35	39	42	93	91	89	29	68	67	66	N	7	W
MERIDA	20	58N	89	38W	72	56	59	61	97	95	94	21	80	79	77	E	11	E
MEXICO CITY	19	24N	99	12W	7575	33	37	39	83	81	79	25	61	60	59	N	8	N
MONTERREY	25	40N	100	18W	1732	31	38	41	98	95	93	20	79	78	77			
TAMPICO	22	17N	97	52W	79	--	48	50	91	89	89	10	83	81	81	N	11	E
VERA CRUZ	19	12N	96	08W	184	55	60	62	91	89	88	12	83	83	82			
MIDWAY ISLAND	28	13N	177	23W	10	--	56	58	87	86	86	9	78	77	76	NNW	9	E
MOROCCO																		
CASABLANCA	33	35N	7	39W	164	36	40	42	94	90	86	50	73	72	70			
NEPAL																		
KATMANDU	27	42N	85	12E	4388	30	33	35	89	87	86	25	78	77	76	W	4	NW
NETHERLANDS																		
AMSTERDAM	52	23N	4	55E	5	17	20	23	79	76	73	10	65	64	63	S	8	E
NEW ZEALAND																		
AUCKLAND	36	51S	174	46E	140	37	40	42	78	77	76	14	67	66	65	W	4	NNW
CHRIST CHURCH	43	32S	172	37E	32	25	28	31	82	79	76	17	68	67	66	NE	6	NNE
WELLINGTON	41	17S	174	46E	394	32	35	37	76	74	72	14	66	65	64			E
NICARAGUA																		
MANAGUA	12	10N	86	15W	135	62	65	67	94	93	92	21	81	80	79	E	9	E

State and Station	Lat. °	'N	Long. °	'W	Elev. feet	Winter, °F Mean of Annual Extremes	99%	97.5%	Summer, °F Design Dry-bulb 1%	2.5%	5%	Mean Daily Range	Design Wet-bulb 1%	2.5%	5%	Prevailing Wind Winter	Summer	Knots
NIGERIA																		
LAGOS	6	27N	3	24E	10	67	70	71	92	91	90	12	82	82	81	WSW	S	8
NORWAY																		
BERGEN	60	24N	5	19E	141	14	17	20	75	74	73	21	67	66	65			
OSLO	59	56N	10	44E	308	-2	0	4	79	77	74	17	67	66	64	N	S	10
PAKISTAN																		
KARACHI	24	48N	66	59E	13	45	49	51	100	98	95	14	82	82	81	N	SSW	4
LAHORE	31	35N	74	20E	702	32	35	37	109	107	105	27	83	82	81	NW	SE	3
PESHWAR	34	01N	71	35E	1164	31	35	37	109	106	103	29	81	80	79	W	NE	5
PANAMA AND CANAL ZONE																		
PANAMA CITY	8	58N	79	33W	21	69	72	73	93	92	91	18	81	81	80			
PAPUA NEW GUINEA																		
PORT MORESBY	9	29S	147	09E	126	62	67	69	92	91	90	14	80	80	79			
PARAGUAY																		
ASCUNCION	25	17S	57	30W	456	35	43	46	100	98	96	24	81	81	80	NE	NE	7
PERU																		
LIMA	12	05S	77	03W	394	51	53	55	86	85	84	17	76	75	74	N	S	10
SAN JUAN DE MARCONA	15	24S	75	10W	197	--	55	57	82	81	79	12	75	73	72	S	S	10
TALARA	4	35S	81	15W	282	--	59	61	90	88	86	15	79	78	76	SSE	S	18
PHILIPPINES																		
MANILA	14	35N	120	59E	47	69	73	74	94	92	91	20	82	81	81	N	ESE	3
POLAND																		
KRAKOW	50	04N	19	57E	723	-2	2	6	84	81	78	19	68	67	66			
WARSAW	52	13N	21	02E	394	-3	3	8	84	81	78	19	71	70	68	E	SE	7
PORTUGAL																		
LISBON	38	43N	9	08W	313	32	37	39	89	86	83	16	69	68	67	ENE	N	5
PUERTO RICO																		
SAN JUAN	18	29N	66	07W	82	65	67	68	89	88	87	11	81	80	79	ENE	E	10

State and Station	Lat. °	Lat. 'N	Long. °	Long. 'W	Elev. feet	Winter, °F Mean of Annual Extremes	99%	97.5%	Summer, °F Design Dry-bulb 1%	2.5%	5%	Mean Daily Range	Design Wet-bulb 1%	2.5%	5%	Prevailing Wind Winter Knots	Prevailing Wind Summer Knots
RUMANIA																	
BUCHAREST	44	25N	26	06E	269	-2	3	8	93	91	89	26	72	71	70		
SAUDI ARABIA																	
DHAHRAN	26	17N	50	09E	80	39	45	48	111	110	108	32	86	85	84	N 8	N
JEDDA	21	28N	39	10E	20	52	57	60	106	103	100	22	85	84	83	N 8	N
RIYADH	24	39N	46	42E	1938	29	37	40	110	108	106	32	78	77	76	N 8	N
SENEGAL																	
DAKAR	14	42N	17	29W	131	58	61	62	95	93	91	13	81	80	80	N 8	NW
SINGAPORE																	
SINGAPORE	1	18N	103	50E	33	69	71	72	92	91	90	14	82	81	80	N 4	SE
SOMALIA																	
MOGADISCIO	2	02N	49	19E	39	67	69	70	91	90	89	12	82	82	81	SSW 16	E
SOUTH AFRICA																	
CAPETOWN	33	56S	18	29E	55	36	40	42	93	90	86	20	72	71	70		
JOHANNESBURG	26	11S	78	03E	5463	26	31	34	85	83	81	24	70	69	69		
PRETORIA	25	45S	28	14E	4491	27	32	35	90	87	85	23	70	69	68	N 4	W
SOUTH YEMEN																	
ADEN	12	50N	45	02E	10	63	68	70	102	100	98	11	83	82	82		
SOVIET UNION (FORMER)																	
ALMA ATA	43	14N	76	53E	2543	-18	-10	-6	88	86	83	21	69	68	67		
ARCHANGEL	64	33N	40	32E	22	-29	-23	-18	75	71	68	13	60	58	57		
KALININGRAD	54	43N	20	30E	23	-3	1	6	83	80	77	17	67	66	65		
KRASNOYARSK	56	01N	92	57E	498	-41	-23	-27	84	80	76	12	64	62	60		
KIEV	50	27N	30	30E	600	-12	-5	1	87	84	81	22	69	68	67		
KHARKOV	50	00N	36	14E	472	-19	-10	-3	87	84	82	23	69	68	67		
SAMARA (KUIBYSHEV)	53	11N	50	06E	190	-23	-19	-13	89	85	81	20	69	67	66		
ST. PETERSBURG (LENINGRAD)	59	56N	30	16E	16	-14	-9	-5	78	75	72	15	65	64	63		
MINSK	53	54N	27	33E	738	-19	-11	-4	80	77	74	16	67	66	65		

State and Station	Lat. °	Lat. 'N	Long. °	Long. 'W	Elev. feet	Winter, °F Mean of Annual Extremes	Winter 99%	Winter 97.5%	Summer Design Dry-bulb 1%	2.5%	5%	Mean Daily Range	Design Wet-bulb 1%	2.5%	5%	Prevailing Wind Winter	Knots	Summer
MOSCOW	55	46N	37	40E	505	-19	-11	-6	84	81	78	21	69	67	65	SW	11	S
ODESSA	46	29N	30	44E	214	-1	4	8	87	84	82	14	70	69	68			
PETROPAVLOVSK	52	53N	158	42E	286	-9	-3	0	70	68	65	13	58	57	56			
ROSTOV ON DON	47	13N	39	43E	159	-9	-2	4	90	87	84	20	70	69	68			
EKATERINBURG (SVERDLOVSK)	56	49N	60	38E	894	-34	-25	-20	80	76	72	16	63	62	60			
TASHKENT	41	20N	69	18E	1569	-4	3	8	95	93	90	29	71	70	69			
TBILISI	41	43N	44	48E	1325	12	18	22	87	85	83	18	68	67	66			
VLADIVOSTOK	43	07N	131	55E	94	-15	-10	-7	80	77	74	11	70	69	68			
VOLGOGRAD	48	42N	44	31E	136	-21	-13	-7	93	89	86	19	71	70	69			
SPAIN																		
BARCELONA	41	24N	2	09E	312	31	33	36	88	86	84	13	75	74	73	N	10	S
MADRID	40	25N	3	41W	2188	22	25	28	93	91	89	25	71	69	67	NNE	5	W
VALENCIA	39	28N	0	23W	79	31	33	37	92	90	88	14	75	74	73	W	7	ESE
SRI LANKA																		
COLOMBO	6	54N	79	52E	24	65	69	70	90	89	88	15	81	80	80	W	6	W
SUDAN																		
KHARTOUM	15	37N	32	33E	1279	47	53	56	109	107	104	30	77	76	75	N	6	NW
SURINAM																		
PARAMARIBO	5	49N	55	09W	12	66	68	70	93	92	90	18	82	82	81	NE	9	E
SWEDEN																		
STOCKHOLM	59	21N	18	04E	146	3	5	8	78	74	72	15	64	62	60	W	4	S
SWITZERLAND																		
ZURICH	47	23N	8	33E	1617	4	9	14	84	81	78	21	68	67	66			
SYRIA																		
DAMASCUS	33	30N	36	20E	2362	25	29	32	102	100	98	35	72	71	70			
TAIWAN																		
TAINAN	22	57N	120	12E	70	40	46	49	92	91	90	14	84	83	82	N	10	W
TAIPEI	25	02N	121	31E	30	41	44	47	94	92	90	16	83	82	81	E	7	E

State and Station	Lat. °	Lat. 'N	Long. °	Long. 'W	Elev. feet	Winter, °F Mean of Annual Extremes	Winter 99%	Winter 97.5%	Summer Design Dry-bulb 1%	2.5%	5%	Mean Daily Range	Design Wet-bulb 1%	2.5%	5%	Prevailing Wind Winter	Summer	Knots
TANZANIA																		
DAR ES SALAAM	6	50S		18E	47	62	64	65	90	89	88	13	82	81	81			
THAILAND																		
BANGKOK	13	44N	100	30E	39	57	61	63	97	95	93	18	82	82	81	N	S	4
TRINIDAD																		
PORT OF SPAIN	10	40N	61	31W	67	61	64	66	91	90	89	16	80	80	79			
TUNISIA																		
TUNIS	36	47N	10	12E	217	35	39	41	102	99	96	22	77	76	74	W	E	10
TURKEY																		
ADANA	36	59N	35	18E	82	25	33	35	100	97	95	22	79	78	77	N	W	8
ANKARA	39	57N	32	53E	2825	2	9	12	94	92	89	28	68	67	66	N	NE	10
ISTANBUL	40	58N	28	50E	59	23	28	30	91	88	86	16	75	74	73	NNE	N	8
IZMIR	38	26N	27	10E	16	24	27	29	98	96	94	23	75	74	73			
UNITED KINGDOM																		
BELFAST	54	36N	5	55W	24	19	23	26	74	72	69	16	65	64	62			
BIRMINGHAM	52	29N	1	56W	535	21	24	27	79	76	73	15	66	64	63			
CARDIFF	51	28N	3	10W	203	21	24	27	79	76	73	14	64	63	62			
EDINBURGH	55	55N	3	11W	441	22	25	28	73	70	68	13	64	62	61	WSW	WSW	6
GLASGOW	55	52N	4	17W	85	17	21	24	74	71	68	13	64	63	61			
LONDON	51	29N	0	0	149	20	24	26	82	79	76	16	68	66	65	W	E	7
URUGUAY																		
MONTEVIDEO	34	51S	56	13W	72	34	37	39	90	88	85	21	73	72	71	N	NNE	11
VENEZUELA																		
CARACAS	10	30N	66	56W	3418	49	52	54	84	83	81	21	70	69	69	E	ENE	8
MARACAIBO	10	39N	71	36W	20	69	72	73	97	96	95	17	84	83	83			
VIETNAM																		
DA NANG	16	04N	108	13E	23	56	60	62	97	95	93	14	86	86	85	NW	N	5
HANOI	21	02N	105	52E	53	46	50	53	99	97	95	16	85	85	84			

State and Station	Lat. °	'N	Long. °	'W	Elev. feet	Winter, °F Mean of Annual Extremes	99%	97.5%	Summer, °F Design Dry-bulb 1%	2.5%	5%	Mean Daily Range	Design Wet-bulb 1%	2.5%	5%	Prevailing Wind Winter Knots	Summer
HO CHI MINH CITY (SAIGON)	10	47N	106	42E	30	62	65	67	93	91	89	16	85	84	83		
YUGOSLAVIA																	
BELGRADE	44	48N	20	28E	453	4	9	13	92	89	86	23	74	73	72	ESE 9	SE
ZAIRE																	
KINSHASA (LEOPOLDVILLE)	4	20S	15	18E	1066	54	60	62	92	91	90	19	81	80	80	NNW 7	W
KISANGANI (STANLEYVILLE)	0	26S	15	14E	1370	65	67	68	92	91	90	19	81	80	80		

INDEX